The Calf

Fifth Edition

Volume 1 **Management of Health**

The Calf Fifth Edition

Volume 1 Management of Health

J. H. B. Roy MA, Dip Agric (Cantab), PhD, DSc, CBiol, FIBiol
Honorary Fellow, University of Reading; formerly, Senior Principal Scientific Officer and Head of Feeding and Metabolism Department, National Institute for Research in Dairying, Shinfield, Reading

Butterworths
London Boston Singapore Sydney Toronto Wellington

First published, 1955, by Farmer and Stock-Breeder Publications Ltd
Reprinted, 1956
Second edition, 1959, by Farmer and Stock-Breeder Publications Ltd
Third edition, 1970, by Iliffe Books Ltd for Farmer and Stock-Breeder Publications Ltd
Reprinted, 1974
Fourth edition, 1980, by Butterworths
Fifth edition (Volume 1), 1990, by Butterworths

British Library Cataloguing in Publication Data

Roy, J. H. B. (James Henry Barstow)
The calf.—5th ed.
Vol. 1: Management of health
1. Livestock. Calves. Rearing
I. Title
636.2′08

ISBN 0-407-00520-X

Library of Congress Cataloging-in-Publication Data

Roy, J. H. B. (James Henry Barstow)
The calf/J.H.B. Roy.—5th ed.
p. cm.
Includes bibliographical references.
Contents: v. 1. Management of health.
ISBN 0-407-00520-X (v. 1):
1. Calves. I. Title.
SF205.R64 1990
636.2′087—dc20 89-22348

Composition by Genesis Typesetting, Borough Green, Sevenoaks, Kent
Printed and bound by Hartnolls Ltd., Bodmin, Cornwall

Preface

Since the fourth edition of *The Calf* was published in 1980, a marked change of emphasis has occurred in cattle production in the developed countries of the world. This has arisen mainly from a saturation of the market in those countries or community of countries, in spite of continuing shortages of animal products in developing countries of the world; a paradox arising from maldistribution due to the developing countries' inability to pay for the surpluses or to increase their own production.

Thus, encouragement of increased production in the developed countries has been superseded by a requirement for the same or reduced output but with increased efficiency of production if incomes are to be maintained.

At the same time, the impact of public opinion has demanded that farmers should show more concern for the welfare of their stock and should search their consciences before adopting systems so intensive that they may be dependent for their success on the increased and continual use of drugs. The worries on this score are reflected in the recent ban of the use of hormones for meat production by the European Community.

Thus, the balance of production has shifted to more extensive systems in the developed countries, although a shift towards more intensive methods in peasant farming communities may still be appropriate. Moreover, demographic changes in the future may necessitate a return to more intensive methods even in the developed countries.

In the last few years also, a very marked reduction in government support in the UK for research in husbandry, nutrition and health of calves has occurred, so this new edition comes at a time of watershed; hopefully it may be considered as a summary of the present state of knowledge from which further steps forward may be taken when financial restraints to progress are less severe.

Since 1980 several other books on calf rearing have been published in the UK, namely:

Calf Management and Disease Notes by A. H. Andrews, published by A. H. Andrews, Harpenden (1983)
Calf Husbandry, Health and Welfare by John Webster, Blackwell Scientific Publications, Oxford (1984)
Calf Rearing by Bill Thickett, Dan Mitchell and Bryan Hallows, Farming Press, Ipswich (1986).

These books are directed at and are more within the restricted financial means of veterinary and agricultural undergraduates and college students.

Excluding the Russian and Spanish translations of *The Calf*, about 60 per cent of the English fourth edition was sold overseas.

This new fifth edition once more aims to present the reader with research findings from throughout the world, so that from often conflicting evidence, the reader can make his own assessment of where the scientific truth lies and use this knowledge for the benefit of food production.

To accelerate the publication of this new edition which has involved the scanning of several thousands of references, it has been decided to publish it in three volumes, each of which will correspond to one of the three parts of the fourth edition.

The first volume is *The Calf: Management of Health*. The information in this volume cannot cover all the hazards to health to which a calf may be exposed, particularly those in subtropical areas of the world, or all the congenital defects that have been reported, but it is hoped that most areas of importance have been covered.

In view of the multifactorial nature of so many calf disorders, in particular enteric and respiratory, even more emphasis than in earlier editions has been given to the predisposing causes that influence the development of these conditions.

Although some information on the more important deficiency conditions relating to mineral, trace mineral and vitamin inadequacy appear in this first volume, these conditions will be considered in more detail in Volume 2.

Whilst scientific facts do not change, but are rather qualified as time progresses, the increased number of references scanned has meant that the book has been completely rewritten and hopefully by extensive and appropriate subdivision will be easier for the reader to find the information that he requires.

J. H. B. Roy
Reading

Acknowledgements

During the course of the preparation of the earlier editions of this book, I was greatly helped by a number of my colleagues at the National Institute for Research in Dairying; this institute unfortunately ceased to exist on 31st March 1985. In particular, I am indebted to Dr K. W. G. Shillam, Dr I. J. F. Stobo, Mr B. F. Bone, Mrs G. Maddox, Mrs P. D. A. McMillan Browse, Mrs S. M. Humphrey, Mrs E. B. Poulter and Mrs N. Gable. In addition, my efforts in the last three editions have been considerably eased by the excellent information retrieval system developed by Mr N. W. Briggs and his staff, which also sadly exists no longer.

I am also especially grateful to the many people who during 35 years have assisted me in the conduct of experiments on calves, the results of many of which are included in this and other parts of this, the fifth edition.

In the preparation of this fifth edition, I have been on my own, but supported in my endeavours by Professor T. R. Morris, Head of the Department of Agriculture, University of Reading. In particular, I would like to thank Mrs V. W. Craig of that department, without whose dedication and accurate interpretation of my often illegible handwriting of text and very large number of references, this edition would not have been possible. Tragically, she has not lived to see the results of her hard work in print.

Contents

Chapter 1

Calf mortality

Definitions

For comparison of mortality rates to be of any value, it is imperative that the time span to which the rates refer should be specified.

Thus, calf mortality can best be subdivided into:

1. Abortions (stillbirths at ≤270 d of gestation).
2. Perinatal mortality (stillbirths at >270 d of gestation and mortality during the first 24 h of life).
3. Neonatal mortality (calves born alive that die between 24 h and 28 d).
4. Older calf mortality (calves born alive that die between 29 and 84 d, or between 29 and 182 d).

Under good management conditions in developed countries, mortality rates in these four classifications are approximately as follows:

1. Abortions, 2–2.5 per cent.
2. Perinatal mortality, 3.5–5.0 per cent.
3. Neonatal mortality, 3 per cent.
4. Older calf mortality, 1 per cent (29–84 d) or 2 per cent (29–182 d).

Abortions

Europe

A survey in Great Britain in 1962–63 showed that abortions accounted for 2.1 per cent of parturitions (including abortions)[5]. In the Republic of Ireland, there were 2.6 per cent abortions in 2598 pregnancies in 1976–77[14] and 2.0 per cent in 8750 pregnancies from 1976–79[15]. Of these abortions in Ireland in 1976–77, 42 per cent were due to brucellosis, 14 per cent to listeriosis, 4 per cent to salmonellosis, 7 per cent to mummified fetuses and 39 per cent were undiagnosed[14], whereas of the abortions from February 1976 to December 1979, 35 per cent were due to brucellosis, 6.5 per cent to listeriosis and 4.5 per cent to salmonellosis[15]. In Yugoslavia, prenatal mortality averaged 2.1 per cent[16].

America

In the USA in 1459 pregnancies at the University of Maine between 1948 and 1972, there were 1.8 per cent abortions[13].

Asia

Significant breed differences in rates of abortion (1.4 to 6.5 per cent) occurred in India for Tharparker, Sahiwal and Red Sindhi breeds and their crosses with Brown Swiss, Holstein-Friesian and Jersey. Crossbreds had higher rates than had purebred Zebus, and three-breed crosses had higher losses than had two-breed crosses. Murrah buffaloes had 4 per cent abortions. A year to year variation in abortion rates was found. More prenatal mortality occurred in the monsoon and winter seasons than in summer[17].

Perinatal mortality

Mortality rates

Europe

In the survey in Great Britain in 1962–63 stillbirths accounted for 3.3 per cent of parturitions[5].

In a detailed study of mortality at calving of calves out of Friesian heifers on 58 farms in Great Britain reported in 1987, the mean mortality for 18 farms, which had the lowest incidence of dystokia and mortality, was only 1.3 per cent, whereas in the other extreme 12 farms, the mean mortality rate at calving was 25.9 per cent[59].

In the Republic of Ireland in 1976–77, an incidence of 4.7 per cent stillbirths in 2598 pregnancies[14] and from 1976–1979 an incidence of 6 per cent in 8750 pregnancies were reported[15]. In S. Belgium, of 1454 births on 47 farms in August–September 1973, the incidence of stillbirths was 3.2 per cent[20]. Of 6285 calves born in 1979 in Norway, perinatal mortality up to 24 h of age was 2.8 per cent[21,22] and in Yugoslavia stillbirths averaged 3.9 per cent[16].

America

In Maine USA from 1948–1972 an incidence of stillbirths of 3.4 per cent was reported with an overall perinatal mortality to 24 h of age of 4.2 per cent[13], whilst in N. Dakota of 9877 cows in 86 herds, 1 per cent of pregnancies resulted in abnormal presentation[19] and in Virginia in 12 300 heifer births, mortality at birth was 1.2 per cent[18].

Asia

In India, the incidence of stillbirths varied from 1.0 to 2.2 per cent for different breeds of Zebu and crosses with European cattle, whilst for Murrah buffaloes the incidence was 1.3 per cent[17].

Comparison with perinatal mortality in infants

These values may be compared with those of perinatal mortality in infants, i.e. stillbirths and deaths within the first week of life, which in the UK in 1986 was 0.99 per cent[58], and in 1987 was 0.89 per cent of births[60].

Causes of death

Rarely septicaemia may be the cause of death in calves within the first 24 h of life.

Europe

A survey of post-mortem findings from Veterinary Investigation Centres in 1971[11] showed that the deaths of about 7 per cent of calves, stillborn or dying during the first 3 d of life, were associated with genetic or prenatal factors. Of these, 22 per cent were due to prematurity; 15 per cent to cardiac abnormalities, mainly a patent foramen ovale (the opening in the septum between the two atria in the heart of the fetus); 14 per cent to thyroid hyperplasia, probably due to a marginal iodine deficiency or to the consumption of goitrogens by the dam; 11 per cent to gastroenteric abnormalities; 11 per cent to nervous conditions; and 14 per cent to agammaglobulinaemia. This latter condition was probably a result of inadequate colostrum ingestion rather than a genetic factor.

In the Republic of Ireland, 85 per cent of perinatal deaths were due to anoxia and the results of prolonged parturition, particularly of heifers[14]. The problem arose mainly from the effects of the sire and dam on birth weight and musculature, but partly to the size of the pelvic opening of the dam[23]. Cows with difficulties at successive calvings had smaller pelvic openings and a strong correlation has been found for dystokia in Holsteins between successive parities[24]. Similarly in Germany, the majority of perinatal losses were due to hypoxia[25].

In the Republic of Ireland in 1976–77, 73 per cent of stillbirths were associated with anoxia, 13 per cent with difficult or prolonged parturition, 7 per cent with septicaemia and 5 per cent were undiagnosed. Anoxia most commonly occurred following parturition outside normal working hours, among heifers' calves and among the progeny of certain bulls[14]. In further work from 1976 to the end of 1979, 75 per cent of stillbirths were associated with anoxia and 10 per cent with serious birth defects[15].

America

In Cuba, examination of 17 stillborn calves revealed fat embolism in the lungs of nine, associated with rib fracture[26].

Genetic factors

Breed

Europe

The slightly higher mortality rate within 12 h of birth of calves sired by a Charolais bull compared with those sired by a Hereford bull, or purebred calves[10], is probably a reflection of calving difficulties rather than of differences in susceptibility to infection.

The Milk Marketing Board (MMB) records in England and Wales for 1986–87 for mortality associated with parturition showed that for bulls of beef breeds used on Friesian-Holstein cows the mortality associated with calving varied from 2.9 per cent for the Hereford to 6.5 per cent for the Chianina and for those used on heifers varied from 5.7 per cent for the Aberdeen Angus to 10.2 per cent for the Limousin[27]. For 1987–88, MMB records showed that calf mortality within 48 h after calving was <3 per cent for calves out of cows mated to a Hereford bull, 3.4 per cent for Limousin-cross calves, 4 per cent for Charolais and Simmental-cross calves and nearly 6 per cent for Blonde d'Aquitaine-cross calves[61].

In Norway, the incidence of stillbirths was in increasing order: Norwegian Red (NRF) × Friesian, NRF, NRF × Swedish Friesian, NRF × Charolais[21,22]. A survey of stillbirths in 163 dairy herds in Hungary in 1978 and 1979 showed that the incidence varied with breed[28].

Asia

In India, the incidence of stillbirths was higher for crossbreds of Tharparker, Sahiwal and Red Sindhi with Brown Swiss, Holstein-Friesian and Jersey than for purebred Zebus; three-breed crosses had higher losses than had two-breed crosses[17].

Sire

The MMB records in England and Wales for 1986–87 for mortality associated with parturition showed a variation within Friesian-Holsteins of 0.6–9.2 per cent for different bulls[27]. For 1987–88, the overall calf mortality to 48 h was 4.2 per cent with a range of 0–18 per cent for different bulls[61]. In the Republic of Ireland a sire influence on perinatal mortality has also been reported[14] and in Iowa perinatal mortality up to 48 h varied from 1 to 16 per cent for different sires that had at least 60 progeny, but the heritability of livability was small[29].

Dam factors

Parity

Stillbirths were much higher in heifers than in cows[20,28,29]. For instance, in Israel, perinatal mortality associated with difficult calvings was 9.1 per cent in heifers compared to 4.1 per cent in cows, but it was not related to age of heifer or cow parity[30]. In Norway, the greatest incidence was found in the first and ≥ sixth parity[21,22].

Age at calving

In the Republic of Ireland, it has been suggested that a young age of the dam at first calving may be important[14].

Gestation length

In spite of the normal gestation length for bull calves, which are more susceptible to perinatal mortality, being longer, gestation length for calves that died before 48 h of age in Iowa was 1.2 d shorter than that for live calves (279.6 d). The highest rate of survival was for gestation lengths that were 2 d longer than the average[29].

Nutrition

It is suspected that the nutritional status of the dam affects the incidence of stillbirths[14]. The incidence of dystokia appears to be greater in overfat animals. Other work has shown that management during the dry period affects the incidence[28].

Hormonal abnormalities and milk yield

In the Republic of Ireland, it was considered that perinatal mortality was related to hormonal abnormalities[14]. Stillbirth incidence in cows in Hungary tended to increase with increasing milk yield[28].

Calf factors

Sex

Stillbirth incidence is higher in bull calves[28]. In Iowa, perinatal mortality up to 48 h in 136 775 Holstein calvings was 7.6 per cent for bull calves and 5.6 per cent for heifer calves[29]. This in part may reflect the greater birth weight of bull calves.

Twinning

In Norway, a greater incidence of stillbirths occurred with twin births[21,22].

Environmental factors

Season

In Israel, perinatal mortality was high in winter and low in summer[30]. Similar findings were obtained in Hungary[28].

Herd size

In New York State, the number of calves born dead or abnormal was unrelated to herd size[52], but in Virginia mortality at birth was positively correlated with herd size in a survey in which 87 per cent of herds contained <120 cows, although the correlation coefficient was only +0.10[18]. In Hungary, no effect of herd size on stillbirths has been found[28].

Supervision and housing

The degree of supervision at calving and the housing conditions did not affect the incidence of stillbirths in Hungary[28].

In the study of 'high' (>20 per cent) and 'low mortality' (<5 per cent) rate farms in Great Britain there was no difference between the two groups of heifers in relation to their measurement of body condition at service or calving, but on the 'high mortality' farms a much greater proportion of the heifers were assisted at parturition and many more difficult calvings were reported.

The 'low mortality' rate farms lost no calves born to heifers with difficult calvings, whereas the 'high mortality' rate farms lost 68.6 per cent. On the 'low mortality' rate farms 4.8 per cent of large calves were born dead or died within 24 h compared with 54.7 per cent of large calves on the 'high mortality' rate farms.

In attempting to elucidate the cause of these large differences, it was found that the frequency of observation during the 3-week period before calving was 4–5 times daily for the 'low mortality' rate farms and only 1–2 times daily for the 'high mortality' rate farms. On the 'low mortality' rate farms, the heifers were calved on a field or paddock and the forage was naturally sparse or deliberately restricted.

The heifers in these herds were usually grazed in a heifer group or run with dry cows rather than with the milking herd. Magnesium supplementation of the cows was the norm on the 'low mortality' rate farms but was unusual on the 'high mortality' rate farms[59].

Neonatal mortality

Mortality rates

Europe

In Great Britain, three surveys during 1936–37[1], 1946–48[2,3] and 1962–63[4,5] have shown very much the same incidence of mortality of young calves. In the first two surveys about 6 per cent of calves born alive in England and Wales did not survive until the age of 6 months, whereas in Scotland the figure was about 12 per cent. In the third survey throughout England, Wales and Scotland it was estimated that, from 3.4 million live births, 140 000 home-bred and 37 000 purchased calves died before 1 year of age. This represented a mortality rate for home-bred calves of 5.0 per cent to 6 months of age and 5.7 per cent to 1 year of age.

In spite of the fact that antibiotics became available to the agricultural industry between the second and third surveys, little reduction in mortality appears to have occurred. It may be argued, therefore, that a mortality of 5 per cent is the economic mortality rate, and that this level reflects the fact that the cost of reducing the mortality rate to about 1.5 per cent, which would appear to be technically feasible and would be similar to that of infant mortality[12], is not at present economically attractive, and that a risk of 5 per cent mortality can be tolerated by the farming community at present levels of input of labour and capital.

In a study on one farm in England overall mortality to 12 weeks was 1.1 per cent[31], and in an intensive calf rearing unit that purchased 1996 artificially reared dairy-bred calves, overall mortality from about 1 week to 6 weeks of age was 4 per cent[57].

In the Republic of Ireland, surveys in 1976–77[14] and to the end of 1979[15] showed mortality rates to 21 d of age of 2.9 and 3.2 per cent respectively. Neonatal mortality to 1 month of age in S. Belgium was 6.5 per cent[20].

In Norway, the mortality of calves born alive was 1.18 per cent to 30 d of age[21,22], and in Yugoslavia to 120 d of age postnatal losses on two farms averaged 11.9 per cent of calves born alive[16].

America

In California, losses averaged 17.3–20.2 per cent in 16 farms during 5.5 years[32–34]. Other work in the USA showed that of calves born alive at the University of Maine, 4.8 per cent died before 24 months of age[13]. In S. Carolina, mortality to 6 months of age of heifer calves in 140 herds in 1978–9, including those born dead, averaged 19.1 per cent with mortality to 1 month of age representing about 84 per cent of the total[35]. In Virginia mortality from birth to 3 months of age was 6.5 per cent[18].

In California, a statistical study showed the following factors to affect the mortality from diarrhoea: vaccination practices, use of dietary supplements for pregnant cattle, control of dehydration, hygienic conditions at birth, housing and bedding, temperature control and farm size[36].

Asia

In India, mortality in 2478 crossbred calves from birth to 6 months at Karnal was 7.9 per cent, with 49 per cent of the deaths occurring in the first month[37]. Mehsana buffaloes in India had a mortality rate to 12 months of 56.4 per cent for males and 27.7 per cent for females, of which 48 and 51 per cent respectively occurred in the first month of life[38].

Africa

In Nigeria, of 1105 calves born alive between 1975 and 1980, 8.7 per cent died before 84 d of age[39].

Microbiological factors

Causes of mortality

In general, in developed countries where milk substitutes are used widely, the main cause of death is enteric infection, whereas in developing countries where whole milk is the normal diet, the main cause of death is often respiratory infections.

Europe

In Great Britain, of 2046 calves examined *post mortem* at Veterinary Investigation Centres in 1959–61, 46 per cent of the deaths to 6 months of age were associated with *E. coli*, 24 per cent with *Salmonella* spp., 10 per cent with poisoning (mainly lead) and 7 per cent with virus infections[6]. In the 1962–63 survey in Great Britain 70 per cent of deaths were due to septicaemia (25 per cent) and gastroenteritis (45 per cent), and 11 per cent due to respiratory disease[4].

In 1971[11] a survey from Veterinary Investigation Centres covered 6042 post-mortem examinations of all ages of cattle, but 74 per cent were for calves under 20 weeks of age. Of diagnoses made on calves that died at less than 1 week of age, 41 per cent were associated with *E. coli*, of which 49 per cent were classified as septicaemia or bacteraemia, and 44 per cent were classified as a digestive condition, presumably an *E. coli* localized intestinal infection. Salmonellosis accounted for 11 per cent of all diagnoses, of which 80 per cent were *S. dublin* and 14 per cent *S. typhimurium*; 50 per cent of the former and 71 per cent of the latter occurred during the first 7 weeks of life, and the incidence for *S. typhimurium* was twice as high in the period 1–3 weeks as in the period 4–7 weeks.

Respiratory infections formed 13 per cent of all diagnoses and were most common during the 8–19th weeks of life. Of these, 22 per cent involved *Pasteurella* spp. and 11 per cent were classified as virus pneumonia.

In a very recent study of one intensive calf rearing unit by the Meat and Livestock Commission (MLC) of 1996 purchased dairy-bred calves, pneumonia was the most common disease, affecting 48 per cent of calves, whereas diarrhoea only affected 14 per cent. More than half the pneumonia cases occurred between 1 and 3 weeks after purchase whereas 67 per cent of the cases of diarrhoea occurred within 1 week of purchase. Mortality rates for both conditions appeared to increase with time from purchase[57]. In a survey in the Republic of Ireland in 1976–77, of the postnatal deaths to 21 d of age, 37 per cent were due to neonatal diarrhoea, 35 per cent to septicaemia, 9 per cent to pleurisy and pneumonia, 7 per cent to birth defects and 12 per cent to miscellaneous injury or infection[14]. In a further survey up to the end of 1979, 46 per cent of deaths to 21 d were associated with diarrhoea, 31 per cent with septicaemia, 78 per cent with birth defects and 16 per cent with miscellaneous infections or accidents[15].

In S. Belgium, diarrhoea was the cause of death in 72 per cent of calves that died up to 1 month of age[20]. Similarly in Norway enteric disease was the main cause of death to 30 d of age[21,22].

Asia

In India, pneumonia was the major cause of loss to 6 months of age, amounting to 30–40 per cent of deaths; diarrhoea and enteritis was of secondary importance[41]. In other work in India, mortality during the first month of life was largely due to respiratory disease, whereas digestive diseases caused more deaths between 3 and 6 months of age. Respiratory disease was responsible for 31 per cent and digestive disorders for 34 per cent of the mortality[37].

Of the mortality to 12 months of age in Mehsana buffaloes, 31 per cent was due to gastroenteritis, 28 per cent to tympany, 24 per cent to pneumonia, 6 per cent to urinary syndromes and 3 per cent to congenital abnormalities[38].

Antibiotic-resistant strains

In the survey in the Republic of Ireland in 1976–77, mortality from neonatal diarrhoea and septicaemia was associated with drug resistant forms of *E. coli* and salmonella[14] (see Chapters 3 and 4).

Immunological factors

Colostrum intake

The importance of colostrum in reducing the incidence of mortality is borne out by the fact that in the survey of 1946–48 in Great Britain[2,3] the mortality rate to 1 month of age for calves that sucked colostrum was 4 per cent, compared with 9 per cent for those that obtained colostrum from the bucket. This latter procedure, often adopted in Scotland, together with the fact that calves of the Ayrshire breed predominated, may in part have accounted for the higher mortality rates found in that country compared with those in England and Wales.

In the Republic of Ireland, the incidence of mortality in 3200 purebred Friesian bull calves was 10.8 per cent when their serum ZST units (see Chapter 2) were <16.2 compared to 3.5 per cent when the serum concentration of Ig was normal[43]. However, workers in Liverpool found that when calves were allowed to suck for a longer period, fewer calves had low ZST values (<10 ZST units) but there was no reduction in the incidence of diarrhoea or mortality[49].

In a further study in the Republic of Ireland of mortality from diarrhoea in the first 3 weeks of life, good standards of hygiene and adequate amounts of colostrum given immediately postpartum reduced morbidity and mortality from 36 and 4.1 per cent respectively when hygiene was poor and colostrum was withheld until 6 h postpartum to 8.3 and 1.2 per cent respectively[40].

In France there was a slightly higher incidence of diarrhoea in calves with serum IgG concentrations of <8 g/l at 48 h, but otherwise IgG values were not correlated with incidence of pneumonia or diarrhoea or with mortality[51]. However in contrast in S. Carolina, mortality was lower when a calf was separated from its dam early in life and was thus less dependent on the mothering ability of the dam[35].

Vaccination of the dam

Herds vaccinated against colibacillosis showed a mortality rate in S. Belgium of 5.8 per cent against 7.6 per cent in unvaccinated herds[20].

Disinfection of umbilicus

In S. Belgium, mortality of calves in which the cord was disinfected was 5 per cent compared to 8.5 per cent when this procedure was not adopted[20].

Calf factors

Age

Neonatal mortality rates decline rapidly with age.

Europe

Thus, in the survey of 1962–63 in Great Britain the percentage probability of death in home-bred calves was about equal to the reciprocal of age and percentage mortality up to age A (in weeks) was equal to $\log_e A + 1.72$[5]. Of these deaths, just over 75 per cent occurred during the first month of life in the surveys of 1936–37[1] and 1946–48[2,3] in Great Britain and 64 per cent during the same period in the 1962–63 survey[4,5], the great majority being associated with the bacterium *Escherichia coli.*

In the Republic of Ireland, mortality rate in 8752 calves from diarrhoea and septicaemia in the first 3 weeks of life was 3.2 per cent[40]. In S. Belgium, 58 per cent of the deaths to 1 month of age occurred in the first 14 d of life[20]. However, in Norway in contrast to other countries mortality from day 2–10 was low but from 21–60 d was relatively high[21,22]. In Yugoslavia, 39 per cent of postnatal losses occurred in the first 10 d of life[16].

America

In Maine USA, of calves that died after 24 h, 59 per cent died before 7 d of age and 76 per cent before 6 months[13]. In California 82 per cent of losses occurred in the first 2 weeks of life[32–34]. Mortality from diarrhoea of 9877 pregnancies in 86 herds in N. Dakota between 1975 and 1977 was 6.8–10.9 per cent with 82 per cent of deaths occurring within the first 56 d of life[19].

Asia

In India losses were greatest in the first month of life and then declined[41].

Africa

In Nigeria, the greatest mortality was in the third week of life and decreased with age after the fourth week. Since mortality has been increasing since 1975, it is suggested that more attention should be given to management of bucket-fed calves and calves from imported breeds during the first 5 weeks of life and especially during the third and fourth week[39].

Sex

Bull calves have a slightly higher rate of mortality than do heifers, a finding similar to that observed in human infant mortality. However, in Norway the difference between the sexes was very large, mortality to 30 d of age being 2.7 per cent for bull calves and only 1.8 per cent for heifer calves[21,22].

In India, no effect of sex on mortality of crossbred calves of Holstein, Jersey and Brown Swiss inheritance was disclosed[37,41].

Birth weight

In the Republic of Ireland, mortality was higher in calves of low birth weight[43], but in S. Belgium, mortality was greater when calves weighed <36 kg or >50 kg[20]. In India no effect of birth weight on mortality rate was found[37].

Breed

Europe

In Great Britain, the Channel Island breeds seem to have a higher incidence than have Ayrshires, which, in turn, have a higher incidence than have Shorthorns and Friesians. Thus, the larger the average birth weight of a breed, the lower seems to be the breed mortality rate. This finding may be related to differences between breeds in digestibility of milk diets which may be exaggerated if the quality of the milk substitute is poor. Moreover, it is much easier to overfeed a small calf than a large one.

In Norway, the highest neonatal mortality for Norwegian Red (NRF) and various crosses was with the NRF × Charolais[21,22].

America

In N. Dakota, mortality was in the order: beef × dairy cross > dairy > beef breed[45].

Asia

In India, mortality was greatest when inheritance from temperate breeds was 0.75 rather than 0.5[37]. Other work from India has shown that mortality rates increase with the amount of inheritance from exotic breeds. Special care should be taken with calves of 0.38 exotic inheritance during the first month of life, particularly during the winter[46]. Two-breed crosses have been shown to have better adaptability to tropical climates than have three-breed crosses with Jersey crosses having the best adaptability to extreme conditions[47].

Results from Karnal in India showed that Brown Swiss × Sahiwal and Brown Swiss × Red Sindhi and Murrah buffaloes had greater mortality to 1 year than had the above pure zebu breeds or the Tharparker[41]. However in other work there was no difference in mortality of crossbred calves of Holstein, Jersey and Brown Swiss inheritance[37].

Africa

In Nigeria, calves from the indigenous breeds N'dama, White Fulani, Sokoto and Gudali had lower mortality than that of calves produced from crossing with imported pure breeds, which in turn had lower mortality than that of imported pure breeds (Friesian and Charolais)[39].

Nutritional factors

Nutrition of the dam

In 47 farms in S. Belgium, mortality was lower when the dams were fed hay (3.9 per cent) rather than silage (maize 7.1 per cent, grass 10.7 per cent, beet pulp 6.4 per cent)[20].

Milk intake

In Scotland, mortality of calves from 5 d of age when reared in a 'climatic house' was 13 per cent when the amount of milk substitute given was 0.8 of their maintenance requirement but 8 per cent when the milk substitute was given at a rate of 1.7 × maintenance[50]. In the other extreme, mortality in the Republic of Ireland was greater for calves given *ad libitum* group feeding than for those given individual bucket feeding of milk[43].

Type of liquid diet

In N. Dakota, mortality was greater when calves received a milk substitute[45]. However in New York State, with its high mortality rate, mortality in calves given milk substitute (16.2 per cent) was similar to that of calves given whole milk (15.1 per cent)[52].

A high concentration of fat in a milk substitute may reduce mortality. In S. Carolina, mortality was 8.7 per cent to 1 month of age in 47 herds when the milk substitute contained 220 g fat per kg compared to 12.8 per cent in 23 herds where a milk substitute containing only 100 g fat per kg was used[35].

Age at weaning

In S. Carolina, mortality in herds in which calves were weaned at 3–4 weeks of age was about 50 per cent higher than that in herds where weaning occurred at 7–8 weeks of age or later[35]. On the other hand, mortality in 12 300 heifer births in Virginia from birth to 3 months of age was positively associated with age at weaning and negatively with average milk and fat production in the herd[18].

Environmental factors

Regional differences

In Great Britain, regional differences in the cause of mortality have been disclosed in the 1962–63 survey, with a low frequency of respiratory disorders in Devon, Cornwall, S.E. England and S.E. Scotland, but a high frequency of alimentary disorders in S.E. Scotland, Devon, Cornwall and Wales[4]. No regional trend in the mortality rate of home-bred calves has been found, but a marked trend, from low mortality in the south to high mortality in the north of Great Britain, has been shown for purchased calves[5].

Season

Europe

In Great Britain, mortality rates are highest in February, March and April, but there seems to be a steady increase from shortly after the beginning of the autumn calving season until the end of the spring calving season. Although the high incidence in the early spring months has been attributed to poor nutrition of the dam[7], the evidence suggests that even with good nutrition of the dam there is a steady increase in the incidence of mortality during the winter months, because of the gradual build-up in the calf house of a dominance of certain strains of *E. coli* against which colostrum is ineffective (Figure 1.1)[8]. In an MLC study of 1996 purchased dairy calves in one unit, mortality was significantly lower in those purchased between April and June than at any other time[57].

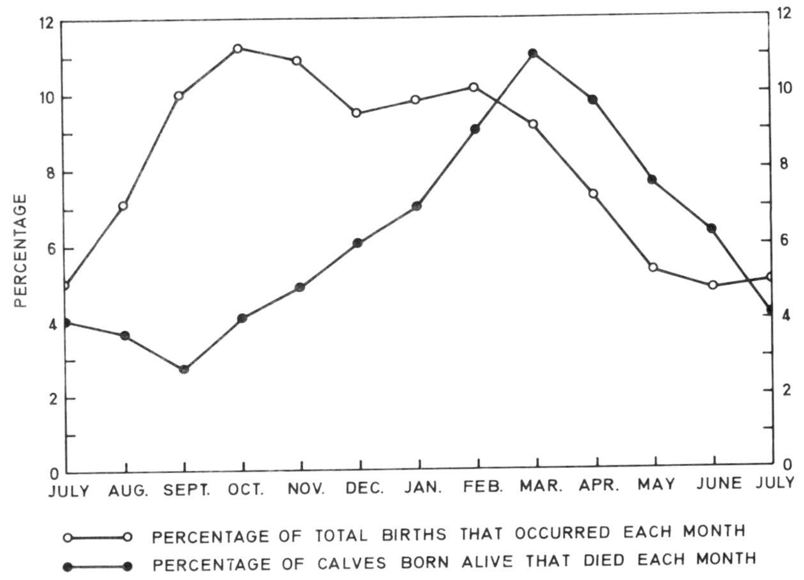

Figure 1.1 Relationship between mortality rate and month of year (based on references 2 and 3): ○——○, percentage of total births that occurred each month; ●——●, percentage of calves born alive that died each month

In Scotland, purchased Friesian and Friesian × Ayrshire calves of >36.3 kg weight born in March had a higher mortality than had those born in September. This was reflected in total Ig concentrations of 11.9 g/l at about 7 d of age in the March-born calves compared to 24.5 g/l in the September-born calves[42]. In the Republic of Ireland, 47 per cent of calves had low ZST values of 0–15 units from March to June, but only 26 per cent had such low values from September to March[43].

America

In N. Dakota, most diarrhoea was associated with cold temperature and snowfall and occurred in March and April[19]. Similarly, the highest mortality in winter in California was at the time of low temperature, large range in temperature[44] and high precipitation[33], but it may also have arisen from an accumulation of microorganisms in the calves' environment[44]. Losses were greatest in mid-summer and mid-winter, but winter losses were 1.26 of those in the summer. Losses in winter were associated with cold, wet and windy weather and losses in summer, to a lesser extent, with hot, dry weather[33].

Asia
In India, the mortality rate of Mehsana buffaloes was affected by season[38]. Other work showed either a tendency for higher losses in winter than in summer or in the monsoon seasons[41], or no effect[37].

Herd size

In S. Belgium, mortality rate increased from 3.9 per cent in herds of <30 cows to 11.1 per cent in herds of >70 cows[20]. Calf losses in New York State were greater in larger herds being 15.8 per cent for herds of <100 cows compared with 27.2 per cent for herds of >200 cows[52]. Although in California size of herd was not related significantly to calf deaths, nevertheless on larger farms there appeared to be a dilution of calf care and there were a greater number of deaths on larger owner-managed farms[32–34].

In contrast in S. Carolina, losses decreased from 21.3 to 16.3 per cent as herd size increased from <100 cows to >200 cows. Since larger herd size was associated with greater milk fat production, mortality decreased from 23 to 12.5 per cent as production increased from 200 kg to >264 kg fat[35].

Housing

Europe
In a survey by the MLC, it was claimed that purchased calves were more likely to die if they were reared in groups 'outdoors' rather than if reared either singly or in groups indoors. The 'outdoor' group had one side of their pen open to the environment. Neither type of penning or housing affected the incidence of pneumonia, but diarrhoea was most common in individually-penned calves. Mortality rates for diarrhoeic calves in group pens were higher than for those in single pens and mortality rates for pneumonia were highest for calves reared in groups 'outdoors'. However, the survey was confounded in that individual penned calves were given restricted feeding, whilst group-fed calves were given *ad libitum* acidified milk from automatic machines. All calves were weaned 5 weeks after purchase, i.e. at 6 weeks. The author admitted that the group-penned calves may not have received such close attention[57].

In the Republic of Ireland, between 1976 and 1977, mortality was reduced from 3.9 to 1.2 per cent by keeping calves in individual pens for the first few days of life, by not mixing calves of different age groups, cleaning and disinfecting the pens after their vacation and giving all diarrhoeic calves only an oral electrolyte solution for 2–3 d[14]. The lowest neonatal mortality in Norway was when calves were reared on elevated and perforated floors rather than on a solid floor[21,22].

In a study involving 5000 calves in E. Germany, the incidence of disease was 75 per cent in calves reared in groups compared to 46 per cent for those in individual pens with automatic milk feeding. Mortality was 17 per cent for calves in groups and 3.7 per cent for those in single pens, but there was no difference in incidence between calves that did or did not have contact with calves in neighbouring pens[56].

America
In New York State, elevated stalls resulted in a lower mortality (13.9 per cent) than free stalls in a barn (18.8 per cent)[52]. However in California[32–34] and in S. Carolina[35], no relationship with calf housing or calving site was found.

Management factors

The person looking after the calves

In S. Belgium, mortality was 6 per cent when the farmer or his wife looked after the calves compared to 13 per cent when an employee did so[20].

Lower calf mortality in New York State, California and S. Carolina has been associated with the calf rearer being either a farmer or a member of the farmer's family rather than an employee[34, 35, 52]. Thus, in New York State, mortality varied from 6.3 per cent when the farmer's wife cared for the calves to 11.7 per cent when it was an employee[52].

Purchase of calves

In Great Britain purchased calves have a higher incidence of mortality than have home-bred calves. A survey[9] of calf losses in herds using artificial insemination gave a value of 6.3 per cent for home-bred calves, compared with 13.5 per cent in purchased calves during the first 2 months of life in 1963–64, but the survey of 1962–63[5] indicated a 60 per cent higher incidence for purchased than for home-bred calves. The main cause of death of purchased calves, however, differs from that of home-bred calves, being shifted towards salmonellosis and away from *E. coli* infection. This is not surprising, as *E. coli* infections mainly afflict calves during the first 10 d of life, during which time a large proportion of the calves will be on farms where they were born.

In a survey of farms which purchased calves, 12 per cent had been treated for enteric disease during the first 2 weeks after arrival[48]. Workers at Liverpool showed that, when buying additional calves was discontinued, mortality decreased from 12 to 2 per cent[49].

In N. Dakota, pneumonia was more important than diarrhoea in imported calves; the younger the age imported, the greater was the risk. Calves bought from farmers or dealers were more at risk than were those bought from sale barns[45].

In complete contrast is the finding that mortality of Friesian male calves in the Republic of Ireland was higher for those purchased direct from farms (7.5 per cent) than for those obtained from markets (4.9 per cent)[43].

Method of feeding

In the survey of 1962–63 in Great Britain[4, 5], mortality was 69 per cent higher for crossbred calves and calves of the Ayrshire and Guernsey breeds given colostrum from the bucket than for those that sucked their dams. However, for the Friesian breed, the method of giving colostrum had no effect on mortality rate.

In N. Dakota, mortality varied from 13.9 per cent for bucket feeding compared to 9.7 per cent for those sucking a nurse cow[45]. Similarly, in Nigeria, suckled calves were found to survive better than bucket-fed calves[39].

Costs of mortality

The cost of mortality should include not only the value of the dead calf, but also the costs of food and treatment before the calf died[54]. Calf rearing units with the lowest gross margin to 85 d had a mortality of 6 per cent compared to 4 per cent for those with the highest gross margin[55].

Long-term survival

A study in 70 000 cows of long-term survival to 3 years after calving in Norway showed an average value of 0.368 corresponding to a culling rate of 0.283 and an average productive life of 3.5 years.

Survival rate was significantly lower (35 per cent) for cows in herds with production >6000 kg milk than for herds with lower yields (37 per cent). Cows calving at 2 years had a greater survival rate than had those calving at younger or older ages. A significant difference between sires in the survival rate of their daughters was found[53].

References

1. LOVELL, R. and HILL, A. B. *J. Dairy Res.* **11**, 225 (1940)
2. WITHERS, F. W. *Br. vet. J.* **108**, 315, 472 (1952)
3. WITHERS, F. W. *Br. vet. J.* **109**, 65, 122 (1953)
4. ANIMAL HEALTH DIVISION, MINISTRY OF AGRICULTURE, FISHERIES AND FOOD. *Vet. Rec.* **78**, 498 (1966)
5. LEECH, F. B., MACRAE, W. D. and MENZIES, D. W. *Calf Wastage and Husbandry in Britain 1962–63.* HMSO, London, (1968)
6. VETERINARY INVESTIGATION SERVICE, MINISTRY OF AGRICULTURE, FISHERIES AND FOOD. *Vet. Rec.* **76**, 1139 (1964)
7. PAYNE, W. J. A. *Br. J. Nutr.* **3**, i (1949)
8. ROY, J. H. B., PALMER, J., SHILLAM, K. W. G., INGRAM, P. L. and WOOD, P. C. *Br. J. Nutr.* **9**, 11 (1955)
9. MILK MARKETING BOARD. *Report of the Breeding and Production Organisation No. 14* (1963–64)
10. MILK MARKETING BOARD. *The Charolais Report* (1966)
11. HUGH JONES, M. E. *Proc. Wld Ass. Buiatric Congr., London,* p.1 (1972)
12. ANONYMOUS. *Br. med. J.* **ii**, 526 (1977)
13. LEONARD, H. A. and GARDNER, D. L. *J. Dairy Sci.* **58**, 1240 (1975)
14. GREENE, H. J. *Ir. J. agric. Res.* **17**, 295 (1978)
15. GREENE, H. J. and BAKHEIT, H. A. *Fm Fd Res.* **11**, 76 (1980)
16. BAČVANSKI, S., COBIĆ, T., DORDEVIĆ, S. and MAGOČ, M. *Zb. Rad. Poljopr. Fak. Inst. Stocarstvo Novi Sad.* **11/12**, 41 (1981)
17. SHARMA, K. N. S. and JAIN, D. K. *Indian J. Anim. Sci.* **54**, 297 (1984)
18. JAMES, R. L., McGILLIARD, M. L. and HARTMAN, D. A. *J. Dairy Sci.* **67**, 908 (1984)
19. HAUGSE, C. N. and STRUM, G. *N. Dak. Res. Rep.* No. 69 (1978)
20. MASSIP, A. and PONDANT, A. *Annls Méd. vét.* **119**, 481 (1975)
21. SIMENSEN, E. *Acta Agric. scand.* **32**, 412 (1982)
22. SIMENSEN, E. *Acta Agric. scand.* **32**, 421 (1982)
23. MENISSIER, F. In *The Early Calving of Heifers and its Impact on Beef Production,* Commission of the European Communities, Brussels, p.81 (1975)
24. THOMPSON, J. R., FREEMAN, A. E. and BERGER, P. J. *J. Dairy Sci.* **64**, 1603 (1981)
25. MULLING, M. *Berl. Münch. tierärztl. Wschr.* **87**, 473 (1974)
26. MERINO, O., CABRERA, J. F. and MUNOZ, M. C. *Revta Salud. Anim.* **5**, 313 (1983)
27. MILK MARKETING BOARD. *Report of the Breeding and Production Organisation No. 37* (1986–87)
28. SZENCI, O. and KISS, M. B. *Acta Vet. hung.* **30**, 85 (1982)
29. MARTINEZ, M. L., FREEMAN, A. E. and BERGER, P. J. *J. Dairy Sci.* **66**, 2400 (1983)
30. BAR-ANAN, R., SOLLER, M. and BOWMAN, J. C. *Anim. Prod.* **22**, 299 (1976)
31. ANDREWS, A. H. and READ, D. J. *Br. vet. J.* **139**, 431 (1983)
32. MARTIN, S. W., SCHWABE, C. W. and FRANTI, C. E. *Am. J. vet. Res.* **36**, 1099 (1975)
33. MARTIN, S. W., SCHWABE, C. W. and FRANTI, C. E. *Am. J. vet. Res.* **36**, 1105 (1975)
34. MARTIN, S. W., SCHWABE, C. W. and FRANTI, C. E. *Am. J. vet. Res.* **36**, 1111 (1975)
35. JENNY, B. F., GRAMLING, G. E. and GLAZE, T. M. *J. Dairy Sci.* **64**, 2284 (1981)
36. FRANTI, C. E., WIGGINS, A. D., LOPEZ-NIETO, E. and CRENSHAW, G. *Am. J. vet. Res.* **35**, 649 (1974)

37. RAO, M. K. and NAGARCENKAR, R. *Trop. Anim. Hlth Prod.* **12**, 137 (1980)
38. TAJANE, K. R., SIDDIQUEE, G. M., RADADIA, N. S. and JHALA, V. M. *Indian J. Anim. Sci.* **53**, 661 (1983)
39. UMOH, J. U. *Br. vet. J.* **138**, 507 (1982)
40. BAKHEIT, H. A. and GREENE, H. J. *Vet. Rec.* **108**, 455 (1981)
41. SHARMA, K. N. S., JAIN, D. K. and NOBLE, D. *Anim. Prod.* **20**, 207 (1975)
42. WILLIAMS, P. E. V., WRIGHT, C. L. and DAY, N. *Br. vet. J.* **136**, 561 (1980)
43. FALLON, R. J. and HARTE, J. F. *Anim. Prod.* **38**, 558 (1984)
44. MARTIN, S. W., SCHWABE, C. W. and FRANTI, C. E. *Can. J. comp. Med.* **39**, 377 (1975)
45. STAPLES, G. E. and HAYGSE, C. N. *Br. vet. J.* **130**, 374 (1974)
46. JAIN, D. K. and SHARMA, K. N. S. *Indian J. Anim. Sci.* **52**, 957 (1982)
47. SHARMA, K. N. S. and JAIN, D. K. *Indian J. Anim. Sci.* **52**, 954 (1982)
48. WEBSTER, A. J. F., SAVILLE, C., CHURCH, B. M., GNANASAKTHY, A. and MOSS, R. *Br. vet. J.* **141**, 472 (1985)
49. BOYD, J. W., BAKER, J. R. and LEYLAND, A. *Vet. Rec.* **95**, 310 (1974)
50. WILLIAMS, P. E. V., DAY, D., RAVEN, A. M. and McLEAN, J. A. *Anim. Prod.* **32**, 133 (1981)
51. LOMBA, F., SPOONER, R. L., MILLAR, P., CHAUVAUX, G. and BIENFET, V. *Annls Méd. vét.* **122**, 101 (1978)
52. HARTMAN, D. A., EVERETT, R. W., SLACK, S. Y. and WARNER, R. G. *J. Dairy Sci.* **57**, 576 (1974)
53. SYRSTAD, O. *Acta Agric. scand.* **29**, 42 (1979)
54. LOWMAN, B. G. *Fm Bldg Prog.* **68**, 15 (1982)
55. MEAT AND LIVESTOCK COMMISSION. In *Beef Handbook* (1985)
56. SCHMOLDT, P., LEMKE, P., BUNGER, U., RIECK, H. and ROTARMUND, H. *Mh. VetMed.* **32**, 11 (1977)
57. PETERS, A. R. *Vet. Rec.* **119**, 355 (1986)
58. ANONYMOUS. *Regional Trends,* HMSO, London (1988)
59. DREW, B. *Proc. Br. Cattle Vet. Ass. 1986–87,* p.142 (1987)
60. ANONYMOUS. *The Times* Nov 9 (1988)
61. MILK MARKETING BOARD. *Report of the Breeding and Production Organisation No. 38* (1987–88)

Chapter 2

Immunity to disease

Classification of immunoglobulins (Ig)

The immunoglobulins have been subdivided into IgG, IgM, IgA, IgD and IgE with further subclasses according to their structure. At first only two classes, IgG and IgM, were recognized but gradually others were specified and their designation is now internationally agreed. Table 2.1 shows the concentration of Ig in various secretions of cattle.

Table 2.1 Immunoglobulins (Ig) in bovine external secretions and serum[24]

	Protein concentration (mg/l)			
	IgA (secretory Ig)	*IgG*$_1$	*IgG*$_2$	*IgM*
Lachrymal secretions	2600	300	120	6
Nasal secretions	1950	40	25	Trace
Saliva	560	30	10	10
Spermatic fluid	130	130	110	Trace
Gastrointestinal secretions	240	250	60	Trace
Bile	80	100	90	50
Urine	0.7	0.8	1	Trace
Colostrum	4400	75 000	1900	4900
Milk	50	350	60	40
Serum	300	10 500	7900	2500

The Ig concentration in dam's serum

A survey of Ayrshire cows showed that the mean plasma concentrations of Ig were 9.8 g IgG_1, 10.9 g IgG_2, 2.7 g IgM and 0.32 g IgA per litre. Whereas the frequency distribution of IgG_1, IgM and IgA concentrations were log normal, there were two populations of IgG_2 with means of 7 and 14 g/l[197].

Transfer of Ig into colostrum

The main Ig transferred from the bloodstream of the dam across the mammary barrier to colostrum is IgG, particularly IgG_1; the passage is made by a specific transport mechanism[195]. Thus, structural differences between the H chains of IgG_1 and IgG_2 seem to be implicated in the selective transport of IgG_1 into colostrum of the goat[199].

IgM and IgA, which are in much smaller amounts in colostrum, are produced locally in the udder by plasmocytes (B lymphocytes that have differentiated into mature antibody-synthesizing plasma cells) adjacent to the secretory epithelium[195]. Thus, the plasma concentrations of Ig (g/l) at 10 d prepartum, parturition and 30 d postpartum were as shown in Table 2.2.

Table 2.2 Plasma concentrations in g/l at 10 d prepartum, parturition and 30 d postpartum[201]

Ig	*−10 d*	*Parturition*	*+30 d*
IgG	8.2	5.8	11.9
IgM	1.35	1.21	1.16
IgA	0.22	0.34	0.18

The main antigenic stimulus for the production of IgG_1 in the dam is probably from the intestinal tract without participation of the mammary gland. However, the local mammary gland immune system also exists, and its artificial stimulation could be of value in practice[196].

One report showed that the decrease in maternal serum IgG concentration began 8 weeks before calving and was correlated with the Ig concentration in colostrum, and that IgM and IgA concentration in serum was unaffected[107]. However, other research showed that both IgG_1 and IgM concentrations in serum were significantly depressed from 4 weeks before until at least 8 weeks after calving, whereas IgG_2 concentration was increased[198].

Factors affecting Ig transfer from blood to colostrum

Age

No effect of age on plasma Ig concentration was found in Ayrshire cows except that cows older than 6 years had significantly higher IgG_2 concentrations[197]. Another report showed that IgG_2 concentrations increased with age in Friesians but decreased with age in Jerseys[198].

Parity

IgG_1 was transferred earlier from serum to colostrum in heifers than in second parity cows[198].

Breed

Since Jersey cows were shown to have a lower serum Ig concentration but a similar colostral Ig concentration than had other breeds, it was postulated that Jersey cows were more efficient in transferring Ig into colostrum[180]. Jersey cows were also shown to have lower serum IgG_2 concentrations than had Friesians[198]. For instance, IgG_2 concentrations of 14 g/l for Friesians and 6.8 g/l for Jerseys have been reported[64]. However, these breed differences in the Ig content of the plasma of adult cattle may have been due to differences in the environment and husbandry[64].

Diet

When cattle were transferred from a balanced complete diet to one of grass or hay, there was a 1.3-fold increase in IgG_1 concentration in serum, a transitory increase in IgM and no change in IgG_2 concentration[198].

The Ig concentration in precolostral (i.e. before sucking) calf serum

Although precolostral calf serum is normally devoid of Ig or contains only small amounts, antibodies against viruses have been found in this serum of calves from vaccinated and unvaccinated dams[194, 235].

Of the precolostral sera of 128 calves in the UK, 16 contained low concentrations of IgG_1, IgG_2 and IgM but none contained any detectable amounts of IgA[232]. However, it has been claimed from the USA that IgM was present in serum in 50 per cent, IgG_1 in 60 per cent and IgG_2 in 30 per cent of calves at birth, a surprisingly high incidence which requires confirmation[193].

Reports of the presence of Ig in precolostral calf serum should be treated with scepticism unless the author has been present at the birth of the calf and removed it from the dam immediately. An attempt to muzzle a calf, unless the muzzle is extremely well designed, is no guarantee that the calf will not obtain some colostrum.

The serum of colostrum-deprived calves does show a bactericidal effect against some strains of *Escherichia coli*, thought to be due to the presence of properdin[5, 6]. The precise importance of this factor relative to that of the antibodies from colostrum has still not been elucidated.

The Ig concentration in colostrum

Although the concentration of Ig in colostrum is the most important factor in protecting a calf from disease, it must be remembered that there are other nutrients present in high concentration that may affect the long-term performance of the calf, e.g. fat-soluble vitamins and many trace elements.

The majority of Ig in cow's colostrum is IgG_1, although natural antibodies to Gram-negative bacteria such as *E. coli* are associated with IgM[23]. In addition, in colostrum there is a much smaller amount of the secretory IgA, which is presumed to be synthesized in the udder tissue rather than to have passed from the bloodstream. The amounts of the various components in bovine colostrum in g/l are IgG_1, 75; IgG_2, 1.9; IgM, 4.9; IgA, 4.4[24].

In W. Germany, mean values per litre for colostrum were 86.4 g whey protein, 53.6 g IgG, 4.8 g IgM and 9.5 g IgA[108], whilst in five samples from the USA, IgG ranged from 28.4 to 46, IgM from 5.6 to 17.4 and IgA from 2.9 to 4.2 g/l[95]. In Japan, the IgA in colostrum varied from 2.5 to 0.4 g/l with one sample of colostrum containing <0.13 g/l[143].

Effect on weight gain

Calves given colostrum with a high Ig concentration (>60 g/l) gained weight from birth to 4 weeks but those given colostrum with a low Ig concentration (<45 g/l) lost weight and had a greater severity and duration of diarrhoea. Serum protein and Ig concentration in the calves were higher for those hand-fed the colostrum containing the high Ig concentration than for those given the colostrum containing the low Ig concentration or allowed to suck their dams for 12 to 24 h[189]. Milk intake as well as live-weight gain to 42 d of age has been correlated with the Ig concentration in colostrum[112].

Estimation of Ig concentration

Since there is a linear relationship between the concentration of Ig and the concentration of solid-not-fat (SNF) in colostrum[236], the concentration of Ig can be obtained from the following formula:

$$Ig(g/kg) = 0.784\ (\pm 0.029)\ SNF(g/kg) - 76.7$$

Measurement of the specific gravity of colostrum using a hydrometer has been advocated as an estimate of the Ig concentration[190]. However, whilst the estimation for first milking colostrum was good, the estimation for the fourth milking gave a value that was only 0.73 of the analytical value for Ig, and at the fifth milking only 0.56 of the value[191].

The use of bovine colostrum for other species

Bovine colostrum can be used for feeding to lambs. Thus, the live-weight gain of lambs given a single feed of bovine colostrum followed by a milk substitute was similar to that of lambs given ovine colostrum followed by the milk substitute or to that of lambs that sucked naturally[192].

However, occasional cases of anaemia in young lambs have been attributed to giving bovine colostrum[233]. It has also been suggested[234] that giving bovine colostrum to unsuckled lambs may not fulfil the same physiological functions as ovine colostrum, since ovine colostrum contains 90 g fat per kg compared to 34–40 g fat per kg in bovine colostrum. The author considered that a high fat concentration is necessary to act as a laxative for the passing of the meconium and suggested supplementing bovine colostrum with 55 g vegetable oil per litre. However, as mentioned below, colostrum in fact appears to have a costive rather than a laxative effect.

The use of excess colostrum

Frequently, a greater quantity of colostrum is produced than the newborn calf can drink. This excess colostrum produced during the first 24 h of life can be diluted in the proportion of two parts of colostrum to one part of warm water, and may then be given in place of whole milk to older calves. Later-drawn colostrum should be diluted with less water. Diluted colostrum should never be fed to newborn calves.

The best results will be obtained if the colostrum is given to older calves soon after it is produced; it will, however, keep for 2–3 d as long as it is kept cold and stored under hygienic conditions. Although older calves given colostrum will tend to have looser faeces than those given whole milk, there is no evidence that this practice will cause diarrhoea[34–39].

Research[40] has shown that colostrum can be stored at ambient temperature and still be suitable for giving calves in place of whole milk or milk substitute diets. After 12 d of storage, the pH becomes stabilized at a value of about 4. Although the nutritional quality of sour colostrum appears to be unaffected, there is about a 50 per cent fall in the total concentration of Ig after 28 d of storage. Sour colostrum must therefore definitely not be used as a source of Ig during the first 2 d after birth.

The first experiment on the use of sour colostrum showed that it could be used continuously or intermittently until weaning without reducing the growth rate to 12 weeks of age. The colostrum was on average 13 d old.

The Ig concentration in milk

In milk the concentrations of Ig per litre were 350 mg IgG_1, 60 mg IgG_2, 40 mg IgM and 50 mg IgA[24]. However, in Israel, the Ig concentration in milk from the second week after calving was constant with mean values per litre of 600 mg IgG, 150 mg IgM and 160 mg IgA. No Ig was detected in a milk substitute[187].

In late lactation IgG_1 concentration increases and is higher in cows of more than third parity. In uninfected quarters the mean IgG_1 concentration was 0.46 g/l, but in *Staphylococcus aureus* and *Corynebacterium bovis* infected quarters the concentration increased to 11.3 g/l. There was no correlation between somatic cell count and IgG_1 concentration in infected quarters but a slight correlation between bovine serum albumin in milk and IgG_1 concentration[188].

Factors affecting Ig concentration in colostrum

Dam factors

Yield

It has been suggested that the concentration of Ig in colostrum will depend on the inherent yield of the cow, higher-yielding cows having lower Ig concentration than, but similar total Ig production to, that of the lower-yielding cows[28]. However, it appears that the inverse relationship between total yield and Ig concentration is far from close. Thus, it seems that there may be some factor associated with high yield which is also associated with an increased transfer of Ig from the blood of the dam to colostrum, possibly some hormonal effect such as from oestrogens, which are known to increase the permeability of the capillaries.

Parity

In Denmark[28, 29], heifers had not only lower yields of colostrum, but also lower yields of Ig than had cows in their second or later lactation. Other work has shown that the Ig concentration is lowest in colostrum from cattle of the first, or first two[107] or three parities[183], and thereafter increases with parity. For instance total Ig in the first 3–4 kg colostrum removed from the udder was 56.8 g/l for first parity colostrum compared with 77.2 g/l for third and fourth parity colostrum[182]. Another report showed that whey protein and IgG concentration increased up to the fourth parity but IgA increased up to the fifth[108].

Quarter of udder

Total Ig concentrations[107], and IgM and IgA concentrations[173], were higher in colostrum obtained from the rear teats.

'Let down'

No difference between the concentration of Ig in colostrum before and after 'let down' has been found[122].

Twin births

No effect on Ig concentration in colostrum has been observed[114].

Abortion

Colostrum of aborted cows is of poor quality and should not be given to newborn calves[185].

Breed

In studies in Denmark on colostrum quantity and quality[83, 84], marked differences occurred between breeds in yield and in concentration of Ig in the first milking after calving. The Black and White Danish breed had a significantly lower colostrum yield but higher IgM concentration than had the Red Danish breed.

In the USA, a comparison of the Ig concentration in the first 3–4 kg colostrum removed after parturition revealed that the total Ig and IgG was highest for the Jersey (90 and 66.5 g/l respectively) and lowest for the Friesian (56 and 41.2 g/l respectively), but IgA and IgM were lowest for the Guernsey (9.0 and 3.9 g/l respectively) and highest for the Jersey (18.6 and 5.3 g/l respectively). It was suggested that the higher calf losses in the Guernsey breed might be related to the low IgA and IgM concentration[182]. In contrast in N. Ireland, the colostrum from Jersey cows did not differ in Ig concentration from that of other breeds[180]. Higher IgG_1 and IgM, but not IgG_2, concentrations have been reported for Blue Grey than for Holstein-Friesian cows[112]. In Italy colostral Ig concentration was significantly lower in the Piedmont than in the Aosta red pied breed[184].

Management factors

Time after parturition

If colostrum is not removed from the udder after parturition, its Ig concentration will not decline for the first 9 h[237]. By 24 h after calving colostrum from unsucked teats contains more Ig than does that from sucked teats[173]. Similarly, in Denmark[28, 29] it was found that an increase in the time interval between calving and milking the cow, even in the absence of its calf, resulted in a reduced Ig concentration in the colostrum.

With a normal milking routine after parturition the concentration of protein, which reflects its Ig concentration, falls exponentially from 140–200 g/l at the first milking to 80–120 g/l at the second, to 50–60 g/l at the third and to 42–44 g/l at the fourth[3,4]. Similarly, the decline in protein and Ig concentration during the first 24 h after calving was from 114 to 58 g/l and from 44 to 8.8 g/l respectively[110]. In Czechoslovakia, the highest concentration of IgG and IgM in colostrum was 69.4 and 7.83 g/l respectively, with the concentrations halving at 12 h after parturition[104].

Length of dry period

A short dry period has a very marked effect in depressing the Ig concentration in colostrum. For instance, four cows that were milked up to 14 d or less before calving had IgG, IgM and IgA concentrations in colostrum of 12–17, 1.7–4.4 and 1.3–4.8 g/l respectively compared with 43.7, 9.4 and 4.5 g/l for 20 normal heifers and cows[180].

Prepartum milking

The composition of colostrum normally changes to that of milk during the first 4 d after calving. If a cow is milked intensively before calving, this change will take place before the calf is born[1]. The extent of this change in composition depends mainly on the total quantity of secretion removed from the udder before calving. If the udder is only partially milked out once or twice before calving, to relieve congestion and discomfort, the secretion after calving differs but little from colostrum, whereas if the udder is completely milked out for a fortnight before calving, the secretion after calving will resemble normal milk.

By intensively pre-milking cows, the concentration of Ig and associated antibodies in the secretion after calving is greatly reduced, and thus the calf will receive a much smaller amount of Ig during the first 24 h of life than when sucking normal colostrum. For instance, the concentration of Ig in the postpartum secretion during the first 24 h after calving may be only 1.6 g/l instead of a value of about 68 g/l. Hence, the secretion produced after calving by cows that have been pre-milked is of less value to the calf than the same volume of normal colostrum[44].

To adequately protect the calves produced, colostrum should be available to the calf from another cow or from a deep-freeze store. When colostrum is for any reason not available for the calf, a tetracycline antibiotic should be given, both by injection and by mouth, as soon after birth as possible. As an alternative, serum from a cow on the farm may be injected, or a substitute for colostrum may be fed (see below) although this method is probably less reliable.

Leakage of colostrum from the udder before calving, depending on its extent, will also reduce the Ig concentration in colostrum[181].

Age at first calving

Independent of parity, increasing age appears to have either no effect[108,114] or to be associated with an increase in Ig concentration[105,110].

Prepartum feeding

It is not possible to influence the concentration of Ig in colostrum by prepartum feeding[114]. High-protein rations before calving lead only to an increase in the non-protein nitrogen fraction[3]. Similarly, a reduction in protein intake during the last 100 d of gestation of beef heifers had no effect on Ig concentration in their serum or colostrum[118].

Season

Colostral Ig concentration is usually unaffected by season, although in W. Germany it has been variously reported as highest in December and January[107] or higher in May, September and October than in other months[105]. Similarly, in W. Germany the concentration of whey protein and IgG in colostrum has been reported to be higher in winter, although the concentration of IgA was unaffected by season[108]. In W. Germany cows calving in the spring had the lowest colostral Ig concentration[110].

Dilution and denaturation

Colostrum from the first milking after calving should be given to the calf at the first feed. This may not happen if a number of cows calve at about the same time and colostrum is bulked. Colostrum bulked from different milkings after calving will have a lower concentration of Ig than will colostrum obtained from the first milking. However, it is not essential for a calf to receive colostrum from its dam, provided that first-milking colostrum is given at the first feed.

Colostrum is sometimes misguidedly diluted with water so that the concentration of Ig is reduced or heated sufficiently to denature the whey proteins including the Ig. This latter may occur if steam is passed into colostrum or some other method is used which allows the temperature to rise above 50°C.

Storage

Colostrum can be stored without reducing its protective value at −18 to −25°C for at least 6 months, provided that it is frozen in small amounts (0.6 litre ≡ 1 pint cartons). Indeed, it has been claimed that colostrum can be kept for 15 years without losing much of its original quality[237]. However, freezing and rethawing colostrum was found to influence its Ig concentration after the fifth repetition of the procedure[108]. From Japan, it was reported that colostrum stored for a year and given at 15, 25 or 35 ml/kg body weight, but not at 5 ml/kg body weight, resulted in an increase in serum Ig concentration in the calf[186].

Addition of either 5 g propionic acid or 5 g lactic acid per litre colostrum stored at 5°C showed that antibody titres to coronavirus, adenovirus (ADV 7) and bovine virus diarrhoea (BVD) virus were stable for 6 weeks but at a concentration of 10 g/l both acids hastened a decline in titres. Not surprisingly with colostrum stored at 20°C, antibody decreased rapidly irrespective of acid addition[186].

It is generally not worth storing colostrum obtained from the second, or possibly, third milkings, since total protein falls from about 150 g/l at the first milking to 90 g/l at the second and 55 g/l at the third.

The quicker the colostrum freezes the better. Since there is a risk of the centre of the colostrum taking some time to freeze, the cartons should not be stacked together until they are frozen. It is imperative to thaw out the colostrum slowly and not allow the temperature to rise above 50°C or the immune globulins and associated antibodies will be denatured. This can be done by standing the cartons in a container of water during the day or overnight. To thaw more rapidly, a porringer, i.e. a saucepan within another containing water, can be used with gentle heat and continuous stirring of the colostrum.

It is not known how much destruction of Ig might occur if a microwave oven is used to warm up colostrum[130].

The only detrimental effect on colostrum of deep-freezing is increased ‘oiling-off’ of the fat, but this does not seem to affect Ig absorption. If ‘oiling-off’ is to be avoided, the colostrum must be homogenized before freezing.

Vaccination

The Ig concentration in colostrum was higher in cows that had received a mixed vaccine at 6 weeks prepartum[107].

Ig intake

This will depend on the intake of colostrum and its Ig concentration. Newborn calves left with their dam ingest about 7–8 kg colostrum on the first day of life, increasing to 10–12 kg by the fourth day; these amounts are much larger than those usually given by bucket feeding of newborn calves removed from their dams at birth[2]. This difference in the amount of colostrum obtained may account partly for the difference in the mortality rate of calves receiving colostrum by these two methods.

In experiments in which colostrum was blended to produce IgG concentrations ranging from 7.3 to 123.8 g/l, IgM concentrations of 0.46–11.1 g/l and IgA concentrations of 0.38–5.53 g/l and in which either 2 litres or 4 litres were given at birth and again at 12 h, the concentration of IgG and IgA in the serum of calves at 24 h was positively linearly related to the concentration in colostrum, whereas IgM concentration in serum was related quadratically. When compared on an equal mass of Ig basis, the amount of colostrum given had less effect on Ig absorption than did the concentration, although it was concluded that Ig concentration, age at first feeding and volume of colostrum intake were the major factors affecting Ig absorption[177]. Other work has suggested that the amount of total γ-globulin or IgG ingested per unit of body weight is the most important factor in determining the Ig concentration in serum[124].

An intake of 4–5 per cent of body weight at 4–6 h after birth, whether given by nipple pail or bucket, gave a serum zinc sulphate turbidity (ZST) value (see p. 39) of 10.3, whilst an additional feed at 8–12 h gave a ZST value of 21.4[55]. When 11.7 kg colostrum, obtained at the first two milkings, was given in three feeds in 24 h followed by 7.8 kg the next day, routine ZST values increased from 8 units before this procedure was adopted to 29 and 21 units respectively for the following 2 years with mortality declining from 8 per cent to 1.8 and 2.3 per cent for the 2 years respectively[103]. Artificially feeding newborn calves with colostrum to appetite at 4–6 h after birth with a second feed 4–6 h later, resulted in 95 per cent of calves having serum values of >15 ZST units[215].

Amount of Ig required

Although as little as 14 g Ig given within 12 h of birth will protect the majority of calves against an *E. coli* septicaemia, and 400 ml of first milking colostrum should protect all calves against this condition[83, 84], such amounts will not protect calves against an *E. coli* localized intestinal infection[219]. Thus, the greater the amount of colostrum ingested, the better the protection afforded to the calf, especially as colostrum is also the main source of fat-soluble vitamins.

To ensure complete protection against a heavy challenge of infection with enteropathogens, including *Salmonella* spp., one would recommend at least 7 kg colostrum, which would supply about 400 g Ig during the first 24–36 h of life (equivalent to 1.7 kg per feed for four feeds from the first two milkings after parturition).

It has been suggested that heifers and cows should produce at least 2 kg colostrum containing not less than 50 g Ig per kg in the first milking (i.e. 100 g Ig) to ensure that a normal Ig concentration is maintained in the calf's blood[28]. It is doubtful whether this amount of Ig is sufficient to protect a calf in an adverse microbiological environment. Indeed, in one survey all calves given <80 g Ig died[178].

It has also been suggested that a minimum quantity of 50 g Ig must be absorbed into the circulation by a 35 kg calf (1.43 g/kg body weight) to protect against neonatal disease since non-diarrhoeic calves had an initial intravascular concentration of 45.7 g Ig and weighed 31.1 kg (1.47 g/kg body weight)[179].

Factors affecting Ig intake

Dystokia or caesarean section

Colostrum intake of dystokia calves or those born by caesarean section, which were showing a respiratory acidosis (plasma pH 6.9–7.15), was lower in the first 8 h of life (3.8 kg) than that of normal calves (5.5 kg) with a plasma pH >7.25[99, 228]. Other work has shown that difficult calvings and weak calves are associated with reduced colostrum ingestion[62].

Diet of dam

Housed beef cows given silage *ad libitum* gave greater yields of colostrum than did outwintered cows receiving no supplementary feeding, but there was no difference in Ig concentration. Of 11 cows on each treatment, seven outwintered and three inwintered gave insufficient colostrum to ensure survival of the calf. Total Ig produced in two milkings was only 76 g for the outwintered cows compared with 144 g for the housed animals[82].

Parity of dam

Because colostrum from heifers had a lower Ig concentration, and calves from heifers were slower to ingest colostrum and also consumed less colostrum than did those from cows, heifers' calves are more at risk of receiving inadequate amounts of Ig[53].

Mothering ability

Reduced mothering ability may result in reduced colostrum ingestion. By reducing the number of suckler beef calves in a building, 'better mothering' and improved supervision at calving were achieved, which, together with giving colostrum by bottle to any calf that had not sucked within 6 h, eliminated hypogammaglobulinaemia[85].

Conformation of dam

Badly shaped udders or very bulbous teats may adversely affect colostrum ingestion.

The absorption of Ig

The Ig ingested by the calf is taken up by the epithelial cells of the small intestine by the process of pinocytosis (engulfing of fluid globules by pseudopodia) and passes into the lymph spaces and then into the blood circulation by way of the thoracic lymph duct[9,10]. Not only the Ig but also the other whey proteins of colostrum can pass into the bloodstream, although the latter appear to be soon excreted by way of the urine[11,12].

Thus, the uptake of macromolecules is in general non-selective, the cessation of transfer occurring at a progressively increased rate after 12 h with a mean closure time at about 24 h. The proportions of the different classes of Ig in calf serum, when absorption is complete, reflect the proportions in the ingested colostrum[124]. However, IgG is preferentially absorbed, and this occurs more quickly than for IgM and IgA. Some hours elapse before Ig appears in the bloodstream, and it is more than 6 h before 50 per cent absorption has been achieved[97].

Taking into account the distribution of Ig in the intravascular and extravascular pools, it was considered that 90 per cent IgG, 59 per cent IgM and 48 per cent IgA were absorbed[56].

However, more recently a computer model has been used to study the transfer from colostrum to serum of IgG_1, IgG_2, IgM, γ-Glutamyl transpeptidase and added D-xylose. γ-Glutamyl transpeptidase (γGT) (EC 2.3.2.2)[238] has been advocated as a test of colostrum intake (see below)[126] and D-xylose as a marker for intestinal absorption of a monosaccharide. It was considered that absorption of IgG_1, IgG_2 and IgM was not selective in the calf and that the transfer efficiencies for IgG_1, IgG_2, IgM, γGT and D-xylose were 46, 49, 47, 18 and 21 per cent respectively[125].

Location of absorption

Uptake of γ-globulin was reported to be greatest in the region that is 15 per cent of the distance from the caecum to the abomasum and decreased progressively towards the abomasum. Uptake was, not surprisingly, greater for intestinal segments exposed to γ-globulin for 1.5 rather than 0.5 h. Uptake, which was a cubic function of the segment position, was 2.14 mg γ-globulin per g intestinal tissue. It is possible that retention of IgG in the tissue may not be the same as absorption[87]. Other work indicated that optimum uptake was in the midgut[88].

Immunofluorescent techniques showed that IgG, IgM and IgA were present in the lamina propria (the layer of connective tissue between the epithelium and the thin layer of smooth muscle), in the villous epithelial cells or on the mucosal surface of the epithelial cells in very young colostrum-fed calves, but by 4 d of age the epithelial cells were devoid of Ig[89].

Using ^{125}I-labelled IgG_1 it was shown that absorption was maximal in the segment represented by 60–80 per cent of the duodeno-caecal length, but there was a significant difference between calves for uptake and absorption (uptake + transfer) but not for transfer. The amount of IgG_1 in the thoracic lymph duct did not reflect the amount that was transferred by the segment, probably due to ^{125}I IgG_1 degradation[90].

Absorption at 6 h occurred in all villous cells, but by 53 h it was restricted to cells at the tip[91]. Thus, pinocytosis of protein by the intestinal epithelium is probably lost afer one or two regenerations of the epithelium.

Transport of Ig

Little is known about the cytological events that occur in the intestinal epithelial cells during Ig absorption. Since there is some selectivity of absorption, this implies that receptors are a necessary component of Ig transport. Selectivity further requires binding of Ig to an endocytic vesicle membrane to ensure transport through the cell, circumvention of intracellular digestion and release at the basolateral cell membrane. The decrease of Ig absorption may occur from competition between intestinal microbes and Ig for a common receptor on the intestinal epithelial cell[220].

Closure time

The globulins with their associated antibodies are only able to pass unchanged into the bloodstream during the first 24 h of the calf's life.

Factors affecting closure time

Ig class

In 1973, it was estimated that the time of closure to absorption of intact Ig from the small intestine varied with class of Ig, from 16 h for IgM, 22 h for IgA to 27 h for IgG. Thus, if colostrum was ingested late, the calf might be deficient in IgM[56]. IgM, which comprises 10 per cent of the total Ig, is of most importance in prevention of a septicaemia[57].

However, more recent work indicates that closure occurs about 24 h for all Ig classes if a calf is not fed. Giving colostrum shortly after birth promotes earlier closure, and a delay in giving colostrum will delay closure up to a maximum of 32 h if calves are first fed at 24 h. Even so, 50 per cent of calves whose first feed was delayed to 24 h were unable to absorb Ig, whereas all calves fed before 12 h absorbed Ig[165].

Intestinal permeability was found to be maximal when age at the initial feed was 13.6 h for IgG, 11.7 h for IgM and 4 h for IgA. The maximum period of absorption lasted for 29.4, 26.3 and 26.0 h for IgG, IgM and IgA respectively[166]. Thus, maximum serum IgG concentrations were reached sooner than for those of IgM, which were in turn reached sooner than for those of IgA. It was suggested that the differences could be due to different intravascular and extravascular distribution of the Ig classes[117].

Effect of epithelial cell replacement

Although the villi are longer and the crypts shorter in newborn than in older calves, epithelial cells produced in the crypts take at least 48 h to reach the tips of the villi in newborn calves. Since calves replace less than half of their villous epithelial cells during the first 48–72 h of life, it is unlikely that cessation of absorption is caused by epithelial replacement in the small intestine, unless absorption occurs near the tips of the villi, which are replaced within the first few hours after birth. In some species, macromolecule uptake continues after 'closure' but transport of the macromolecules into the circulation is prevented[169].

Intestinal flora

The presence and multiplication of bacteria in the small intestine may affect absorption by accelerating cell migration along the villi and reduce closure time[49,168,206]. Microorganisms multiplying within the digestive tract may also degrade Ig[49]. They may also cause thickening of the walls of the intestines. When duodenal fluid was given before the first feed of colostrum, the serum Ig concentration at 24 h was 5.3 g/l instead of 12.4 g/l in the uninoculated calf[49].

Giving a feed before colostrum is given

The suggestion that feeding calves with other foods before colostrum blocks absorption of Ig is unproven. Thus, the feeding of 1 litre of saliva from healthy cows before colostrum was fed gave equivocal results[167].

Season

It has been suggested that the period of permeability is shorter in winter than in summer, and may be as low as 6 h in some calves[7].

The meconium

Once the meconium has been passed completely, no further absorption appears to occur[33].

Meconium, the first intestinal discharge of the newborn calf, differs in appearance and properties from the faeces of later life. It has a viscid sticky consistency and its blackish-green colour is thought to be due to bile pigments. Meconium is considered to be an accumulation of debris that has occurred during fetal life, and it has been shown to have a high concentration of polysaccharides[32]. In the infant's meconium, these polysaccharides were found to be similar to blood-group substances and, in fact, to have blood-group activity.

The length of time after birth that the meconium is present in the alimentary tract of the calf is very similar to that during which the colostral globulins can be transferred unchanged into the bloodstream. The time interval between birth and expulsion of the meconium depends on the quantity of colostrum given, and whether the calf is bucket-fed or sucking its dam. The greater the amount of colostrum given by bucket feeding, the greater the time interval before the meconium is expelled, a finding in contrast to the usually suggested purgative action of colostrum. When a milk substitute containing skim milk that had received a severe heat treatment was first given 4–5 h after birth, the meconium was expelled, on average, by 23 h after birth; when whole milk was given, the period was 33 h. If colostrum from the first two milkings was given throughout the period, expulsion did not occur until 42 h. However, calves that sucked their dams for the first 4 d of life expelled their meconium by 28 h after birth, which might have been due to the greater volume ingested at each nursing, but more probably was due to stimulation by the dam[33]; it is a well-known phenomenon that a cow tends to lick the anal region of her newborn calf and this may stimulate defaecation.

Hormonal status
It has been suggested that calves which are hyperthyroid at birth have a reduced period of absorption of Ig[31]. Other aspects of the hormonal balance of the calf may also be important[11].

Development of digestion
The low acidity in the abomasum, caused by the under-development up to 24 h after birth of the parietal cells of the abomasum that produce hydrochloric acid[13], the absence of enzymes that would normally degrade the proteins[14], the presence of an antitryptic enzyme in colostrum[15] (but see below), and the presence of a non-heat-coagulable protein in colostrum that may act in a manner analogous to that of a surface-active agent[16], may all affect the period of permeability.

Management factors affecting absorption of Ig

Time between birth and first feed

The maximum Ig concentration in serum is linearly related to the amount of Ig ingested and the age at the first feed; the lower the age and the greater the amount of colostrum, at least up to 2 litres, the higher the serum concentration.

Studies in Denmark[28,29] showed that the increase in serum Ig concentration during the first 24 h after feeding was a function of the mass of Ig given to the calf, the time after birth that the colostrum was given and the birth weight, the first two factors being the predominant ones. The efficiency of absorption, assuming a plasma volume of 5 per cent of body weight and a transfer into the extravascular space of 40 per cent of the Ig mass present in the intravascular space during 24 h after birth, was almost entirely a reflection of the age of the calf at first feed, and was unaffected either by the Ig concentration in colostrum or the quantity of Ig given to the calf.

The transfer of Ig was found to be most rapid during the first 4 h after feeding, and then decreased linearly[120]. A delay from 2 h to 20 h in giving colostrum halved the absorption coefficient[29].

With increasing age, there are a decreasing number of intestinal epithelial cells capable of pinocytotic activity and transmission of colostral constituents into the circulation. Since with increasing age there are fewer cells to saturate for maximum absorption, maximum absorption is represented by progressively lower Ig concentrations in serum[170]. Some work has claimed that the quantity of Ig given in the first feed is of greater influence on serum Ig than the time of the first feed[171], whereas other work considers the reverse to be true[106]. Highest serum IgA, IgG and IgM occurred when first sucking was within 3 h of birth and when colostrum contained more than 185 g IgG, 13 g IgM and 15 g IgA per litre[173].

Time to first sucking, 28–650 min (mean 3.8 h), was shortest for multiparous cows. Calves from heifers are slower to ingest colostrum; the shortest time was for heifers calving in December and January[107,173]. In theory, calves should ingest more colostrum sucking their dams than is likely to be offered from a bucket, but in practice 32 per cent failed to suck within 6 h of birth in a study in the UK[175], whilst in the USA, 42 per cent failed to suck within 12–26 h of birth[176]. Of these latter calves, 13 per cent were agammaglobulinaemic but 87 per cent of these survived, possibly because of the local prophylactic effect of colostral Ig in the intestine even though closure had occurred[176].

Delays in sucking due to abnormal behaviour of the dam were found to be restricted to heifers. Offspring of older animals with pendulous udders and those that had difficult calvings were likely to fail to suck. Differences in time of first sucking occurred between the offspring of different sires and low temperatures reduced the calf's desire to suck[175].

For a 12 h delay in feeding colostrum, absorption rates of IgA were reduced by 0.4 per cent, IgG by 1.6 per cent and IgM by 2.3 per cent[172]; it is presumed that these reductions are of percentage units.

In France, on three farms where calves were separated from their dams at birth, many calves did not receive colostrum for 5–6 h. Over 70 per cent of calves showed serum Ig concentrations at 48 h of <15 g/l and this was linked with the delay in feeding colostrum and the small number of daily feeds, so that colostrum intake was much lower than the amount normally ingested by sucking calves[174].

Calves born at night are assumed to have sucked and are removed from the dam the following morning and then may not be given colostrum until the afternoon. As a result the calf may absorb very little Ig. However, time of birth did not affect the transfer of immunity in a study of 360 Friesian calves born on three farms in France[111].

Purchased calves

Purchased calves often have low serum Ig concentrations. Of 1100 calves purchased at auction in the Republic of Ireland, 45 per cent had Ig levels of <16 ZST units[55]; a further report for 1250 purchased calves showed that 40 per cent had serum Ig values of <15 ZST units[215].

Stress

Eight stressors are known to affect the immune system of animals and alter their susceptibility to infectious and non-infectious diseases. They are heat, cold, crowding, mixing, weaning, restricted feeding, noise and restraint[151]. Although ambient temperature, humidity, pain, fear and apprehension have been claimed to affect Ig absorption, it has not been possible to prove that 'stress' is responsible for reducing colostrum Ig absorption or curtailing intestinal permeability.

Early work suggested that increased serum corticosteroid levels in the calf as a result of pain or fear reduced Ig absorption[30]. More recently it is considered that neither exogenous corticosteroids nor induction of endogenous secretion by naturally occurring stressors have caused any reduction in Ig absorption[213]. This is in contrast to neonatal rodents in which stress causes cessation of absorption of macromolecules.

Diarrhoea, on the other hand, does affect serum corticosteroid concentration.

Factors affecting plasma corticosteroid concentration

Age

Bound and free plasma cortisol (hydrocortisone), which is the major glucocorticoid in calves, is high at birth and drops rapidly during the first 5 d of life and then more gradually during the following 10 d to reach a concentration of 2.2 μg/l[202, 209, 210]. The decline occurs in both colostrum-fed and colostrum-deprived calves[203]. Levels in diarrhoeic calves decrease with age, except during the actual diarrhoeal episodes[202].

Dystokia

The concentrations of serum cortisol at birth tend to be lower in dystokia calves[212].

Diarrhoea

Plasma cortisol is higher in diarrhoeic calves[208, 211]. Moreover, when calves were given orally enteropathogenic *E. coli*, whether or not the calves were treated with electrolytes, the plasma corticosterone and aldosterone increased before death, showing that newborn calves were competent to increase cortisol output in response to severe stress and water and electrolyte loss[209, 210].

Cortisol concentrations during the first 4 d of life were higher in those calves that subsequently developed diarrhoea than in those that remained healthy[202].

Effect of plasma corticosteroid concentration

On Ig absorption

In earlier research, the concentration of cortisol at birth did not appear to affect Ig absorption[212]. However, later work showed that increasing serum glucocorticoid concentrations at birth by injections of ACTH tended to increase rather than reduce serum Ig concentration at 48 h[205, 217]. Similarly a higher uptake of ^{125}I-labelled globulin in calves occurred with higher serum corticosteroid concentrations[206]. In other work a negative correlation was found between serum cortisol in the calf and serum Ig. Moreover if the normal rapid decline in serum cortisol with age in newborn calves was delayed by injecting ACTH, then transfer of Ig as measured by the concentration at 12 and 24 h was increased[204].

These findings would conflict with the view that cortisol, being a metabolic decelerator, would cause aberrant synthesis or recycling of the cell membrane receptors for Ig in the small intestines. Failure to recycle receptors would decrease efficiency of absorption of Ig[220].

On cell-mediated immunity

It is thought that high cortisol levels at birth could affect neutrophil function and phagocytosis[203].

On 'weak calf syndrome'

Pathological hypersecretion of cortisol at birth can produce symptoms similar to those of 'weak calf syndrome' (see below)[207].

Induction of parturition

Induction of parturition by long-acting synthetic corticosteroids, given 6–8 weeks before parturition, has been shown to result in reduced IgG_1 transfer to colostrum[60], with no characteristic decrease in serum IgG_1 in the dam before parturition and to reduce the efficiency of absorption of Ig by the calf born of such cows[61] by decreasing the permeability of the small intestine.

No significant difference in Ig uptake by calves was found between quick-acting (62 h before parturition) betamethasone given between 260 d and full term and slow-acting (14 d before parturition) dexamethasone trimethylacetate. Corticosteroids were considered to suppress all aspects of reaction to bacterial invasion, including antibody production. Moreover, there was less opportunity for Ig to accumulate in the udder, and the premature calf being weaker was likely to suck less effectively[216].

However, in contrast, other work showed that induction of parturition by corticosteroids given 3–22 d and full term did not adversely affect colostrum or calf serum Ig compared with those from full-term cows and calves[218].

Sucking versus bucket feeding

The rate of absorption and maximum absorption of Ig was found to be superior in calves that sucked their dams regardless of age at time of sucking or of the amount of colostrum ingested[8]. It is suggested that this is not due to the mothering effect (see above) since calves were only with the dams for short interrupted periods, but rather to the transfer of some labile messenger to stimulate pinocytotic activity in the absorptive cells and rapid absorption; this would also have the advantage of producing early closure of the endocytes to invasion by pathogenic organisms. It is suggested that the factor could be a hormone or an enzyme, but why it should not be transferred in colostrum given by bucket is not explained. With 2.4 litres of colostrum sucked (initial feed of 1.4 litres), serum concentrations of IgG, IgM and IgA were 22, 5.8 and 3.3 g/l respectively compared with concentrations of 15, 3.6 and 2.9 g/l when 4 litres were given by bucket (initial feed 2 litres)[121].

The greater absorption of colostral Ig in calves sucking their dams is not due to a higher concentration of Ig in the teat and gland cistern compared to that in alveolar tissue and collecting ducts since the composition, in respect of Ig, is the same before and after 'let-down'[122].

In spite of the fact that suckled calves normally have higher Ig concentrations at 48 h[106, 123], it has been shown that giving colostrum from a bucket was more effective in ensuring adequate Ig intake than leaving cow and calf together for 2 d postpartum[103].

Early assisted sucking of colostrum to saturation produced a consistently high serum concentration of absorbed Ig with a mean ZST of 27.2 units for 100 calves[181].

Giving colostrum by stomach tube

It has been suggested that if newborn calves refuse to suck, colostrum should be given by stomach tube. Unless the procedure is done carefully, serum Ig will be lower than that achieved with a nipple feeder, since radiography has shown that colostrum given by stomach tube goes straight into the rumen[99]. Even so, adequate absorption of Ig appears to occur, so the preruminant rumen must empty itself efficiently[100]. If the colostrum is to flow down the oesophageal groove to the abomasum, the end of the tube must be located in the oesophagus. If the end of the tube is caudal to the cardia, colostrum will pass into the rumen. On occasions it may be possible to pass the tube down the oesophageal groove into the abomasum. It is suggested that the stomach tube should be inserted to its full length to ensure oesophageal rather than tracheal intubation, and should then be withdrawn slightly, so that it delivers into the thoracic oesophagus[101].

Temperature of colostrum

Neither colostrum intake nor serum Ig concentration in the calf were significantly reduced by giving colostrum at a temperature of 14°C[215]. Thus, two pail feeds of colostrum at 4–6 h and 8–12 h of age at body temperature resulted in slightly higher ZST values (23.9) than when given at an ambient temperature of 14°C (21.9)[55].

Pasteurization of colostrum

Colostrum can be effectively pasteurized at not less than 63°C but not more than 66°C for 30 min without greatly impairing its protective action, although about 15 per cent of the whey proteins will be denatured. However, it has been reported that calves given pasteurized colostrum had higher IgG, but not IgM, concentrations in serum at 12 h but not at subsequent 4 h intervals to 36 h. This possibly arose because bacteria that might interfere with the uptake of Ig were destroyed[115].

Fermented colostrum

Fermented colostrum should certainly not be used as a source of Ig for the young calf, since it results in consistently lower serum IgG levels[77]. Even when allowed to suck colostrum from their dams for 24 h, calves reared on fermented colostrum have gained much less weight during the first 28 d of life than have those given frozen colostrum[78]. Adjustment of the pH of fermented colostrum to that of fresh colostrum by addition of 7 g $NaHCO_3$/kg fermented colostrum improved Ig absorption[116].

Supplementation with potassium isobutyrate

The suggestion[80] that potassium isobutyrate given with colostral whey increased absorption of Ig has not been confirmed. In fact, it was shown to reduce efficiency of absorption from 36 to 25 per cent[81].

Dam factors affecting absorption of Ig

Concentration of Ig in colostrum

Although it has been reported that the concentration of Ig in pooled colostrum had no influence on the rate of absorption and that 2 litres colostrum given at one feed to Holstein calves may be the optimum for maximum pinocytotic activity of the absorptive cells and maximum rate of absorption[120], other work disputes this. In W. Germany as colostral IgG, IgM and IgA concentrations increased from <145, <8.5 and <9 to >185, >13 and >15 g/l respectively, absorption increased from 33, 23 and 39 per cent to 57, 28 and 53 per cent respectively[107].

Class of Ig

The rate of absorption of the different classes of Ig varies, and it is considered that IgM, because of its greater molecular size, is absorbed less efficiently than are IgG and IgA and that the proportion of IgM absorbed, unlike that of IgG and IgA, increases as the amount of IgM ingested decreases. With 1–5 g IgM, absorption exceeded 90 per cent. However there was a wide variation in absorption of Ig from different samples of colostrum of similar Ig concentration[119]. Other work has confirmed that the efficiency of IgM absorption decreases as its intake increases[124].

Trypsin inhibitor

The serum Ig concentration at 24 h was not found to be related to the trypsin inhibitor concentration in colostrum[217].

Age

The IgG_1 concentration in calf serum at 24–48 h increased with the age of the dam[113].

Dystokia

Dystokia at parturition in Herefords[113] or caesarean section[105] was associated with a lower IgG_1 concentration in the calf. However, in other work dystokia did not appear to cause a lower absorption of Ig[212].

Parity

In W. Germany, serum Ig did not differ between calves suckled by heifers or cows[105]. However, another report from the UK showed lower serum Ig in calves from heifers[204]. Moreover, there was a tendency for cows of greater than eight parities to have a higher percentage of calves with low serum Ig levels than was found for younger cows[69]. Whilst a more recent study has shown that serum IgG_1 and IgM at 48 h decreased as the parity of the dam increased, there was no effect on serum IgG_2 and IgA[232].

Conformation

Serum Ig in calves was lower from cows with pendulous udders[204].

Presence of the dam

Calves given colostrum by bottle or bucket but left in the presence of their dams seem to absorb more Ig than do calves given the same amount of colostrum but reared in isolation[53–55, 215]. In the presence of the dam, two pail feeds at 4–6 h and at 8–12 h of age gave a ZST value of 26.1, whereas in the absence of the dam the value was 20.5[55].

Prepartum feeding

In one year, but not in another, an increase in energy intake during late pregnancy resulted in higher serum Ig concentrations in sucking calves[112]. Moreover, reduction of protein intake during the last 100 d of gestation of beef heifers had a marked effect on the absorption of IgG_1 and IgG_2 by the calves. The mean serum IgG_1 concentration for the calves of dams given high and low protein diets was 6.0 and 0.78 g/l respectively at 36 h, but there was no relationship between the serum

IgM concentration of the calf and the protein intake of the dam. It is suggested that the selective decrease in absorption of IgG_1 and IgG_2 might be due to underdevelopment of the jejunal absorptive cells or to a lack of the transport component for absorption[118].

Calf factors affecting absorption of Ig

Intestinal flora

To be effective prophylactically and prevent bacterial adherence in the small intestine, colostrum must be fed before the establishment of the microflora[221], since there is competition between the intestinal microbes and Ig for the common receptor on the intestinal epithelial cell[220]. Thus, lower uptakes of ^{125}I-labelled globulin were associated with high bacterial counts in the intestine[206]. Bacteria that are established in the small intestine before the first colostrum feed can be absorbed by pinocytosis in the same manner as for Ig[222]. Thus, absorption is drastically reduced by concurrent enteric disease, and calves infected at birth will be unable to utilize colostrum properly[97].

Thyroid status

It has been claimed in France that a hyperthyroid state at birth, and probably also during fetal development, reduced the period of intestinal absorption of Ig, since the plasma thyroxine at birth was negatively correlated with IgG_1 absorption. However, thyroxine injections at birth, which caused illness in all calves, did not alter IgG_1 absorption[31]. Moreover, no correlation was found in the UK between thyroxine concentration in the serum of the dam or its calf and the Ig concentration in the serum of the calf[204].

Birth weight

Birth weight has either been shown to have no effect on serum Ig concentration at 24 h[170,217] or for the concentration of serum Ig to be greater in smaller calves[105].

Sex

Serum Ig was found to be higher in heifer than in bull calves. This might possibly be associated with the lower birth weight of the heifer calves[105].

Genetic factors affecting absorption of Ig

Breed

There appear to be breed differences in the absorption of Ig. After feeding Holstein colostrum, containing 42.6 g γ-globulin per litre, at an intake of 2.12 g γ-globulin per kg body weight at 0 and 12 h after birth, serum levels at 24 h for Holstein calves were 10.1 g/l, compared with 7.0 g/l for Ayrshire calves. The efficiency was calculated using a plasma volume of 93 ml/kg body weight[63]; before the first feed of life, the plasma volume is about 64 ml/kg body weight. Similarly, in the UK, Friesian × Ayrshire calves appeared to have a higher efficiency of absorption than did pure Ayrshire calves[54], and in Denmark the Black and White Danish Breed (SDM), the Danish equivalent of the Friesian, is claimed to absorb Ig more efficiently than the Red Danish[29].

In the USA, serum IgG_1 concentration at 24–48 h was greater in Aberdeen Angus than in Red Poll calves which in turn was greater than that in Hereford calves[113].

Higher serum Ig concentrations in suckled calves of the Blue-Grey crossbred than in Holstein-Friesian calves were due to a higher IgG_1 and IgM, but not IgG_2, concentration in the colostrum of the Blue-Grey dams[112]. Serum Ig concentration at 24 h was found to be positively correlated with Ig concentration in colostrum for Hereford × Friesian bull calves, but not for Hereford × Friesian heifer calves nor for Friesian calves of either sex[217].

Genotype of dam and calf within a breed

There is a high degree of repeatability in the Ig produced by the same cow at successive parities[114] and in the transfer of the Ig to her calf[112]. Thus, serum Ig was low in calves born of cows with a history of poor colostral transfer and in calves sired by certain bulls[204].

Calves from a Hereford line that had been selected for high weaning weight, yearling weight and muscle score had lower serum IgG_1 at 24–48 h than did Hereford calves from a randomly selected control line[113].

Environmental factors affecting absorption of Ig

High temperature

Calves exposed to a hot environment with low air movement were found to have a higher mortality rate, higher serum corticoid concentrations and lower serum IgG_1 level at 2 and 10 d of age than had calves in a well-ventilated building. At 2 d of age the IgG_1 serum levels were 25.5 and 18.6 g/l for calves in the well-ventilated and poorly-ventilated building, respectively[30]. In France, the maximum IgG_1 concentration attained was negatively correlated with the environmental temperature at birth[31].

Low temperature

Very low temperature, i.e. in which core body temperature was lowered by 10°C, may decrease the rate of absorption (based on serum concentrations) of all classes of Ig for up to 15 h after first giving colostrum in spite of the body temperature returning to normal within 8 h of the first feed. However by 21 h the concentrations of Ig did not differ[214].

Season

Serum Ig concentration at 1–3 d of age is normally highest in mid-summer and lowest in winter[103,104,109,217], with a positive correlation with mean weekly temperature[217]. A decline in concentration from December to April occurred in W. Germany[102], and from September to February followed by a gradual increase to

April in the UK[103]. The lowest values in Czechoslovakia occurred from February to April[104]. In other work in the UK, serum Ig concentration at 48 h decreased with month of calving from August to January[232]. The highest values in W. Germany occurred in May, September and October[105] and in Czechoslovakia from May to October[104]. However in France, no effect of season on serum Ig was found[106, 111].

Since the Ig concentration in colostrum is usually not affected by season, this gradual reduction in absorption of Ig during the winter calving season may be related to the 'build-up' of infection that occurs in calf houses during the winter months and the effect of the intestinal microflora (see above).

Site of calving

Transfer of Ig appeared to be lower for calves born indoors in winter than for those born outside in summer[204]. If calves are born in a crowded maternity area rather than in a box, serum Ig concentration in the calves will be depressed[217].

Degradation of Ig

It would appear that microorganisms multiplying in the digestive tract may degrade the immunoglobulins from colostrum[49]. Hypercatabolism of IgG_1 in diarrhoeic calves suggests that IgG may be playing a role in detoxification[50].

Half-life

Determination of the half-lives of IgG, IgM and IgA have given varied results, but IgG has the longest. One report gives normal half-lives of 20–21.5 d for IgG, 4 d for IgM and 2–2.8 d for IgA[51] or 18 d for IgG_1[52]. These half-lives were found to be only 11, 3.4 and 3 d for IgG, IgM and IgA, respectively, after calves had been given *E. coli* endotoxin[133]. However, these normal half-lives may be exaggerated, since the half-life of ^{125}I-labelled IgG_1 was found to be 11.5 d, compared with 19.9 d when calculated from the decrease in plasma concentration, a value that is affected by endogenous production of IgG_1[65].

Other work estimated the half-lives of IgG, IgM and IgA to be 19, 2 and 4 d respectively. The short half-life of IgM was compensated by its early endogenous secretion[131, 132]. Since IgM is important for the prevention of septicaemia, this condition is not likely to occur after 1 week of age. A deficiency of IgA could occur between 1 and 2 weeks of age, since the 1-week-old calf is unable to synthesize IgA[131].

Digestion

Although colostral whey passes out rapidly from the abomasum and acid secretion in the neonate is limited, little is known about conditions required for Ig degradation. Although all three main classes of Ig appear in the faeces within 48 h of birth[58], the amounts that survive have not been determined. The IgG_1 is digested easily by pepsin[134, 135] but not by pancreatic enzymes[137] whereas the reverse is true for IgG_2[136].

The Ig concentration in calf serum

This is a measure of the immune status of the calf. Even in suckled beef calves about 23 per cent were found to be hypogammaglobulinaemic[62]. A similar proportion of dairy calves was also found to be hypogammaglobulinaemic[86].

A comparison of the protein components of calf sera before and after a calf has ingested colostrum is shown in Table 2.3. It will be seen that the concentration of the γ-globulin fraction, which differs only slightly from the immune lactoglobulins of colostrum, is negligible or very low before a calf has ingested colostrum, but rises very markedly afterwards.

Table 2.3 Concentration of serum proteins before and after ingestion of colostrum (percentage of total serum proteins)

	Albumin	*Globulin*					*Reference no.*
		α_1	α_2	α_3	β	γ	
At birth (before ingestion of colostrum)	52	1	20	10	16	0.8	20
	57		37		6	0	20
	63		27		8	1.4	21
	49		41		9	1.3	22
24 h postpartum (approx. 2.3 kg colostrum bucket-fed)	38		30		10	22	22
36 h postpartum (suckled)	30		22		7	42	20

The increase in the concentration of γ-globulin is largely at the expense of the albumin. In absolute terms, the γ-globulin increases during the first 24 h of life by about 17 g protein/l serum, the albumin decreases from a value of about 22 g protein/l to 16 g during the same period, and the α- and β-globulins are largely unaffected[21].

In Scotland, at 24 h of age IgG, IgM and IgA serum levels in suckled beef calves were 38, 3.1 and 3.0 g/l respectively, the values declining to 15, 0.8 and 0.2 g/l at 5 weeks of age[62]. In the USA, normal concentrations for Holstein-Friesian calves left with their dams for the first 24 h of life were only 10.6 g IgG_1, 0.93 g IgG_2, 1.71 g IgM and 1.25 g IgA per litre[230]. In Czechoslovakia, the values for serum obtained 24 h after birth were 23.3 ZST units, 17.6 g IgG and 1.8 g IgM per litre[104], values considerably higher than those from the USA[230]. In the UK, no calf sucking unaided within 6 h of birth or assisted to suck soon after this time failed to achieve adequate serum Ig concentrations at 48 h, namely 30–50 g IgG_1, 0.6–0.8 g IgG_2, 1.2–2.6 g IgM and 1.5–2.3 g IgA per litre[232].

Tests for adequate Ig intake

There are several simple tests that can be used to verify that a calf has received adequate colostrum. The most popular test is the zinc sulphate turbidity test.

Zinc sulphate turbidity (ZST) test

The routine for this test for the presence of adequate γ-globulin and associated antibodies in the sera of young calves is as follows: 5 ml solution containing 250 mg zinc sulphate ($ZnSO_4.7H_2O$)/l is added to 0.1 ml calf serum in 1 ml distilled water at pH 7.0. An immediate marked turbidity, followed by precipitation reaching its peak about 1 h later, will indicate that the calf has had adequate colostrum. The solution containing the serum from a calf deprived of colostrum will remain completely clear[45].

Turbidity tests for Ig are a measure of the intravascular concentration of total Ig. It is questionable whether the level of circulating Ig in peripheral blood completely reflects the calf's Ig status. While this is probably so for low intakes of Ig, it is not necessarily true for higher levels of Ig intake, where blood levels may have reached a maximum but Ig is still passing into the tissues. In man 40 per cent IgG and IgA and 80 per cent IgM are intravascular. In cows around parturition IgG_1 and IgG_2 are divided equally between the intravascular and extravascular pools[66].

Since there is a linear relationship between the sum of (IgG + IgM + IgA) and ZST units up to an Ig level of 30 g/l, and since the relationship[68] is:

$$\text{total Ig(g/l)} = 0.04 + 0.98 \text{ ZST units}$$

20 ZST units is equivalent to about 20 g total Ig/l serum or 16 g IgG_1/l serum[129].

Since (a) IgG_1 concentration in whole blood = IgG_1 concentration in plasma (l-haematocrit) and (b) the mean haematocrit of 453 market calves was found to be 0.36, then 16 g IgG_1/l serum is equivalent to 10 g IgG_1/l whole blood[129].

Sodium sulphite turbidity test

The sodium sulphite turbidity test is an alternative to the zinc sulphate turbidity test, and can be used for haemolysed samples[48] and those from dead calves. It involves adding 19 parts 360 g $Na_2SO_3.7H_2O$/l to one part serum[70].

Refractometer

Another alternative is the use of a refractometer to measure the refractive index, which depends on the concentration of protein in the solution. As the albumin concentration is fairly constant, the refractive index can be used as a measure of globulin concentration[71,230].

Immune assay

A simple immune assay system using whole blood has been developed to evaluate the colostrum status in calves and was tested on 650 calves on 10 farms. Two people could test 40 calves per hour. By this test it was found that more than 50 per cent of 7 d old market calves were colostrum-deficient[223].

γGT activity

Plasma γ-glutamyl transferase (γGT) (EC 2.3.2.2) activity, with a concentration in colostrum three times that in milk and 300 times that in serum, can be used as a test for colostrum intake. Colostral activity, which is at a maximum on the first day after calving (2900–27 200 U/l), decreased for the first 6 d after calving. Before sucking, calf γGT was low (10–31 U/l) and reached a maximum on the first and second day (370–5000 U/l) and then decreased slowly[224].

Latex agglutination test

A quantitative latex agglutination test, in which polystyrene latex beads are coated with antibodies against bovine IgG_1, is commercially available. It is based on the relationship between the concentration of IgG_1 and the time taken for the first signs of agglutination to appear[128]. For concentrations of IgG_1 up to 8 g/l there was a highly significant correlation with radial immunodiffusion (0.93) and with ZST test (0.74), although a later report gives a correlation coefficient with the ZST test of 0.64[127]. In addition a good correlation was obtained between the latex agglutination test done on whole blood on the farm and in the laboratory and on plasma in the laboratory[128].

A qualitative test to give positive or negative results on whether IgG_1 concentration was > or <5 g/l whole blood related poorly to the quantitative version, with 65 per cent false positives and 11 per cent false negatives[127]. Using the test, no significant relationship was found between plasma IgG_1 concentration and initial live weight, live-weight gain or disease incidence. Calves treated for infectious disease, particularly respiratory disease after weaning, had a significantly lower live-weight gain than did healthy calves[127].

Comparison of tests

A comparison by the French of the ZST, glutaraldehyde coagulation and sodium sulphite precipitation tests claimed that only the sodium sulphite gave fully satisfactory results. They stated that turbidity occurred with ZST irrespective of γ-globulin concentration[225]. However, it is possible that these workers did not ensure that the distilled water used in the ZST test was at pH 7, which is essential for an accurate result.

Serum Ig concentration in the calf in relation to health

In a survey of published work on the serum Ig concentrations in calves purchased from markets, involving 6566 calves in the UK and USA during the last 20 years, calves with inadequate IgG_1 concentration were four times more likely to die and twice as likely to suffer disease as calves with adequate IgG_1 in their circulation[129].

It has been suggested that the minimum limits for IgG, IgM and IgA in calf serum for health are 5.0, 0.45 and 0.5 g/l respectively[107]. Similarly, calves were at greater risk if the Ig concentration was <5 g/l[226]. However, since the immune status required for protection depends on the level of infection in the environment, it is not possible to be dogmatic about the number of ZST units necessary in calf serum for protection. To give good protection under adverse conditions, the aim should be serum levels of 30 ZST units, since several workers have shown that, on occasions, mortality rates can be very high when serum levels are below 20 ZST units[72–75]. For instance, in a survey[79] of neonatal diarrhoea in home-bred calves on a farm in which 50 per cent of rectal swabs contained pathogenic *E. coli* the post-colostral serum (2–8 d) of healthy calves averaged 23.6 ZST units; those that suffered from non-fatal diarrhoea, 19.3 ZST units; and those that succumbed to diarrhoea, 16.1 ZST units. When calves were left to suck their dams for a longer period, there were fewer calves with low ZST units (<10 units), but there was no reduction in mortality until purchase of calves was discontinued, after which mortality decreased from 12 to 2 per cent.

In contrast, under good environmental conditions, as few as 10 ZST units in the serum have been associated with less than 1 per cent mortality and with beef calves 20 ZST units gave reasonable protection[67]. With suckled beef calves born outside, 26 per cent of calves in eight herds had ZST values of fewer than 20 units, but mortality was low[69].

In Czechoslovakia it was found that, in 76 dead calves aged 3 d to 1 month, those that died from septicaemia or pyaemia had IgG and IgM concentrations of 2.3 and 0.09 g/l respectively, whilst those that died from pneumonia or due to nutritional disorders had concentrations of 8.3 and 0.4 g/l[227]. Similarly, calves with enteritis or omphalitis (inflammation of the umbilicus) had lower serum Ig concentrations[102]. From W. Germany it was reported that whilst sick calves from cows had lower serum Ig concentrations than did healthy calves, there was no difference in concentration between healthy and sick calves from heifers[105].

Healthy calves had significantly higher serum ZST units at 24 h (21.4) than those that developed non-fatal diarrhoea between 0 and 5 weeks (17.7) or died from gastroenteritis or septicaemia (6.5). Calves treated for pneumonia at 0–3 months had lower ZST at 24 h (17.9), but those treated at 3–5 months had higher values (26.2) than those not treated (21.2). Thus, there was some protection against pneumonia and delay in its onset[217].

Herds with mortality rates above average in the USA were shown to have a greater proportion of calves with IgG of <2.5 g/l serum at 1 week of age[226]. However, French workers were unable to correlate Ig concentrations with incidence of diarrhoea or pneumonia or with viability of calves, although calves with <8 g IgG_1/l serum showed a slightly higher incidence of diarrhoea[106]. In W. Germany, calves that developed disease in the first week of life showed a mean serum Ig at 24 h of 19.8 g/l compared with 33.5 g/l for those that remained healthy[228]. Calves with higher body temperatures were found to have lower Ig concentrations[102].

In Chile, a positive correlation between total Ig, IgM and IgA concentrations in the serum of suckled calves and their health was observed. In particular calves that remained healthy had significantly higher IgM concentrations than did those that developed diarrhoea, which occurred at a mean age of 15 d with a mean duration of 3.4 d[231]. However, in other work in the Republic of Ireland, whereas there was a negative relationship between initial serum Ig concentrations and subsequent mortality, there was no relationship with subsequent live-weight gain or incidence of diarrhoea[215].

High blood Ig levels are also important in protection from arthritis, since Ig appears in synovial fluid within 4–8 h of colostrum ingestion[47], although with a lower antibody titre than that in serum. In the lamb[46] both IgG_1 and IgG_2 are transferred with equal facility to the fluid bathing the mucosa of the respiratory tract, and colostrum may thus have a protective effect against respiratory infections.

Serum Ig concentration in relation to live-weight gain

With very high Ig concentrations (30–50 g IgG_1/l), no correlation was found between concentration of any class of Ig and live-weight gain to weaning[232]. However, at the NIRD the greater the serum ZST units after colostrum ingestion the greater the growth rate to 8 weeks of age of early weaned calves. This increased growth rate could largely be accounted for by increased concentrate intake.

Factors affecting Ig concentration in calf serum

Age

The highest serum IgA, which occurred at 24 h, was 44 per cent of the colostral value. Peak values of IgG and IgM were reached at 36 h and were 35 and 57 per cent of that in colostrum. The lowest IgA and IgM values occurred at 4 weeks of age, whilst IgG continued to fall for 3 months. At 3 months, concentrations for IgA and IgG were similar to those for cows but the IgM was only a third of that for a mature animal[172]. IgG concentrations at 1 week were correlated with those at 2–6 weeks but the correlation became poorer with time[226]. In Czechoslovakia, minimum concentrations of IgG and IgM occurred at 4–5 weeks followed by a marked rise. For calves with a low initial Ig concentration, an increase was observed after the first week, but for calves with a high initial Ig, the increase only occurred after the fifth week[104].

In Japan calves allowed to suck their dam during the first week of life had their peak IgM and IgA serum concentrations at 18 h after birth and their peak IgG_1 and IgG_2 concentrations at 30 h after birth. Ig concentration declined from 1–5 weeks and then gradually increased. No correlation was found between the Ig concentration after sucking, incidence of disease or live-weight gain[200].

Breed

Breed differences were found in total serum protein which reached a maximum at 3–4 months of age. The concentrations were significantly higher in Jerseys than in Holsteins[229].

The role of Ig in the intestine

The role of Ig in the lumen of the alimentary tract is not clear. It has been suggested that in calves of <10 d of age, immunological resistance to enteric pathogens is mainly derived from antibodies in the intestinal mucosa[144, 145].

Experiments have shown that if a calf had not initially received sufficient colostrum to protect it from a septicaemia, no benefit occurred from feeding colostrum after the intestine was impermeable to the transfer of antibodies. However, if the calf had been thus protected, additional colostrum after the period of permeability tended to reduce mortality and diarrhoea associated with an *E. coli* infection[33].

It has been suggested that the prophylactic role of colostrum in the lumen of the intestine is exerted by IgM and IgA[122]. Other work has suggested that IgG and IgM are of comparable activity in prevention but not of curing enteric infection, with IgA being less effective[58, 59]. However, since all three classes of Ig can be found adhering to the luminal surface of epithelial cells, they may play a synergistic role in prevention of enteric disease[97].

In Australia, calves that were allowed to suck their dams for 2 d were thereafter given either whole milk, whole milk supplemented with 10 per cent of normal colostrum or colostrum from cows vaccinated with rotavirus. The calves were affected by a natural outbreak of diarrhoea but feeding the immune colostrum delayed the onset of diarrhoea and reduced its incidence, duration and severity, and improved live-weight gain to a significantly greater extent than did whole milk and tended to be greater than for normal colostrum. The calves with diarrhoea were found to be excreting rotavirus, enteropathogenic *E. coli* and cryptosporidia[140].

IgA has been shown to have a localized action in the intestinal tract in man in preventing multiplication of bacteria in the duodenum[25,26], but the role of IgA in the alimentary tract of the calf has not yet been established. In the newborn piglet there is selectivity against the absorption of IgA from colostrum, and the antibody function of this molecule is confined to the lumen of the intestine, but in the calf there is no selectivity and the IgA after absorption apparently passes to the gastrointestinal tract through external secretions; these may thus have a protective action[27].

However, unlike in man and a number of other species, the secretory immune system in the calf intestine seems to consist largely of IgG_1 with relatively small amounts of IgA, whereas in the respiratory tract it seems that IgA may provide surface protection and IgG_1 in plasma cells beneath the respiratory epithelium may form a second line of defence.

IgA in faeces steadily declined over the first 2 weeks of life[132]. The necessity to prevent interference with rumen fermentation was considered the reason for the low IgA concentration in colostrum, but rumen fermentation does not start before 7–10 d of age. Moreover, IgA deficiency in secretions is overcome partly by enhanced local synthesis and secretion of IgM which has been associated with a secretory component[143].

Faecal excretion of IgG has been shown to be inversely related to serum IgG and IgA concentrations and to be greatest in those calves that die. It is considered that this is either due to malabsorption of Ig because of early bacterial establishment in the intestine or possibly to intraluminal leakage of Ig[142].

Although high serum Ig concentrations have been associated with reduced susceptibility of calves to rotavirus infection[138], absorbed antibodies apparently do not give protection whereas maternal antibodies remaining in the intestine are effective. Thus, for this condition in particular, there is benefit from continued feeding of colostrum even after closure[138,139,141].

Ig synthesis in the calf

Age

Whereas hypogammaglobulinaemic calves begin synthesizing serum Ig within a week of birth, calves with high Ig levels after birth do not begin to produce their own autogenous Ig until 4 weeks of age[62,76]. If only one class of Ig is fed, then the autogenous appearance of only that class will be delayed[58]. Other workers have also found endogenous antibody production to occur much earlier in colostrum-deprived than in colostrum-fed calves, the latter being markedly unresponsive to antigens injected at birth if maternal antibody specific for that antigen was present in the circulation[76].

Thus, calves that survive after being deprived of passive immunity from colostral γ-globulins begin producing their own autogenous globulins to *E. coli* from about 10 d of age, the serum γ-globulins reaching normal levels at about 8 weeks of age[17–19]. More recently the synthesis of IgA and IgM has been detected as early as 4–7 d of age[92,200].

Using ^{125}I-labelled IgG_1, it was found that no intestinal absorption occurred from 2 d of age. The origin of new IgG_1 in the calf after about 36 h was from endogenous production at the rate of about 1 g IgG_1/d. The amount produced was considered dependent on the antigenic stress to which the calf had been exposed[95].

Site

In colostrum-deprived calves, intestinal Ig is synthesized in the Peyer's patches (a collection of many lymphoid nodules packed together to form an oblong elevation on the mucous membrane of the small intestine) and crypt region of the small intestine mucosa, being found in the crypt epithelial cells and in the plasma cells of the surrounding lamina propria[89]. The immunocytes (plasma cells) which infiltrate the lamina propria are mainly IgM and IgA-synthesizing cells, but there are also a few IgG, mainly IgG_2, cells. However, in one calf IgG_1 activity was found in the ileum and colon[92]. The plasma cells did not begin to infiltrate the lamina propria of the intestinal mucosa until 3–4 d of age[92].

These findings are at variance with the views of others who consider the IgM-producing cells in the lamina propria of the intestine and free IgM molecules in the intestinal tissues and secretions are important components of the early development of the local immune system prior to or during the eventual changeover to the IgA secretory system[96]. Meanwhile the role of IgA in mucosal defence is assumed by IgG_1[96].

Stimulation

Stimulation of the intestinal mucosa with heat-killed Gram-negative bacteria produced secretory antibodies over a period of 3 weeks. A second administration produced a response indicating that a continuous stimulus was necessary to maintain antibody secretion[92]. By oral administration of antigens to colostrum-fed calves from 5–8 d of age, faecal antibodies were produced. This indicated that intestinal synthesis of antibodies could be successfully interrelated with declining passive immunity to maintain a continuous level of intestinal antibody in early life[92]. Similarly, calves quickly produce an immune response to rotavirus, antibodies appearing in the faeces within a few days[97].

In fact, calf fetuses have been stimulated to produce Ig by prenatal vaccination. By depositing bacterial antigen (killed *E. coli*) or viral antigen (live reovirus) in amniotic fluid, an immune response occurred in 10–14 d for bacterial antigen and at 8–10 d for viral antigen. Calves that were vaccinated with *E. coli* antigen from 9–102 d before birth and were then colostrum-deprived, survived oral doses of viable *E. coli* that killed calves not vaccinated. Immunofluorescence showed IgG, IgM and anti-*E. coli* antibody in the duodenum, jejunum and ileum as well as in the jejunal lymph nodes. The procedure, however, was not entirely successful as some stillbirths and premature births occurred[93].

Beef calves vaccinated *in utero* with *E. coli* O26:K60:NM in the last 6 weeks of gestation were found to be devoid of IgA at birth, but IgA was present after 9 d[98].

Non-specific inhibitors to bovine enteroviruses have been found in serum, allantoic and amniotic fluid of 53 bovine fetuses of 90–240 d. Serum IgG was mostly within normal values. No IgG or IgM was detected in amniotic fluid, but two samples of allantoic fluid contained traces of IgG[94].

Cell-mediated immunity

The lymphocytes present in colostrum may be of considerable importance to the newborn calf[43]; they are certainly considered to have significant clinical implications in man[146]. A marked increase occurs in blood neutrophils between 6 and 12 h after birth in colostrum-fed but not in colostrum-deprived calves and phagocytosis is much more efficient in the former[203]. Colostrum-fed compared with colostrum-deprived calves had an improved ability to respond to intravenous *E. coli* endotoxin by increases in neutrophil and lymphocyte counts and partial thromboplastin time[149]. In fact, leucocyte and differential neutrophil counts are correlated with serum IgG_1 concentrations at 24 h of age[217]. However, the concentration of B lymphocytes (lymphocytes with surface antibodies) in the circulation in 6-day-old calves was only 33 per cent of that in adult cattle[150].

The titre of haemolytic complement activity in calves immediately after birth was 99 and one day later dropped to 39. (Complement is a thermolabile substance in normal serum which is destructive to bacteria and other cells when it is brought into contact with antibody, which has an affinity to both complement and to cells.) After about 4 weeks the titre reached the precolostral value which was much lower than that of the adult (200). A similar pattern occurred in the third component of complement (C_3) whose concentration at birth was 28 per cent of the adult value[147]. Colostral whey, without the presence of complement from unheated cow's milk or precolostral calf serum, had no bactericidal effect against *E. coli* O111[148].

The effect of the physical environment on cell-mediated immunity

The metabolism of lymphoid tissue and immune function may be affected by the thermal environment as well as by endocrine changes and the psychological and physical well-being of the calf[152]. Survival of microorganisms may also be affected by the thermal environment, for instance airborne particles of bovine rotavirus survived longer at 10°C than at 20°C or 30°C[153].

Exposure of calves to 1°C rather than to 23°C for 3 d did not affect the bactericidal effect of serum, neutrophils or lymphocytes[157–159]. However, whilst exposure of calves to −5°C had no effect on the Ig concentration in plasma, the proportion of neutrophils was increased, and the proportion of lymphocytes and cell-mediated immune function were decreased[154, 155]. Further work has shown that exposure of calves to such severe cold that the colon temperature was decreased by 10°C caused a decreased concentration of lymphocytes associated with a neutropenia[156].

Non-specific antimicrobial factors in colostrum and milk

The lactoperoxidase system

The lactoperoxidase system, which prevents attachment of *E. coli* possessing K88 and K99 to the brush border of the small intestine consists of three components, the enzyme lactoperoxidase (LP) (EC 1.11.1.7)[238], thiocyanate and hydrogen peroxide[162]. The lactoperoxidase is derived from cow's milk and from calf saliva, potassium thiocyanate is secreted by the abomasum and hydrogen peroxide is produced by species of lactobacillus. The optimum pH for the bactericidal effect of the lactoperoxidase system is 5, or possibly somewhat less because of the synergistic bactericidal effect of hydrochloric acid[163, 164]. The LP reacts with hydrogen peroxide and oxidizes thiocyanate to a short-lived intermediary oxidation product, hypothiocyanite, which is antibacterial. To activate the system, 25 ml of a solution containing 1 g potassium thiocyanate and 300 g/l glucose and 20 ml of a solution containing 0.5 g glucose oxidase to produce the hydrogen peroxide were used.

In Sweden and in the UK, activating the LP system increased live-weight gain during the first 3 weeks of life of calves given whole milk but not of those given a milk substitute diet, because processing of skim milk destroys the LP. No significant effect on the incidence of diarrhoea and mortality was found[160].

Lysozyme (EC 3.2.1.17)[238] and lactoferrin

The role of these factors in colostrum has yet to be elucidated. The bactericidal effect of precolostral calf serum is considered to be due to the presence of both transferrin and bicarbonate as well as a low level of citrate. Citrate competes with iron-binding proteins for iron and makes iron available to bacteria. Bicarbonate is required for the binding of iron by transferrin and lactoferrin and can overcome the effect of citrate. The lack of inhibition of bacteria by undiluted whey is considered to be due to the high concentration of citrate[161].

Substitutes for colostrum

If colostrum is not available from the calf's dam or from another cow in the same herd, it has been suggested that the following treatment will assist in keeping a calf alive:

'Give the following mixture three times per day for the first 3–4 d of life: one whipped egg in 0.3 litres water to which is added half a teaspoonful of castor oil and 0.6 litres whole milk.'

The scientific basis for this treatment may lie in the finding that egg white has a marked antibacterial action on at least certain strains of *E. coli*[41], and it has also been shown that egg albumin, like the globulins of colostrum, can pass unchanged into the bloodstream of the calf during the first 24 h of life[11]. This entry into the bloodstream via the lymphatic system[9] is possibly helped by the large amount of the emulsifying agent, lecithin, in egg yolk.

An experiment at Shinfield[42] gave a strong indication that one whole egg given daily for the first 10 d of life to calves deprived of colostrum had a protective action against the septicaemic form of infection by *E. coli*. As it is doubtful whether it will be successful in every case, it is probably wiser, if possible, to obtain some colostrum from a neighbouring farm or to use an antibiotic as soon after birth as possible.

References

1. ROWLAND, S. J., ROY, J. H. B., SEARS, H. J. and THOMPSON, S. Y. *J. Dairy Res.* **20**, 16 (1953)
2. WALKER, D. M. *Bull. Anim. Behav.* **8**, 5 (1950)
3. PARRISH, D. B., WISE, G. H., HUGHES, J. S. and ATKESON, F. W. *J. Dairy Sci.* **31**, 889 (1948)
4. PARRISH, D. B., WISE, G. H., HUGHES, J. S. and ATKESON, F. W. *J. Dairy Sci.* **33**, 457 (1950)
5. GLANTZ, P. J., DUNNE, H. W., HEIST, C. E. and HOKANSON, J. F. *Bull. Pa agric. Exp. Stn* No. 645 (1959)
6. SMITH, H. W. *J. Path. Bact.* **84**, 147 (1962)
7. FISHER, E. W. *Vet. Rec.* **77**, 1482 (1965)

8. SMITH, H. W., O'NEIL, J. A. and SIMMONS, E. J. *Vet. Rec.* **80**, 664 (1967)
9. COMLINE, R. S., ROBERTS, H. E. and TITCHEN, D. A. *Nature, Lond.* **167**, 561 (1951)
10. COMLINE, R. S., ROBERTS, H. E. and TITCHEN, D. A. *Nature, Lond.* **168**, 84 (1951)
11. DEUTSCH, H. F. and SMITH, V. R. *Am. J. Physiol.* **191**, 271 (1957)
12. BANGHAM, D. R., INGRAM, P. L., ROY, J. H. B., SHILLAM, K. W. G. and TERRY, R. J. *Proc. R. Soc.* **B149**, 184 (1958)
13. HILL, K. J. *Q. Jl exp. Physiol.* **41**, 421 (1956)
14. SMITH, V. R. and ERWIN, E. S. *J. Dairy Sci.* **42**, 364 (1959)
15. LASKOWSKI, M. Jr. and LASKOWSKI, M. *Fedn Proc. Fedn Am. Socs exp. Biol.* **9**, 194 (1950)
16. BALFOUR, W. E. and COMLINE, R. S. *J. Physiol., Lond.* **160**, 234 (1962)
17. HOWE, P. E. *J. biol. Chem.* **53**, 479 (1922)
18. HANSEN, R. G., PHILLIPS, P. H., WILLIAMS, J. W. and SMITH, V. R. *J. Dairy Sci.* **29**, 521 (1946)
19. PIERCE, A. E. *Colston Pap.* **13**, 189 (1961)
20. JAMESON, E., ALVAREZ TOSTADO, C. and SORTOR, H. H. *Proc. Soc. exp. Biol. Med.* **51**, 163 (1942)
21. PIERCE, A. E. *J. Hyg., Camb.* **53**, 247 (1955)
22. HANSEN, R. G. and PHILLIPS, P. H. *J. biol. Chem.* **171**, 223 (1947)
23. MICHAEL, J. G. and ROSEN, F. S. *J. exp. Med.* **118**, 619 (1963)
24. MACH, J-P. and PAHUD, J.-J. *J. Immun.* **106**, 552 (1971)
25. HERSH, T., FLOCH, M. H., BINDER, H. J., CONN, H. O., PRIZONT, R. and SPIRO, H. M. *Am. J. clin. Nutr.* **23**, 1595 (1970)
26. PRIZONT, R., HERSH, T. and FLOCH, M. H. *Am. J. clin. Nutr.* **23**, 1602 (1970)
27. PORTER, P. *Proc. Nutr. Soc.* **32**, 217 (1973)
28. KRUSE, V. *Anim. Prod.* **12**, 619 (1970)
29. KRUSE, V. *Anim. Prod.* **12**, 627 (1970)
30. STOTT, G. H., WIERSMA, F., MENEFEE, B. E. and RADWANSKI, F. R. *J. Dairy Sci.* **59**, 1306 (1976)
31. CABELLO, G. and LEVIEUX, D. *Annls Rech. vét.* **9**, 309 (1978)
32. RAPOPORT, S. and BUCHANAN, D. J. *Science, N.Y.* **112**, 150 (1950)
33. ROY, J. H. B. Studies in calf nutrition with special reference to the protective action of colostrum. *Ph.D. Thesis,* University of Reading (1956)
34. ALLEN, N. N. *Bull. Vt agric. Exp. Stn* No. 520, p.18 (1944)
35. JACOBSON, W. C., CONVERSE, H. T., WISEMAN, H. G. and MOORE, L. A. *J. Dairy Sci.* **34**, 905 (1951)
36. PAYNE, W. J. A. *Scott. Agric.* **32**, 186 (1953)
37. KEYES, E. A., PEACE, E. J. and BRENCE, J. L. *J. Dairy Sci.* **37**, 655 (1954)
38. GAUNYA, W. S., MOCHRIE, R. D., EATON, H. D. and JOHNSON, R. E. *J. Dairy Sci.* **37**, 655 (1954)
39. WING, J. M. *J. Dairy Sci.* **41**, 1434 (1958)
40. SWANNACK, K. P. *Anim. Prod.* **13**, 381 (1971)
41. FLEMING, A. and ALLISON, V. D. *Lancet* **i**, 1303 (1924)
42. ROY, J. H. B., SHILLAM, K. W. G. and HAWKINS, G. M. *Rep. natn. Inst. Res. Dairy,* p. 51 (1956)
43. PARMELY, M. J. and BEER, A. E. *J. Dairy Sci.* **60**, 655 (1977)
44. ASCHAFFENBURG, R., BARTLETT, S., KON, S. K. *et al. Br. J. Nutr.* **5**, 343 (1951)
45. ASCHAFFENBURG, R. *Br. J. Nutr.* **3**, 200 (1949)
46. WELLS, P. W., DAWSON, A. McL., SMITH, W. D. and SMITH, B. S. W. *Vet. Rec.* **97**, 455 (1975)
47. STONE, S. S. and DEYOE, B. L. *Am. J. vet. Res.* **23**, 1259 (1974)
48. FISHER, E. W. and MARTINEZ, A. A. *Vet. Rec.* **96**, 113 (1975)
49. JAMES, R. E., POLAN, C. E., BIBB, T. L. and LAUGHON, B. E. *J. Dairy Sci.* **59**, 1495 (1976)
50. McDOUGALL, D. F. The metabolism of plasma proteins in the young calf. *PhD Thesis,* University of Glasgow (1973)
51. LOGAN, E. F., PENHALE, W. J. and JONES, R. A. *Res. vet. Sci.* **14**, 394 (1973)
52. McDOUGALL, D. F. and MULLIGAN, W. *J. Physiol., Lond.* **201**, 77P (1969)
53. SELMAN, I. E., McEWAN, A. D. and FISHER, E. W. *Res. vet. Sci.* **12**, 1 (1971)
54. SELMAN, I. E., McEWAN, A. D. and FISHER, E. W. *Res. vet. Sci.* **12**, 205 (1971)
55. FALLON, R. J. *Anim. Prod.* **24**, 142 (1977)
56. PENHALE, W. J., LOGAN, E. F., SELMAN, I. E., FISHER, E. W. and McEWAN, A. D. *Annls Rech. vét.* **4**, 223 (1973)
57. BYWATER, R. J. and LOGAN, E. F. *J. comp. Path.* **84**, 599 (1974)

58. LOGAN, E. F., STENHOUSE, A. and ORMROD, D. J. *Res. vet. Sci.* **17**, 290 (1974)
59. LOGAN, E. F., STENHOUSE, A., PENHALE, W. J. and ARMISHAW, M. *Vet. Rec.* **94**, 386 (1974)
60. BRANDON, M. R., HUSBAND, A. J. and LASCELLES, A. K. *Aust. J. exp. Biol. med. Sci.* **53**, 43 (1975)
61. HUSBAND, A. J., BRANDON, M. R. and LASCELLES, A. K. *Aust. J. exp. Biol. med. Sci.* **51**, 707 (1973)
62. LOGAN, E. F., McBEATH, D. G. and LOWMAN, B. G. *Vet. Rec.* **94**, 367 (1974)
63. McEWAN, A. D., FISHER, E. W. and SELMAN, I. E. *Res. vet. Sci.* **9**, 284 (1968)
64. PENHALE, W. J. and CHRISTIE, G. *Res. vet. Sci.* **10**, 493 (1969)
65. SASAKI, M., DAVIS, C. L. and LARSON, B. L. *J. Dairy Sci.* **60**, 623 (1977)
66. SASAKI, M., DAVIS, C. L. and LARSON, B. L. *J. Dairy Sci.* **59**, 2046 (1976)
67. BARBER, D. M. L. *Vet. Rec.* **98**, 121 (1976)
68. FISHER, E. W. and MARTINEZ, A. A. *Vet. Rec.* **98**, 31 (1976)
69. LOGAN, E. F and GIBSON, T. *Vet. Rec.* **97**, 229 (1975)
70. GITTER, M. *Vet. Rec.* **96**, 255 (1975)
71. REID, J. F. S. and MARTINEZ, A. A. *Vet. Rec.* **96**, 177 (1975)
72. GAY, G. C., ANDERSON, N., FISHER, E. W. and McEWAN, A. D. *Vet. Rec.* **77**, 148 (1965)
73. McEWAN, A. D., FISHER, E. W. and SELMAN, I. E. *J. comp. Path.* **80**, 259 (1970)
74. FISHER, E. W. *Ann. N.Y. Acad. Sci.* **176**, 64 (1971)
75. THOMAS, L. H. and SWANN, R. G. *Vet. Rec.* **92**, 454 (1973)
76. HUSBAND, A. J. and LASCELLES, A. K. *Res. vet. Sci.* **18**, 201 (1975)
77. SNYDER, A. C., SCHUH, J. D., WEGNER, T. N. and GEBERT, J. R. *J. Dairy Sci.* **57**, 641 (1974)
78. PLOG, J., HUBER, J. T. and OXENDER, W. *J. Dairy Sci.* **57**, 642 (1974)
79. BOYD, J. W., BAKER, J. R. and LEYLAND, A. *Vet. Rec.* **95**, 310 (1974)
80. HARDY, R. D. *J. Physiol., Lond.* **204**, 607 (1969)
81. BAUMWART, A. L., BUSH, L. J. and MUNGLE, M. *J. Dairy Sci.* **60**, 759 (1977)
82. LOGAN, E. F. *Br. vet. J.* **133**, 120 (1977)
83. ASCHAFFENBURG, R., BARTLETT, S., KON, S. K. *et al. Br. J. Nutr.* **3**, 196 (1949)
84. ASCHAFFENBURG, R., BARTLETT, S., KON, S. K., *et al. Br. J. Nutr.* **5**, 171 (1951)
85. McBEATH, D. G. and LOGAN, E. F. *Vet. Rec.* **95**, 466 (1974)
86. McBEATH, D. G., PENHALE, W. J. and LOGAN, E. F. *Vet. Rec.* **88**, 266 (1971)
87. JAMES, R. E., POLAN, C. E. and McGILLIARD, M. L. *J. Dairy Sci.* **62**, 1415 (1979)
88. EL NAGEH, M. M. *Annls Méd. vét.* **111**, 384 (1967)
89. LOGAN, E. F. and PEARSON, C. R. *Annls Rech. vét.* **9**, 319 (1978)
90. FLETCHER, A., GAY, C. C., McGUIRE, T. C., BARBEE, D. D. and PARISH, S. M. *Am. J. vet. Res.* **44**, 2149 (1983)
91. EL NAGEH, M. M. *Annls Méd. vét.* **111**, 400 (1967)
92. ALLEN, W. D. and PORTER, P. *Clin. exp. Immun.* **21**, 407 (1976)
93. SCHMOLDT, P., LEMKE, P., BUNGER, U., RIECK, H. and ROTARMUND, H. *Mh. VetMed.* **32**, 11 (1977)
94. ROSSI, C. R., KIESEL, G. K. and HUBBERT, W. T. *Cornell Vet.* **66**, 381 (1976)
95. DEVERY, J. E., DAVIS, C. L. and LARSON, B. L. *J. Dairy Sci.* **62**, 1814 (1979)
96. OLSON, D. P. and WAXLER, G. L. *Am. J. vet. Res.* **38**, 1177 (1977)
97. DEPARTMENT OF AGRICULTURE N. IRELAND. *Ann. Rep. Res. Tech. Work.* **176** (1978)
98. YAMINI, B. *Diss. Abstr. Int., Sect. B.* **37**, 709 (1976)
99. ZAREMBA, W. *Prakt Tierarzt* **64**, 977 (1983)
100. MOLLA, A. *Vet. Rec.* **103**, 377 (1978)
101. WENHAM, G. and ROBINSON, J. J. *Vet. Rec.* **104**, 199 (1979)
102. FRERAING, H. and AEIKENS, T. *Annls Rech. vét.* **9**, 361 (1978)
103. BARBER, D. M. L. *Vet. Rec.* **104**, 385 (1979)
104. VAJDA, V. and SLANINA, L. *Vet. Med., Praha* **25**, 527 (1980)
105. BERGER, W. Comparison of the immunoglobulin contents of colostrum and calf blood. *Thesis*, Tierärztl. Hochsch., Hannover (1979)
106. LOMBA, F., FUMIERE, I., TSHIBANGU, M., CHAUVAUX, G. and BIENFET, V. *Annls Rech. vét.* **9**, 353 (1978)
107. SCHMIDT, F. W., KIM, J. W., DERENBACH, J. and LANGHOLZ, H. J. *Tierärztl. Umsch.* **37**, 485 (1982)
108. LAMBRECHT, G. Immunoglobulin and milk protein content in the first colostrum of cows with particular reference to environmental conditions and heredity. *Thesis*, Tierärztl. Hochsch., Hannover (1980)

109. WILLIAMS, P. E. V., WRIGHT, C. L. and DAY, N. *Br. vet. J.* **136**, 561 (1980)
110. KRAMER, U. Immunoglobulin content of colostrum ingested by calves on the first day after birth during October 1974–April 1975 with special reference to the health of calves. *Thesis*, Tierärztl. Hochsch., Hannover (1977)
111. TSHIBANGU, M. L., FUMIERE, I., CHAUVAUX, G., LOMBA, F. and BIENFET, V. *Revue Agric., Brux.* **35**, 3207 (1982)
112. HALLIDAY, R., RUSSEL, A. J. F., WILLIAMS, M. R. and PEART, J. H. *Res. vet. Sci.* **24**, 26 (1978)
113. MUGGLI, N. E., HOHENBOKEN, W. D., CUNDIFF, L. V. and KELLEY, K. W. *J. Anim. Sci.* **59**, 39 (1984)
114. DARDILLAT, J., TRILLAT, G. and LARVOR, P. *Annls Rech. vét.* **9**, 375 (1978)
115. BUSH, L. J., CONTRERAS, R., STALEY, T. E. and ADAMS, G. D. *Okla. Agric. Exp. Stn Anim. Sci. Res. Rep.* MP-112 p. 246 (1982)
116. FOLEY, J. A., HUNTER, A. G. and OTTERBY, D. E. *J. Dairy Sci.* **61**, 1450 (1978)
117. LOGAN, E. F., McMURRAY, C. H., O'NEILL, D. G., McPARLAND, P. J. and McRORY, F. J. *Br. vet. J.* **134**, 258 (1978)
118. BLECHA, F., BULL, R. C., OLSON, D. P., ROSS, R. H. and CURTIS, S. *J. Anim. Sci.* **53**, 1174 (1981)
119. STOTT, G. H. and MENEFEE, B. E. *J. Dairy Sci.* **61**, 461 (1978)
120. STOTT, G. H., MARX, D. B., MENEFEE, B. E. and NIGHTENGALE, G. T. *J. Dairy Sci.* **62**, 1766 (1979)
121. STOTT, G. H., MARX, D. B., MENEFEE, B. E. and NIGHTENGALE, G. T. *J. Dairy Sci.* **62**, 1908 (1979)
122. STOTT, G. H., FLEENOR, W. A. and KLEESE, W. C. *J. Dairy Sci.* **64**, 459 (1981)
123. LOMBA, F., SPOONER, R. L., MILLAR, P., CHAUVAUX, G. and BIENFET, V. *Annls Méd. vét.* **122**, 101 (1978)
124. BUSH, L. J. and STALEY, T. E. *J. Dairy Sci.* **63**, 672 (1980)
125. BOYD, J. W. and BOYD, A. J. *Res. vet. Sci.* **43**, 291 (1987)
126. BRAUN, J. P., TAINTURIER, D., LAUGIER, C., BENARD, P., THOUVENOT, J. P. and RICO, A. G. *J. Dairy Sci.* **65**, 2178 (1982)
127. CALDOW, G. L., WHITE, D. G., KELSEY, M., PETERS, A. R. and SOLLY, K. J. *Vet. Rec.* **122**, 63 (1988)
128. WHITE, D. G. *Vet. Rec.* **118**, 68 (1986)
129. WHITE, D. G. and ANDREWS, A. H. *Vet. Rec.* **119**, 112 (1986)
130. SENIOR, J. *Vet. Rec.* **116**, 576 (1985)
131. FISHER, E. W. and MARTINEZ, A. A. *Br. vet. J.* **134**, 231 (1978)
132. PORTER, P. *Immunology* **23**, 225 (1972)
133. FISHER, E. W. and MARTINEZ, A. A. *Vet. Rec.* **96**, 527 (1975)
134. KUMANO, Y., KANAMARU, Y., NIKI, R. and ARIMA, S. *Jap. J. zootech. Sci.* **47**, 554 (1976)
135. KUMANO, Y., KANAMARU, Y., NIKI, R. and ARIMA, S. *Jap. J. zootech. Sci.* **47**, 543 (1976)
136. KANAMARU, Y., NIKI, R. and ARIMA, S. *Agric. biol. Chem.* **41**, 1203 (1977)
137. BROCK, J. H., PINEIRO, A. and LAMPREAVE, F. *Annls Rech. vét.* **9**, 287 (1978)
138. WOODE, G. N. *Vet. Rec.* **103**, 44 (1978)
139. SNODGRASS, D. R. and WELLS, P. W. *Annls Rech. vét.* **9**, 335 (1978)
140. SNODGRASS, D. R., STEWART, J., TAYLOR, J., KRAUTIL, F. L. and SMITH, M. L. *Res. vet. Sci.* **32**, 70 (1982)
141. HENDERSON, W. M. *Br. vet. J.* **132**, 136 (1976)
142. FISHER, E. W., MARTINEZ, A. A., TRAININ, Z. and MEIROM, R. *Br. vet. J.* **131**, 402 (1975)
143. SETO, A., OKABE, T. and ITO, Y. *Am. J. vet. Res.* **38**, 1895 (1977)
144. ACRES, S. D. *J. Dairy Sci.* **68**, 229 (1985)
145. SAIF, L. J. and SMITH, K. L. *J. Dairy Sci.* **68**, 206 (1985)
146. OGRA, S. S., WEINTRAUB, D. and OGRA, P. L. *J. Immun.* **119**, 245 (1977)
147. MUELLER, R., BOOTHBY, J. T., CARROLL, E. J. and PANICO, L. *Am. J. vet. Res.* **44**, 747 (1983)
148. REITER, B. and BROCK, J. H. *Immunology* **28**, 71 (1975)
149. DELDAR, A., NAYLOR, J. M. and BLOOM, J. C. *Am. J. vet. Res.* **45**, 670 (1984)
150. SENOGLES, D. R., MUSCOPLAT, C. C., PAUL, P. S. and JOHNSON, D. W. *Res. vet. Sci.* **25**, 34 (1978)
151. KELLEY, K. W. *Annls Rech. vét.* **11**, 445 (1980)
152. HUDSON, R. J., SABEN, H. S. and EMSLIE, D. *Vet. Bull., Weybridge* **44**, 119 (1974)
153. MOE, K. and HARPER, G. J. *Archs Virol.* **76**, 211 (1983)
154. KELLEY, K. W., OSBORNE, C. A., EVERMANN, J. F., PARISH, S. M. and GASKINS, C. T. *J. Dairy Sci.* **65**, 1514 (1982)
155. KELLEY, K. W., GREENFIELD, R. E., EVERMANN, J. F., PARISH, S. M. and PERRYMAN, L. E. *Am. J. vet. Res.* **43**, 775 (1982)

156. OLSON, D. P., SOUTH, P. J. and HENDRIX, K. *Am. J. vet. Res.* **44**, 572 (1983)
157. WOODARD, L. F., ECKBLAD, W. P., OLSON, D. P., BULL, R. C. and EVERSON, D. O. *Am. J. vet. Res.* **41**, 561 (1980)
158. WOODARD, L. F., ECKBLAD, W. P., OLSON, D. P., BULL, R. C. and EVERSON, D. O. *Am. J. vet. Res.* **41**, 1208 (1980)
159. WOODARD, L. F., ECKBLAD, W. P., OLSON, D. P., BULL, R. C. and EVERSON, D. O. *Cornell Vet.* **70**, 266 (1980)
160. REITER, B., FULFORD, R. J., MARSHALL, V. M., YARROW, N., DUCKER, M. J. and KNUTSSON, M. *Anim. Prod.* **32**, 297 (1981)
161. REITER, B., BROCK, J. H. and STEEL, E. D. *Immunology* **28**, 83 (1975)
162. REITER, B. *Annls Rech. vét.* **9**, 205 (1978)
163. REITER, B., MARSHALL, V. M. E., BJÖRCK, L. and ROSEN, C. G. *Infect. Immun.* **13**, 800 (1976)
164. REITER, B., MARSHALL, V. M. and PHILLIPS, S. M. *Res. vet. Sci.* **28**, 116 (1980)
165. STOTT, G. H., MARX, D. B., MENEFEE, B. E. and NIGHTENGALE, G. T. *J. Dairy Sci.* **62**, 1632 (1979)
166. MARX, D. B. and STOTT, G. D. *J. Dairy Sci.* **62**, 1819 (1979)
167. BALBIERZ, H., NIKOLAJCZUK, M. and SAWICKI, T. *Arch. Immun. Ther. Exp.* **24**, 55 (1976)
168. JAMES, R. E. and POLAN, C. E. *J. Dairy Sci.* **61**, 1444 (1978)
169. MOON, H. W. and JOEL, D. D. *Am. J. vet. Res.* **36**, 187 (1975)
170. STOTT, G. H., MARX, D. B., MENEFEE, B. E. and NIGHTENGALE, G. T. *J. Dairy Sci.* **62**, 1902 (1979)
171. MENSIK, J., DRESSLER, J., FRANZ, J. and POKORNY, J. *Vet. Med., Praha* **22**, 449 (1977)
172. KIM, J. W. and SCHMIDT, F. W. *Dt. Tierärztl. Wschr.* **90**, 283 (1983)
173. KIM, J. W., SCHMIDT, F. W., LANGHOLZ, H. J. and DERENBACH, J. Z. *Tierzücht. ZuchtBiol.* **100**, 187 (1983)
174. TSHIBANGU, M. L., FUMIERE, I., CHAUVAUX, G., LOMBA, F. and BIENFET, V. *Revue Agric., Brux.* **35**, 3295 (1982)
175. EDWARDS, S. A. *Anim. Prod.* **34**, 339 (1982)
176. BRIGNOLE, T. J. and STOTT, G. H. *J. Dairy Sci.* **63**, 451 (1980)
177. STOTT, G. H. and FELLAH, A. *J. Dairy Sci.* **66**, 1319 (1983)
178. MEYER, H. and STEINBACH, G. *Mh. VetMed.* **20**, 84 (1965)
179. McEWAN, A. D., FISHER, E. W. and SELMAN, I. E. *Res. vet. Sci.* **11**, 239 (1970)
180. LOGAN, E. F., MENEELY, D. J. and LINDSAY, A. *Br. vet. J.* **137**, 279 (1981)
181. PETRIE, L. *Vet. Rec.* **114**, 157 (1984)
182. MULLER, L. D. and ELLINGER, D. K. *J. Dairy Sci.* **64**, 1727 (1981)
183. OYENIYI, O. O. and HUNTER, A. G. *J. Dairy Sci.* **61**, 44 (1978)
184. GHIONNA de MARIA, C. and AUXILIA, M. T. *Annali Ist. sper. Zootec.* **13**, 79 (1980)
185. MENSIK, J., SALAJKA, E., STEPANEK, J., ULMANN, L., PROCHAZKA, Z. and DRESSLER, J. *Annls Rech. vét.* **9**, 255 (1978)
186. HASHIGUCHI, Y. and YADA, T. *J. Jap. vet. med. Ass.* **34**, 166 (1981)
187. MEIROM, R., COHEN, R., BARNEA, A., RATNER, D. and TRAININ, A. *Refuah vet.* **34**, 51 (1977)
188. CAFFIN, J. P., POUTREL, B. and RAINARD, P. *J. Dairy Sci.* **66**, 2161 (1983)
189. NOCEK, J. E., BRAUND, D. G. and WARNER, R. G. *J. Dairy Sci.* **67**, 319 (1984)
190. FLEENOR, W. A. and STOTT, G. H. *J. Dairy Sci.* **63**, 973 (1980)
191. SKRIVAN, M., SKRIVANOVA, V. and HORYNA, B. *Sb. čsl. Akad. zemed. věd Zivocisna Vyroba* **29**, 299 (1984)
192. LOGAN, E. F., FOSTER, W. H. and IRWIN, D. *Anim. Prod.* **26**, 93 (1978)
193. LAMOTTE, G. B. *Am. J. vet. Res.* **38**, 263 (1977)
194. MENSIK, J., DRESSLER, J., FRANZ, J. and POKORNY, J. *Vet. Med., Praha* **22**, 463 (1977)
195. LARSON, B. L., HEARY, H. L. and DEVERY, J. E. *J. Dairy Sci.* **63**, 665 (1980)
196. BOURNE, F. J. *Vet. Sci. Commun.* **1**, 141 (1977)
197. WILLIAMS, M. R., MAXWELL, D. A. G. and SPOONER, R. L. *Res. vet. Sci.* **18**, 314 (1975)
198. WILLIAMS, R. and MILLAR, P. *Res. vet. Sci.* **26**, 81 (1979)
199. MICUSAN, V. V. and BORDUAS, A. G. *Res. vet. Sci.* **21**, 150 (1976)
200. ISHIKAWA, H. and KONISHI, T. *Jap. J. vet. Sci.* **44**, 555 (1982)
201. PAAPE, M. J. and PEARSON, R. E. *J. Dairy Sci.* **58**, 1242 (1975)
202. HUDSON, S., MULLORD, M., WHITTLESTONE, W. G. and PAYNE, E. *Br. vet. J.* **132**, 551 (1976)
203. LAMOTTE, G. B. and EBERHART, R. J. *Am. J. vet. Res.* **37**, 1189 (1976)

204. BOYD, J. W. and HOGG, R. A. *J. comp. Path.* **91**, 193 (1981)
205. JOHNSTON, N. E. and OXENDER, W. D. *Am. J. vet. Res.* **40**, 32 (1979)
206. JAMES, R. E., POLAN, C. E. and CUMMINS, K. A. *J. Dairy Sci.* **64**, 52 (1981)
207. FARWELL, S. O., KAGEL, R. A., GUTENBERGER, S. K. and OLSON, D. P. *Analyt. Chem.* **55**, 985A (1983)
208. DVORAK, M. *Br. vet. J.* **127**, 372 (1971)
209. LOPEZ, G. A. *Diss. Abstr. Int., Sect. B.* **35**, 1007B (1974)
210. LOPEZ, G. A., PHILLIPS, R. W. and LEWIS, L. D. *Am. J. vet. Res.* **36**, 1245 (1975)
211. CABELLO, G. *Br. vet. J.* **136**, 160 (1980)
212. STOTT, G. H. and REINHARD, E. J. *J. Dairy Sci.* **61**, 1457 (1978)
213. STOTT, G. H. *J. Dairy Sci.* **63**, 681 (1980)
214. OLSON, D. P., PAPASIAN, C. J. and RITTER, R. C. *Can. J. comp. Med.* **44**, 19 (1980)
215. FALLON, R. J. *Annls Rech. vét.* **9**, 347 (1978)
216. LANGLEY, O. H. and O'FARRELL, K. J. *Vet. Rec.* **99**, 187 (1976)
217. FORD, E. J. H ., BOYD, J. W. and HOGG, R. A. *Index Res. Meat and Livestock Comm.* p. 4 (1978)
218. HOERLEIN, A. B. and JONES, D. L. *J. Am. vet. med. Ass.* **170**, 325 (1977)
219. LANG, J. M., ROY, J. H. B., SHILLAM, K. W. G. and INGRAM, P. L. *Br. J. Nutr.* **13**, 463 (1959)
220. STALEY, T. E. and BUSH, L. J. *J. Dairy Sci.* **68**, 184 (1985)
221. LOGAN, E. F., PEARSON, G. R. and McNULTY, M. S. *Vet. Rec.* **101**, 443 (1977)
222. CORLEY, L. D., STALEY, T. E., BUSH, L. J and JONES, E. W. *J. Dairy Sci.* **60**, 1416 (1977)
223. PRIOR, M. E. and PORTER, P. *Vet. Rec.* **107**, 220 (1980)
224. BRAUN, J. P., TAINTURIER, D., LAUGIER, C., BENARD, P., THOUVENOT, J. P. and RICO, A. G. *Proc. XIIth Wld Congr. Dis. Cattle, Amsterdam, Netherlands,* p. 1222 (1982)
225. PIVONT, P., ANTOINE, H. and GREGOIRE, R. *Annls Méd. vét.* **126**, 621 (1982)
226. HANCOCK, D. D. *J. Dairy Sci.* **68**, 163 (1985)
227. HORYNA, B. and DRAGANOV, D. *Vet. Med., Praha* **25**, 537 (1980)
228. EIGENMANN, U. J. E., ZAREMBA, W., LUETGEBRUNE, K. and GRUNERT, E. *Berl. Munch. tierärztl. Wschr.* **96**, 109 (1983)
229. MASSIP, A. and FUMIERE, I. *Recl. Méd. vét., Éc. Alfort* **151**, 363 (1975)
230. NAYLOR, J. M. and KRONFELD, D. S. *Am. J. vet. Res.* **38**, 1331 (1977)
231. VILLOUTA, G., GONZALEZ, M. and RUDOLPH, W. *Br. vet. J.* **136**, 394 (1980)
232. EDWARDS, S. A., BROOM, D. M. and COLLIS, S. C. *Br. vet. J.* **138**, 233 (1982)
233. WINTER, A. C. *Vet. Rec.* **118**, 344 (1986)
234. HEATH, S. E. *Vet. Rec.* **116**, 647 (1985)
235. FRANZ, J., KREJČÍ, J. and MENŠÍK, J. *Zentbl. VetMed.* **B21**, 540 (1974)
236. ROY, J. H. B. *Jl R. agric. Soc.* **132**, 81 (1971)
237. STRAUB, C. C. and MATTHAEUS, W. *Annls Rech. vét.* **9**, 269 (1978)
238. INTERNATIONAL UNION OF BIOCHEMISTRY. *Enzyme Nomenclature,* Academic Press, New York and London (1984)

Chapter 3

Neonatal calf diarrhoea (excluding salmonellosis)

No one would dispute that enteric and respiratory disorders are the two main hazards to calf health, but many would argue about the relative importance of specific pathogens and other predisposing factors in the aetiology of these disorders.

The diversity of the microorganisms that may be involved emphasizes the importance of prevention of enteric disease by eliminating the predisposing causes. It would seem that for enteric infections as well as for respiratory infections in calves (see Chapter 5) the saying attributed to Claude Bernard 'Le microbe n'est rien, le milieu est tout', is very appropriate. Thus, calf diarrhoea results from the interaction of potentially pathogenic enteric microorganisms with the calf's immunity, nutrition and environment[212].

However, enteric and respiratory disorders occur at different ages, mortality from enteric disorders being limited largely to the first 14 d of life and respiratory infections being most important from 6–8 weeks of age onwards. Both conditions are aggravated by the domestication of cattle.

Classification

The classification of neonatal diarrhoea, i.e. faeces of <120 g dry matter per kg, has been attempted at least twice. First, it has been classified by the substrate on which the dominant microorganisms thrive.

In infants and, presumably, in calves there is a balance (eubacteriosis) between the saccharoproteolytic organisms, including *E. coli*, whose main activity is proteolytic, and the saccharolytic organisms, mainly lactobacilli, whose activity is associated with the fermentation of sugars, mainly to lactic acid and acetic acid. If the balance is swung too far in either direction (dysbacteriosis), the animal is likely to be affected with diarrhoea, which has been designated putrefactive in the case of the dominance of the saccharoproteolytic organisms and fermentative when the saccharolytic organisms are dominant[14].

The dominance of either group of organisms can result from infection with highly invasive strains of the organism, from changes in the composition of the chyme passing from the abomasum or from the use of antibiotics.

Secondly, diarrhoea has been classified as nutritional or infectious, the former usually being synonymous with fermentative and the latter with putrefactive diarrhoea.

The risks of research workers being at cross purposes when discussing diarrhoea is shown by the early work of the late Williams Smith. Contrary to the views of most other workers on the subject, he suggested that lactobacilli rather than *E. coli* were associated with calf diarrhoea and that high counts of *E. coli* were due to agonal invasion, i.e. at the time of death[418]. However, it now seems likely that his studies were confined mainly to calves that had fermentative diarrhoea, whose body weight loss was about 60 g/kg compared to a loss of up to 170 or even 240 g/kg in the much more severe putrefactive diarrhoea[13].

The definitions of putrefactive and fermentative diarrhoea are not mutually exclusive; damage caused to the villi of the duodenal mucosa by mixed infections of rotavirus and *E. coli* may reduce enzyme activity, e. g. lactase[27], and this may result in fermentation in the large intestine with production of organic acids of low molecular weight[26].

Thus a primary putrefactive diarrhoea may be associated with a secondary fermentative diarrhoea. Moreover, subdivision into infectious or nutritional diarrhoea cannot be upheld since nutritional diarrhoea arising from gastric dysfunction may develop into an infectious diarrhoea if enteropathogenic strains of *E. coli* become dominant in the environment.

The relative importance of different microorganisms in the aetiology of calf diarrhoea

A number of different organisms can be associated with diarrhoea[433]. The most common organisms are *E. coli*, *Salmonella* sp., rotavirus (reo-like virus)[15–17, 20, 21], coronavirus and *Cryptosporidium*[149] (a coccidium). The relative importance of the presence of these viruses[16] and of enteropathogenic strains of *E. coli* in the alimentary tract as a cause of localized intestinal infection and diarrhoea in the neonatal calf has yet to be elucidated.

In immunologically susceptible calves, the organism may be the primary pathogen able alone to cause diarrhoea, or the diarrhoea may be caused by a combination of pathogens and other predisposing factors. The pathogens may be present normally in the calf's intestinal tract or environment, or be brought into the environment by purchased calves[148].

The importance of an organism is usually based on the proportion of diarrhoeic calves from which a particular organism is isolated, irrespective of whether the organisms may also be isolated from the faeces of healthy calves or whether one organism causes more serious pathological consequences than another. There is thus a danger of associating proportion of isolations with proportion of causes of diarrhoea[138].

Moreover, whilst it is generally agreed that in the UK the isolation of enterotoxigenic *E. coli* (ETEC) from diarrhoeic calves is not as common as that of other enteropathogens, the illness is more severe[137].

England

A study of faeces samples from diarrhoeic calves in 45 outbreaks showed the presence of rotavirus in 42 per cent and coronavirus in 14 per cent of 490 calves.

Calici-like virus was found in 11 per cent, *Cryptosporidium* in 23 per cent, salmonella in 12 per cent and ETEC with K99 adhesin in 3 per cent of 310 calves. In 20 per cent of calves with diarrhoea more than one enteropathogen was detected, whilst in 31 per cent no enteropathogen was isolated.

From a study of the faeces from 385 healthy calves in the same outbreaks, a significant statistical association with disease was found for rotavirus, coronavirus, *Cryptosporidium* and salmonella spp. Healthy calves were not examined for calici-like virus, and the incidence of $K99^+$ *E. coli* was too small for statistical analysis[150]. The incidence of $K99^+$ *E. coli* seems now to be lower in the UK than in other countries[150].

Association of enteropathogen with calf rearing system

Rotavirus was most common in dairy herds and single-suckled beef herds, *Cryptosporidium* was most common in single- and multiple-suckler beef herds, and salmonella in calf rearing units. ETEC was found in one dairy herd and one multiple-suckler beef herd with unhygienic calving accommodation. No association between coronavirus and calf rearing system was noted[150].

Scotland

In a study of the faeces of normal and untreated diarrhoeic calves on 20 farms, diarrhoea was significantly associated with the presence of rotavirus, ETEC and *Cryptosporidium* and nearly significantly with coronavirus. Rotavirus was isolated on nine farms, ETEC on two, *Cryptosporidium* on one, rotavirus + coronavirus on two, rotavirus + *Cryptosporidium* on one, rotavirus + coronavirus + *Cryptosporidium* on one and none on four farms[140].

Further work showed that ETEC was isolated from 23 of 306 diarrhoeic calves (7.5 per cent) in 8 of 70 (11.4 per cent) farms. Whereas ETEC was not isolated from clinically normal calves, rotavirus was detected in 10.7 per cent of normal calves as well as in 29 per cent of diarrhoeic calves; even so these percentages differed significantly. Isolations of coronavirus and *Cryptosporidium* did not differ between faeces of diarrhoeic calves and those clinically normal. Mixed infections were found in 21 diarrhoeic calves (7 per cent). Including ETEC isolations, at least one enteric pathogen was detected in 48 per cent of diarrhoeic and 25 per cent of clinically normal calves[121].

Samples of faeces from 302 untreated calves obtained on the day of onset of diarrhoea and from 49 healthy calves located on 32 farms that were experiencing outbreaks of diarrhoea showed that rotavirus, coronavirus, *Cryptosporidium*, ETEC and salmonella spp. were all excreted more frequently by diarrhoeic than by healthy calves but the difference was only statistically significant for rotavirus. Rotavirus was excreted by 18 per cent of healthy calves, coronavirus by 4 per cent, *Cryptosporidium* by 14 per cent but no ETEC or salmonella were detected. The most common enteropathogen in diarrhoeic calves was rotavirus excreted by >50 per cent of such calves on 18 farms. Coronavirus was also excreted at the same rate on one farm. *Cryptosporidium* was excreted on five farms and ETEC on three farms. Concurrent infection with two or more microorganisms occurred in 15 per cent of diarrhoeic calves[149].

France

In Limousin, there was a decline in isolation of *E. coli* with K99 antigen from 8 per cent of diarrhoeic samples in 1982 to 3 per cent in 1983, whereas isolation of rotavirus increased from 8 per cent in 1982 to 36 per cent in 1983. The decline in *E. coli* K99 was attributed to a vaccination programme[147].

Czechoslovakia

In Czechoslovakia, the most important pathogens were considered to be rotavirus and coronavirus followed by enteropathogenic *E. coli* and *Cryptosporidium*[144].

Canada

In a study of 40 calves, of which 32 developed diarrhoea before 10 d of age, 22 enterotoxaemic strains of *E. coli* (i.e. those that cause dilation of ligated intestinal loops in calves as a result of passage of water and electrolytes from the tissues into the lumen of the intestine) were isolated from 11 diarrhoeic and one normal calf. Rotavirus was isolated from 15 calves before, during and after the onset of diarrhoea. Four calves had both enterotoxaemic *E. coli* and rotavirus, but in all cases the *E. coli* was isolated first. No potentially enteropathogenic agent was isolated from 11 of the 32 calves with diarrhoea. These workers did not think that rotavirus was the initiating factor in enteric colibacillosis[18]. A further study of 55 cases of acute neonatal calf diarrhoea revealed the following infectious agents: one, rotavirus alone; four, rotavirus + *E. coli*; two, rotavirus + *Cryptosporidium*; five, rotavirus + corona-like virus; three, rotavirus + corona-like virus + *Cryptosporidium*; one, rotavirus + corona-like virus + IBR virus; two, corona-like virus alone; one, corona-like virus + mycotic abomasitis; one, corona-like virus + *Cryptosporidium*; six, *E. coli* alone; five, *Cryptosporidium* alone; three, mycotic abomasitis; one, mycotic rumenitis; and 20, undetermined[22].

In 59 spring-calving single-suckled herds, the incidence of organisms isolated from faecal, nasal and blood samples was 31 per cent for ETEC and 37 per cent for rotavirus with both agents present in 14 per cent of herds[139].

Another report estimated that only 20 per cent of cases of calf diarrhoea could be attributed to *E. coli*[142]. Similarly, of a total of 190 cases of diarrhoea, enteropathogenic *E. coli* were not responsible for more than 20 per cent of cases. However, no pathogen was recovered in more than 25 per cent of the cases, possibly due to previous therapy[146]. Coronavirus and rotavirus could be demonstrated in 30 per cent of 134 cases of diarrhoea, 5–6 h after its onset. In 56 per cent of cases both viruses were present, in 53 per cent coronavirus alone and in 14 per cent rotavirus alone[143]. To complicate matters, there is a great difference in the detection rate of rotavirus in calf faeces according to the method used, immunodiffusion giving higher detection rates than electron microscopy or fluorescent antibody techniques[145].

Escherichia coli infections (colibacillosis)

It was in 1885 that Theobald Escherich, a paediatrician and bacteriologist, first described a colon bacillus and called it *Bacterium coli commune*. It was recommended in 1920 that the name should be changed to *E. coli* but it was not until 1939 that the name was changed in Bergey's *Manual of Determinative Bacteriology*[304, 417].

It is sobering to look back some 65 years to the classical work of Theobald Smith and his colleagues at the Rockefeller Institute for Medical Research at Princeton, USA who in the period after World War I established the basic knowledge on *E. coli* infections in calves, which has been confirmed at 20–25 year intervals by successive generations of research workers in their attempts to make further progress. Even in 1925, Theobald Smith commented that calf diarrhoea had been the subject of extensive investigations, but it was he who indicated that *E. coli* diarrhoea in calves resembled cholera in man, and he therefore referred to it as calf cholera[11]. The similarity with cholera and the dissimilarity from fermentative diarrhoea in infants in which there is carbohydrate malabsorption and excessive colonic fermentation has been reaffirmed more recently[28, 377].

Colibacillosis is still considered to be the most important disease of young calves and occurs mainly during the first 3 weeks of life[1–3]. The incidence of mortality is highest in the first fortnight of life; affected calves that survive this period often recover. From the results of experiments in the 1950s there appeared to be two distinct conditions[4] (see Figure 3.1): (a) an *E. coli* septicaemia and (b) an *E. coli* localized intestinal infection or enterotoxaemia, although intermediate forms were found on occasions. Moreover, in calves that appeared to show some resistance to an *E. coli* septicaemia, death might be associated with pleurisy and peritonitis[5]. In a localized intestinal infection certain specific strains tended to become dominant in the intestine, although they did not usually invade the tissues further than the mesenteric lymph nodes.

The main harmful products of the multiplication of *E. coli* are considered to be toxins and amines, and it has been suggested that in man the degree of damage to the intestinal walls depends on the activity of the amine oxidases and the amount of antibody present in the cell walls of the colon[14]. The feeding of very large amounts of colostrum may thus saturate the calf's tissues with antibodies and so protect calves from an *E. coli* localized intestinal infection as well as from an *E. coli* septicaemia, even though the calves may be given subsequently a poor-quality milk substitute.

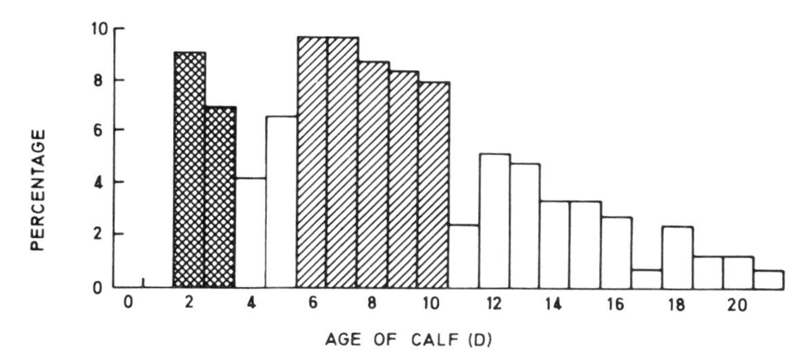

Figure 3.1 Frequency distribution of diarrhoea during the first 3 weeks of life showing age at which calves usually die from an *Escherichia coli* septicaemia ⊠ and an *E. coli* localized intestinal infection ▨

In 1974, *E. coli* infections in calves were classified into four separate diseases[307], as follows:

1. Enterotoxic – proliferation of *E. coli* in the small intestine with release of enterotoxin which causes intestine to produce fluid.
2. Enterotoxaemic – proliferation of *E. coli* in the small intestine with the production of a toxin which is absorbed and acts elsewhere.
3. Local-invasive – invasion by *E. coli* of the walls of the small intestine with destruction of the epithelium.
4. Septicaemic – *E. coli* causes bacteraemia or is localized outside the intestinal tract.

More recently, *E. coli* have been classified into three types[305]:

1. Strains that invade the intestinal epithelium (enteroinvasive *E. coli*)[306].
2. Strains that are non-invasive and produce gut adhesive antigens and enterotoxins (ETEC)[307].
3. Strains consistently associated with diarrhoea in man which are neither enteroinvasive nor enterotoxigenic, but elaborate a Vero cell cytotoxin. This has been shown to be identical with shiga toxin produced by *Shigella dysenteriae*. Thirteen of such strains have been isolated from nine of 306 diarrhoeic calves (3 per cent). Serogroups O26 and O111 which were isolated from five calves in the study were associated with shiga-like toxin production in *E. coli* recovered from diarrhoeic man. Hence, calves may harbour *E. coli* capable of producing enteric infection in man.

E. coli septicaemia

In acute cases there is a septicaemia caused by *E. coli* and the calf is usually found dead, or prostrated with death ensuing within a few hours. The acute cases usually occur very early in life, sometimes on the day after birth.

There is general agreement that *E. coli* is the responsible agent, as a bacteraemia occurs before death[5]. The microorganisms appear to invade the tissues during the first 24–36 h after birth, when the intestine is permeable to the transfer of intact antibodies into the bloodstream. Indeed, *E. coli* have been found to enter the intestinal epithelial cells by invagination in the same manner as for macromolecular proteins. However, *E. coli* were only found in large numbers in the mesenteric lymph nodes if they had been ingested before colostrum was fed[24].

Thus, when calves aged 27 h were given 1.5×10^5 *E. coli* O26:K60 NM all had diarrhoea within 8 h. Of those given 2 litres colostrum every 12 h from birth one out of five died, whereas all died when given milk replacer or polyvinylpyrrolidone to imitate Ig and cause closure, and four out of five died when given isotonic saline[420].

Certain serological types of *E. coli* – in particular 'O' groups 15, 78, 86, 115 and 117 – are associated with the disease in calves. These types accounted for 57 per cent of the cases of septicaemia, with O78:K80 (B) being the predominant type and being associated with 36 per cent of the infections. Agammaglobulinaemia (lack of γ-globulin in the blood serum) was found in 96 per cent of calves with septicaemia, and 11 per cent of normal calves had low levels of serum γ-globulin in spite of having received colostrum[6–9]. Other 'O' groups encountered are 26, 35, 119 and 137[213].

In a further study of 2356 calves suffering from neonatal diarrhoea 288 had an *E. coli* bacteraemia, based on isolation from the liver. Of these, 24 per cent were associated with 'O' group 78, 22 per cent with O15, 14 per cent with O101 and 5 per cent with O8[23]. A common feature of *E. coli* strains that cause bacteraemia in calves is their ability to produce colicine V[25]. Colicine V is controlled by a transmissible plasmid (a structure of extrachromosomal DNA[304] in the cell cytoplasm which is capable of reproducing autonomously), and gives greater virulence and ability of the organisms to resist the defence mechanisms of the calf and thus invade the tissues.

Colostrum-deprived calves injected intravenously with endotoxin from strains O101 and O137 showed shock symptoms similar to those of the natural disease[118].

In France, it has been suggested that septicaemic *E. coli* strains with adhesive properties could be responsible for mucoid enteritis in calves[421].

It seems that *E. coli* septicaemia may be more prevalent in developing countries. In Egypt, enteritis was responsible for 42 per cent of deaths of buffalo calves and 34 per cent of deaths of Friesian calves, with a mortality rate of 11 and 2.7 per cent of buffalo and Friesian calves born respectively during an 8-year period. Of 100 dead calves examined, 62 had died of colibacillosis, with *E. coli* isolation from intestines, kidney, liver, blood, spleen and lungs and less frequently from bone marrow and muscles. In calves dying of symptoms of meningitis, *E. coli* could be isolated from the cerebrospinal fluid. A total of 492 strains of *E. coli* were isolated, of which 368 could be typed into the following 'O' groups: 8, 9, 11, 15, 26, 35, 55, 78, 86, 101, 111, 115, 119, 125, 126 and 137[422]. In Iraq, non-enterotoxigenic serotypes of *E. coli* isolated from diarrhoeic calves were quite divergent and belonged to the following 'O' groups: 5, 8, 9, 19, 21, 28, 45, 70, 71, 77, 103, 105, 131 and 141 and were associated with K30, 35 and 85[122].

In colostrum samples obtained within 16 miles of Reading, Berkshire, UK, all contained antibodies to most of the strains isolated from cases of *E. coli* septicaemia. However, no sample of colostrum contained antibodies against O78:K? (B)[10]. Thus, unless a virulent strain of *E. coli* is present in the environment, the feeding of adequate colostrum during the first 24 h of life should protect calves against an *E. coli* septicaemia.

E. coli localized intestinal infection or enterotoxaemia

The diarrhoea caused by enterotoxigenic *E. coli* (ETEC) is an infectious disease that begins during the first few days of life, although peak mortality occurs about 6–7 d of age. The *E. coli* that cause the disease possess special attributes of virulence that allow them to colonize the small intestine, proliferate rapidly and produce an enterotoxin that causes hypersecretion of fluid into the intestinal lumen.

These ETEC are shed into the environment by infected animals in the herd and are ingested by newborn calves soon after birth. There is some natural immunity to ETEC but it often fails to protect calves that are born and raised under modern intensive methods[137].

Symptoms

The obvious symptom of an *E. coli* localized intestinal infection is diarrhoea, and the tail becomes soiled with faeces. Diarrhoeic faeces may contain as little as 50 g dry matter per kg. The colour of the faeces seems to vary in different outbreaks of the disease; in some it is yellow and watery, whereas in others it is white and pasty.

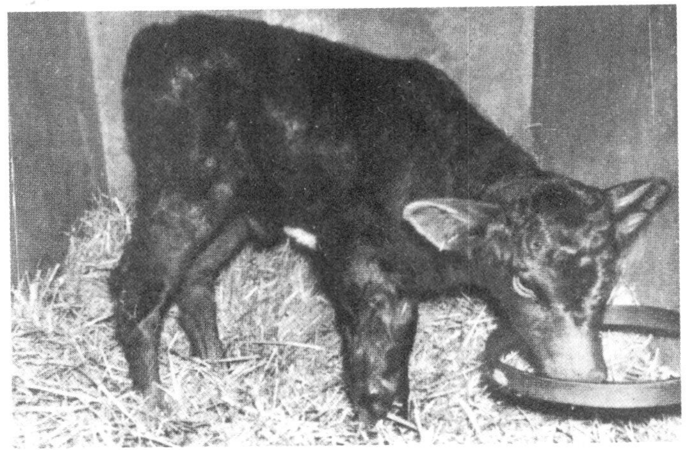

Figure 3.2 Calf suffering from *Escherichia coli* localized intestinal infection. Note the sunken eyes and unhealthy stance

In succeeding days the appetite is variable, the diarrhoea increases and the calf becomes dehydrated, the sinking of the eyes being the most obvious symptom (Figure 3.2). Decreased skin elasticity associated with dehydration is considered to occur when live-weight loss has reached 70 g/kg[218].

If no treatment is given, the calf becomes prostrated, hypothermic[214] and usually dies. This condition is usually restricted to calves that have received colostrum, although often in inadequate amounts, and starts within the first week of life. It is associated with the presence of large numbers of *E. coli*, which form a carpet on the wall of the intestinal tract from the lower end of the small intestine up to the duodenum[11]. A similar condition may be found in older veal calves if they have been allowed to become anaemic.

Age incidence

Workers in Edinburgh concluded that in Scotland and N. Ireland, enterotoxigenic colibacillosis was a major cause of diarrhoea in calves <3 d old, but was not associated with the typical diarrhoea problem seen on most farms in calves >3 d old[121]. Similarly, in Canada ETEC was considered primarily a disease of <5 d of age, whereas rotavirus could infect calves from a few days to 1 month of age, probably depending on host susceptibility[139]. However, certain strains of *E. coli* (see below) have been isolated from diarrhoeic calves between the age of 3 and 21 d[413].

Differential diagnosis

It has been claimed that it is possible to differentiate between diarrhoea caused by ETEC and by rotavirus. ETEC infection occurs at 1–5 d, mostly 1 or 2 d, and the faeces are watery without clots, fibrin, mucus or blood, whereas with rotavirus the age varied but all calves were >6 d and the faeces were watery or pulpy and always contained lumps. In 50 per cent of cases, faeces contained blood. The rotavirus-infected calves were clinically in better shape but often had severe metabolic acidosis[366].

An indirect fluorescent antibody technique involving a suitable pooled 'OK' antiserum has been found to be a satisfactory rapid method for identification of enteropathogenic types of *E. coli* in the faeces of calves with diarrhoea[128].

Establishment of the microflora

At birth there are generally no bacteria present in the contents of the alimentary tract. The first organisms to colonize the alimentary tract of the normal calf are *E. coli*, streptococci and *Clostridium perfringens* (*Cl. welchii*). These are followed by the lactobacilli, which become the commonest organisms in the stomach and small intestine. *Bacteroides* spp. are slow to colonize and are mainly restricted to the large intestine. The temporary dominance of the first organisms to colonize has been attributed to the high pH of the abomasal contents at this time, and the decline of these organisms has been associated with the subsequent drop in pH[12].

Within 5.5 h after birth *E. coli* are mainly restricted to the ileum and the large intestine, but by 8.5 h after birth they are found in the abomasum at a level of 10^3/ml. In the duodenum the viable *E. coli* count reaches a peak at 1–4 d of age, and in normal healthy calves it declines to a very low level at 10 d of age (10–10^2/ml), presumably because of the dominance of lactobacilli. In contrast, calves whose deaths are associated with an *E. coli* localized intestinal infection have viable counts of 10^7–10^8/ml in the duodenum at this latter age[13].

As long ago as 1925, Theobald Smith[11] showed that coli adhered to the mucous membrane of the ileum and spread towards the duodenum (Figure 3.3) and it was he who suggested that *E. coli* became dominant in the alimentary tract as a result of

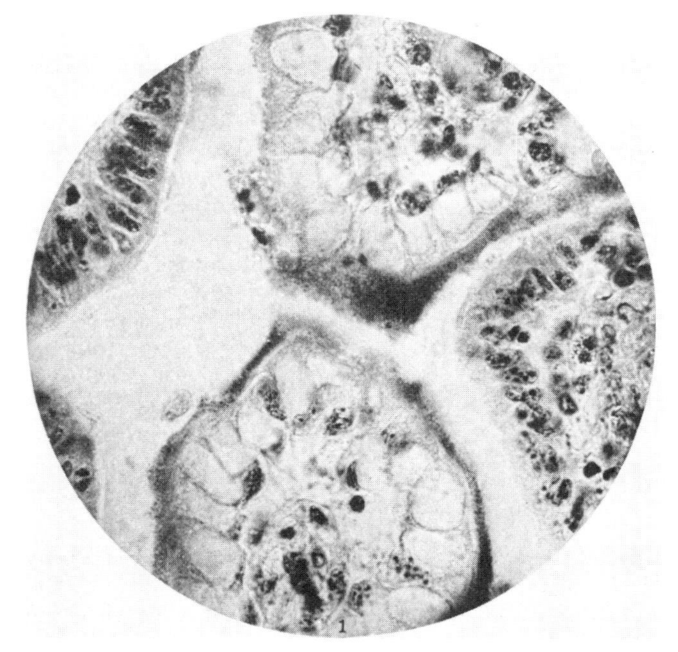

Figure 3.3 The year 1925. Section of the unopened small intestine above the ileum of a calf with *E. coli* localized intestinal infection. The tips of four villi are in view, two of which show the hydropic condition of the epithelium[141]. Dense masses of *E. coli* are attached to the top of these and are also present on the unchanged villi[11]. (Reproduced by copyright permission of the Rockefeller University Press)

failures in peptic digestion. He showed that *E. coli* reached the caecum within 12 h of birth in calves that were not fed and that even in calves killed and promptly examined, coli were in the duodenum in large numbers. He pointed out that in the last few hours before death, coli could spread rapidly throughout the intestine. In contrast more recently, it has been claimed that a challenge organism could only be found adhering to the proximal small intestine in calves that died and in which post mortem examination had been delayed[119].

Although the location of *E. coli* is of interest, it is perhaps irrelevant in relation to its effect since the enterotoxin or a mediator, possibly cyclic AMP (3′,5′-adenosine monophosphate), produced at one site results in excessive secretion throughout the rest of the small intestine[423].

Enterotoxigenic E. coli

Certain strains of *E. coli* are enterotoxigenic for the calf (ETEC). The multiplication and adherence of these specific enteropathogenic strains of *E. coli* which cause profuse diarrhoea in the calf are usually associated with an antigen designated K99. They are also associated with a heat-stable toxin (ST)[216, 426].

Adherence factors

The adherence of *E. coli* is associated with the presence of filamentous proteinaceous surface antigens, fimbriae, that mediate the attachment of the bacteria to the epithelial cells. The principal adherence antigen in calves is designated K99, but another fimbrial antigen, F41, can occur with K99 on *E. coli* of the O9 and O101 groups or occasionally on its own[123–125, 407, 408]. The *in vitro* biosynthesis of K99 antigen is inhibited by alanine[217].

Most K99 strains produce ST toxin. However, expression of K99 depends on the 'O' group of the strain[433]. Group O101 produces 10 times more K99 than do groups O8, O9 and O20[432]. K99 is considered to cause the stunting and fusion of the cells. K99 is controlled by a plasmid that can be transmitted from one organism to another by conjugation[29]. In Canada, the most common ETEC strains belonged to 'O' groups 8, 9 and 101[139]. Of 56 enterotoxigenic isolates, 32 belonged to 'O' group 9, 19 to group O101, three to group O8 and one to group O17. However, only 14 of the isolates had K99 antigen, and it was suggested that either the antigen had been lost on storage or some enterotoxigenic strains do not have K99 antigen[30]. In Czechoslovakia the most frequently occurring enteropathogenic strains were O9 and O101, followed by O8, O8 + O141 and O117[23]. It has been claimed that the ileal site has a specificity for adherence of the K99 antigen[426].

A monoclonal antibody (an antibody produced from the multiplication of a single B lymphocyte) to K99 has been shown to be as effective as conventional absorbed rabbit antiserum for detecting K99 *E. coli*[412].

More recently, an adhesive antigen designated F(Y) has been isolated in France[409] and one designated Att 25 in Belgium[410, 430] and these have also been isolated in the UK[411]. These two antigens are now considered to be the same. *E. coli* producing adhesive antigen FY (Att 25) were isolated from 46 of 1341 (3.4 per cent) calves on 20 of 164 (12.1 per cent) farms in England and Scotland. Of the 46 calves, 20 had diarrhoea and nine of these calves had mixed infections with rotavirus, coronavirus, *Cryptosporidium* and *S. typhimurium*. The F41 fimbrial antigen was found on one of the FY(Att 25)$^+$ *E. coli*. This strain also produced a heat-stable enterotoxin. The remaining FY(Att 25)$^+$ isolates did not produce either adhesins, enterotoxins or Vero cell cytotoxins. The FY(Att 25) antigen was not detected on 109 pathogenic *E. coli* isolated from calves, chickens, lambs, pigs or man.

Almost 50 per cent of calves from which FY(Att 25)$^+$ *E. coli* was isolated presented no clinical signs of diarrhoea. Whereas ETEC diarrhoea in calves is severe in the first 3–4 d of life and is responsible for 50–60 per cent of calf diarrhoea in this age group in Canada[137], FY(Att 25)$^+$ *E. coli* was isolated from diarrhoeic calves between the ages of 3 and 21 d. K99 was not detected on any of the FY(Att 25)$^+$ *E. coli*[413].

Enterotoxin

Enteropathogenic *E. coli* are capable of elaborating one or both of two main types of enterotoxin. Of 345 *E. coli* strains from calves with diarrhoea, none were positive for heat-labile (LT) enterotoxin but 13 per cent of strains representing 23 per cent of herds were positive for heat-stable (ST) enterotoxin[425]. The heat-labile toxin is inactivated by heating at 60°C, whereas the heat-stable toxin can resist this treatment. The heat stable enterotoxin has been divided into two types, STa and STb, with different biological activities[427].

In Belgium, none of 721 isolates of *E. coli* from calf diarrhoea, whether enteropathogenic or not, produced LT toxins. Thus, enteropathogenic *E. coli* strains from calves produce only ST toxin, which is not antigenic[428].

In the USA, of 1004 isolates of *E. coli* obtained in spring 1975 from calves with diarrhoea in seven different states, 124 isolates from six states were enterotoxigenic, i.e. they had the ability to cause distension of the calf ligated intestinal segment[429].

To complicate the issue, it was found that in Belgium *E. coli* strains that produced no STa type toxin and had no adhesive factor (K99, Att 25 and others) nevertheless caused fluid accumulation in the ligated intestine of the calf[430]. This is in contrast to the view that *E. coli*, to be enteropathogenic, must have an adherence factor and at least one enterotoxigenic factor[431].

The enterotoxin is presumed to act by increasing the concentration of 3′-5′ guanosine monophosphate (cyclic GMP) which stimulates increased secretion by decreased absorption of water and sodium in the small intestine; these workers found that cAMP concentration was unaffected by enterotoxin[386].

However, the action of the toxin of ETEC is usually considered to be similar to that of cholera. The cholera toxin binds to receptors on the intestinal epithelial cells and stimulates adenylate cyclase activity[424]. The resulting increase in cAMP causes secretory diarrhoea by inhibiting sodium chloride and water uptake by the villi and by stimulating chloride secretion by the crypt cells. Since the formation of cAMP is inhibited by α_2-agonists (e.g. clonidine), i.e. a drug possessing both receptor affinity and intrinsic activity, and its action is blocked by neuroplegic drugs (e.g. chlorpromazine), i.e. those that block synaptic transmission in the central nervous system, and since certain opiates (e.g. loperamide) have an antisecretory effect, it has been suggested that these drugs should be used in combination with oral rehydration in the treatment of *E. coli* enteritis in calves[357].

Placed in Thira-vella loops, the enterotoxin increased the flux of water into the small intestine, particularly in the jejunum and ileum, but did not affect the absorption of glucose[39, 43, 387].

The increased peristalsis of diarrhoea may help to decrease the concentration of enterotoxin and also the duration of exposure of the mucosa to the enterotoxin; decreased and disorganized intestinal EMG in calves with diarrhoea has been reported[458].

In Canada, it was found that of 200 strains of *E. coli* obtained from calves with diarrhoea under 14 d of age, 36 per cent were enterotoxigenic in piglet intestine and 28 per cent in calf intestine[30].

Incidence of ETEC

Of 1529 isolates of *E. coli* from diarrhoeic and clinically normal calves in Scotland and N. Ireland, 88 (5.7 per cent) were found to possess K99 pilus antigen ($K99^+$). However, in contrast to the results above, complete correlation was found between possession of $K99^+$ antigen and ST endotoxin production and ability to dilate intestinal loops. Thus, detection of K99 alone could be used for diagnosis of ETEC infections in calves. Between 3 and 4 per cent of the sera from calves and cows contained antibodies to $K99^+$ antigen[121]. In France, of *E. coli* strains isolated from 89 diarrhoeic calves under 4 d of age, 45 per cent had ST and 38 per cent were $K99^+$ [120].

In the USA, during the spring beef-calving seasons in 1974 and 1975 in Montana, ETEC was detected in 118 of 355 herds. Of 124 isolates, 114 could be allocated to six different 'O' groups; O90:K35, O101:K30, O8:K85, O20:K?, O8:K25 and O101:K28. Of 35 ETEC isolates serotyped, 28 had K99 antigen, whereas the antigen was not detected in any of the 10 isolates of non-ETEC studied[429].

In Iraq, 40 per cent of 103 diarrhoeic calves were excreting ETEC. Of these 74 per cent were 1–3 d old and 71 per cent died. ETEC strains belonged to the following 'O' groups; 3, 9, 15, 17, 20, 21, 28, 45, 86, 117, 125 and 127 and were usually associated with capsular antigens K30, 35, 80 and 85. The predominant 'O' groups were 9 and 20 and to a lesser extent 17, 21 and 45. Of the 46 ETEC strains, 12 produced enterotoxin as well as $K99^+$, whereas 34 produced enterotoxin only. Furthermore four $K99^+$ strains were non-enterotoxin producers and K88 was found in three non-enterotoxigenic strains although there are no pigs in Iraq. Thus, calves may be one of the hosts that maintain and propagate *E. coli*. All toxin-producing strains, except for two, and 28 of 60 non-toxin-producing strains (including four $K99^+$ and three $K88^+$) were found to exhibit pellicle formation indicating the possibility of the presence of some kind of somatic pili[122].

Atypical E. coli

An atypical strain of *E. coli* ($O5:K^-:H^-$) that colonized the colon and rectum in the same manner that a $K99^+$ *E. coli* colonizes the ileal mucosa was found to cause dysentery in gnotobiotic calves. Neither infections with $O101:K^-$, $K99^+$ or with O9:K30, $K99^+$ or with *E. coli* that cause a septicaemia produced dysentery or the pattern of colonization of the large intestine seen with this atypical *E. coli*[127].

Association with abortion

It has been claimed that *E. coli* was isolated from an aborted bovine fetus at 4 months of gestation from a Boran cow[415].

'Collapse syndrome' (see also 'Weak calf syndrome', Chapter 7)

In the winter of 1971–2, a collapse syndrome was reported in Scotland in 25 of 91 diarrhoeic single-suckled beef calves on hill farms. The condition was characterized by dilation of the abomasum and small intestine with fluid and the presence of

ETEC[238]. The seven strains able to dilate the gut all belonged to O8, O9 and O101 serotypes[239]. The highest morbidity or mortality was with the O8 serotype.

In a further study, 13 calves died after acute collapse with a terminal rise in blood potassium, magnesium, inorganic phosphorus and total protein, whilst 10 died from diarrhoea that persisted for several days and resulted in a gradual rise in chloride and urea and in packed cell volume (PCV). Calves that died had lower blood glucose concentrations before the onset of clinical symptoms than did those that survived[240].

The syndrome occurred mainly within the first 10 d of life with a range of 1–21 d. It was least common from June to September and most common in calves of heifers. The mortality rates increased in wet and windy weather with losses on occasion exceeding 25 per cent. It was considered to be a clinical entity distinct from other forms of diarrhoea. No association was found with the type of feeding of the dam or with the Ig concentration in colostrum or in the serum of the calves[241]. Some success in treating the collapse syndrome by giving injections of 400 ml dextrose solution (400 g/l) providing the body temperature of the calves was above 98°F (a body temperature indicative of severe hypothermia) has been claimed[242].

Campylobacter infection

These Gram-negative curved organisms, whose old name was *Vibrio*, consist of a number of species and subspecies. Although *C. fetus* subspecies *jejuni* has been associated with human[243] and calf enteritis[245], in calves it was considered to be endemic with its incidence no higher in diarrhoeic than in healthy calves[140, 149].

However, it has been claimed that *Campylobacter* have been obtained from diarrhoeic calves, from which no other enteric pathogen has been isolated[244]. Moreover, gnotobiotic calves infected orally with *C. jejuni* produced mucoid faeces. The lesions were mainly in the large intestine and consisted of scattered, dilated or damaged crypts containing abundant neutrophils with associated inflammatory infiltrates in the surrounding lamina propria[247].

Indeed, *C. jejuni* must be considered a zoonosis, as it has been isolated from a naturally occurring case of bovine mastitis in the herd of a small producer-retailer. The same serotype was isolated from 18 people suffering from gastrointestinal illness, who had drunk unpasteurized milk from that source[250].

Cultures of *C. fetus* subspecies *intestinalis* isolated from the congested mucosa of the small intestine of a 2-week-old calf were used to infect both preruminant and ruminant calves. All developed fever and diarrhoea with clear mucus and occasional spots of blood. The ileum, which was most markedly affected, became thickened and slightly inflamed with accumulation of lymphoid cells and crypts filled with inflammatory cells. The mesenteric lymph nodes were pale and enlarged. It was claimed that the syndrome was similar to that produced by subspecies *jejuni*[246].

Many of the *Campylobacter* species have been isolated from bovine fetal material, and are associated with abortions and infertility. Concurrent bovine virus diarrhoea (BVD) and *C. fetus fetus* infection has been reported in an aborted fetus[248]. Vaccination of Jamaican heifers with *C. fetus* subspecies *venerealis* cleared 44 per cent of them of infection and improved the fertility of infected heifers threefold[249].

Clostridium infection

Cl. perfringens

No evidence has been found of any relationship between the presence of *Cl. perfringens* (*Cl. welchii*) in the gut or faeces of calves and the incidence of diarrhoea or gastroenteritis. Comparisons between *E. coli* and *Clostridium* concentrations per g gave no indication of any direct effect of one organism on the other in healthy or diseased calves[252].

Cows vaccinated during pregnancy with a multicomponent clostridial vaccine produced antitoxin to *Cl. perfringens* type D in blood and colostrum; it was suggested that the colostrum could be given to lambs to provide passive immunity against the infection[256].

Cl. sordelii

Oral inoculation of calves with *Cl. sordelii*, isolated from a 2-year-old animal with haemorrhagic enteritis, produced a slight rise in rectal temperature, the presence of soft faeces containing blood and mucus within 1–6 d. Although the organism was isolated from faeces, it was only found in the large intestine at slaughter. The effect may have been due to a toxin, since a filtrate of the culture produced mild congestion and inflammatory changes in the jejunal and ileal mucosa, dilation of the crypts and capillary dilation in the large intestine mucosa[251].

Cl. botulinum

Cl. botulinum type D toxin and *Cl. perfringens* enterotoxins were isolated from a bull calf that was unable to stand but was able to feed and ruminate normally[254]. Botulism in cows as a result of their grazing an area of stubble, on to which poultry broiler litter containing carcasses had been spread, has been reported[255].

Proteus and *Pseudomonas* infection

In Scotland, a diarrhoeic syndrome in neonatal calves has been associated with calves excreting faeces heavily contaminated with *Pseudomonas* or *Proteus vulgaris*[19]. Neonatal calves with a mixed infection of *E. coli* and *Pseudomonas* or *Proteus vulgaris* in their faeces died quickly with little weight loss[19]. In Belgium, of 277 calves with diarrhoea in 66 farms, 107 calves in 46 farms yielded *P. vulgaris* or *P. mirabilis*. Both were generally resistant to several antibiotics[419].

Cryptosporidium infection

Cryptosporidium, although a coccidium, is considered here rather than in Chapter 8 because of its isolation in and association with neonatal diarrhoea.

Cryptosporidia have a marked predilection for the follicle-associated epithelium over the ileal Peyer's patch. Cryptosporidial antigen was found in the subepithelial tissue, both in the domes over the ileal Peyer's patch and in villi, apparently in macrophages (large mononuclear phagocytic leucocytes, usually immobile in connective tissue but stimulated into mobility by inflammation) where the parasites seemed to be progressively degraded. The follicle-associated epithelium showed long tightly-spaced microvilli, replacing the normal low folds, and protrusions, particularly in late infection[257].

Anti-*Cryptosporidium* antibodies of IgG, IgM and IgA classes were produced in the faeces within 5–6 d after infection, reached a peak at 8–14 d after infection and then declined. Similarly in serum, IgG was detected 6 d after infection and remained elevated throughout infection, but IgM and IgA in serum showed little change[258].

Enteric viruses

The organisms

At least eight enteric viruses have been identified within the last seven years, e.g. calicivirus (Newbury agent), astrovirus, bredavirus, rotavirus-like agent, that produce diarrhoea and lesions in the intestine of experimentally inoculated calves; the aetiological role of these newly identified viruses is uncertain. On the other hand, coronavirus and rotavirus are frequently associated with outbreaks of neonatal diarrhoea[15]. All diarrhoeal viruses replicate within the epithelial cells of the small intestine resulting in variable degrees of villous atrophy[291]. If the virus is virulent or present with another virus or bacteria, severe diarrhoea may occur.

When enteropathogenic viruses (bovine coronavirus, bovine calici-like virus and bovine rotavirus) were used to infect gnotobiotic calves, there was increased faecal output and elevated concentrations of the two ileal hormones, enteroglucagon and neurotensin. The calici-like virus and the rotavirus, but not the coronavirus, caused xylose malabsorption. Infection with bovine astrovirus and parainfluenza type 3 (PI3) virus, which are not enteropathogenic, had no effect on faecal output or the concentration of the two hormones[321].

Rotavirus

Site of infection

Rotavirus is confined to the alimentary tract and attacks the epithelial cells (enterocytes) of the villi, particularly in the duodenum, causing the cells either to die or slough off or to become cuboidal and squamous so that the villi are stunted; as a result diarrhoea develops. There is disagreement as to how far the rotavirus spreads towards the ileum. Infection in gnotobiotic calves is similar to that in the field with diarrhoea, weight loss, atrophy of villi accompanied by impaired D-xylose absorption and a mortality of 5–30 per cent[351]. Strains of rotavirus vary in virulence[313] and subclinical infection is probably more common than clinical infection.

Incidence of infection

Europe

In a longitudinal study in England, 83 per cent of 48 unweaned calves between 3 and 35 d of age excreted rotavirus in their faeces but only 50 per cent had any clinical signs of disease[318].

In a similar survey of rotavirus in faecal samples from a self-contained herd in N. Ireland during two successive calving seasons, rotavirus was detected in the faeces of 79 per cent of the calves with the first detection at a mean age of 6.1 d. It is suggested that elimination of antibody from the gut at about 6 d of age renders a calf susceptible to infection. All calves had 2 litres colostrum by bottle and teat and were left with their dam for 1 d, given whole milk for 1 week and then milk substitute. The clinical disease was of mild to moderate severity, only one calf requiring intravenous fluid therapy. Diarrhoea or excretion of abnormal faeces was associated with 58 per cent of the infected calves, whilst in the remaining 42 per cent the infection was subclinical[313].

In a study of neonatal diarrhoea in one herd in the Netherlands, either a mild diarrhoea occurred within the first 3 d of life when only one out of 34 calves excreted rotavirus, or a more severe diarrhoea occurred from 4–14 d of age when 34 out of 45 calves excreted rotavirus. The diarrhoea at the older age occurred exclusively in the later part of the year. Rotavirus excretion was observed in 11 calves that remained healthy or had recovered from diarrhoea before the virus was first detected.

A further study in five dairy herds in the Netherlands, with a history of severe diarrhoea in calves under 5 d of age, showed that in 22 diarrhoeic calves between 1 and 3 d, six excreted rotavirus, seven $K99^+$ *E. coli* and one both organisms, whereas of 20 diarrhoeic calves between 4 and 14 d of age, 16 excreted rotavirus, one $K99^+$ *E. coli* and one both[184].

In Wallonia, of 52 faecal samples from diarrhoeic calves 13 contained rotavirus and three contained ETEC strains, but no samples showed the presence of both pathogens[187]. A reo-like virus, which was probably rotavirus, was isolated in Denmark in 1974[317].

America

In Argentina, rotavirus was found in 53 per cent of 177 diarrhoeic calves but only in 7 per cent of 40 healthy calves from 19 farms[319].

Africa

In S. Africa, the faeces of calves experimentally infected with rotavirus contained the virus for at least 3 d[300].

Asia

In 1977, the wide prevalence of diarrhoea associated with rotavirus was reported in Japan[194].

Australasia

In New Zealand, six viruses were isolated from the faeces of diarrhoeic calves and yearling cattle; five were enteroviruses and the remaining one was probably a rotavirus[195]. The incidence of rotavirus in calves on the Taeri Plain, New Zealand, in the first 2 weeks of life during the calving periods in April and September–October was 13 per cent. Of those calves with symptoms of disease, only 27 per cent were positive to rotavirus[199].

Effect of breed

A breed difference in susceptibility to rotavirus has been postulated[197]. In Bulgaria, rotavirus has been isolated from buffalo calves affected with acute diarrhoea[320].

Interaction with E. coli

Although the primary infection may be rotaviral, *E. coli* may complicate the situation and frequently increase the severity so that rapid dehydration and death occurs within 12–24 h after the onset of diarrhoea. The viral infection alters normal digestion and absorption, and the resultant increase in partially digested milk enables bacterial proliferation to occur and the damaged epithelium allows the bacteria to enter the body[197, 300, 345].

In general, although there are exceptions[18], rotavirus becomes established some time before *E. coli* and increases the pathogenicity of *E. coli*. Other work has shown that nearly all outbreaks of rotavirus are associated with two or more agents[446].

Human rotavirus

Diarrhoea in gnotobiotic calves has been caused by a rotavirus from human infantile gastroenteritis, but no illness occurred until the second to fourth serial passage when the diarrhoea began 15–30.5 h later, but lasted less than 24 h[193].

Antibodies to rotavirus in adult cattle

Most adult cattle appear to have antibodies against rotavirus in their serum and colostrum, so rotavirus infection occurs mainly in the second week of life when passive immunity from colostrum is waning. Antibodies to rotavirus were found in 73–76 per cent of cattle tested in Saskatchewan[196]. However in other research, although nearly all cows excreted rotavirus-specific antibodies in colostrum, no relationship was found between the colostral titre and the development of rotavirus diarrhoea in the calf[184].

Excretion of rotavirus by adult cattle

Since rotavirus was found in the faeces of cows at parturition, adults may be a major source of infection[313].

Serum Ig concentration in calves and susceptibility

Whereas calves with $\geq$20 zinc sulphate turbidity (ZST) units in their sera were susceptible to diarrhoea and body weight loss from rotavirus but recovered, those with $\leq$10 ZST units developed severe diarrhoea and died[446].

A survey of 654 cattle of all ages in three herds showed that between 2 and 37 per cent had low rotavirus antibody titres, whether or not they had suffered a recent rotavirus infection[446].

Cell-mediated immunity

Rotavirus was found to cause accumulation of fluid, which contained leucocytes as well as an antiviral substance, in ligated intestinal loops. The antiviral substance was detected in the sera of orally-infected calves as well as those infected by placing virus in the loops. *In vitro* rotavirus was inhibited by interferon[196].

Identification

A comparison of immunofluorescence and immune electron microscopy of diarrhoeal faeces showed that 88 per cent of the results of the two tests agreed in diagnosing the presence or absence of rotavirus[189]. Similarly good agreement was obtained in an independent study in N. Ireland[191].

Other work showed a much higher detection rate of rotavirus in calf faeces with immunodiffusion than with electron microscopy or fluorescent antibody techniques[145]. However, rotavirus and mycoplasma-like particles have been observed in the faeces of calves, whether they had diarrhoea or not[16].

Coronavirus

Site of infection

This virus infects both the small and large intestine, but infection in the small intestine is in the ileum rather than in the anterior small intestine, where rotavirus occurs.

In Quebec, variation was found in coronavirus with two distinctive precipitating antigens[185].

Incidence of infection

In Belgium, coronavirus antigen was detected by fluorescent microscopy in the rectal mucosa of 20 per cent of 458 calves with digestive disorders. The reaction was most distinct in calves that died shortly after the onset of symptoms and weakest in calves kept alive by therapy for more than 10 d. Of the 91 cases diagnosed, 90 occurred between November and April[183].

In a study of diarrhoea in one herd in the Netherlands, coronavirus was isolated from four out of 45 calves with severe diarrhoea at 4–11 d of age and also from three healthy calves[184].

In S. Dakota, coronavirus was detected in 16 per cent of 689 calf enteric disease specimens. Most cases occurred between 1 and 7 d with the upper age limit of 3 months[192].

Of four outbreaks of diarrhoea between 2 and 15 d of age in beef herds in the Po valley in Italy, two were associated with rotavirus, one with coronavirus and one with both types of virus; clinically it was not possible to distinguish the outbreaks according to the virus involved[447].

Duration of infection

Whereas rotavirus produced a diarrhoea that lasted only 5–6 h, coronavirus induced a diarrhoea that lasted 5–6 d[349].

Excretion of coronavirus by adult cattle

Coronavirus has also been isolated from the faeces of cows with profuse diarrhoea, but it is not known whether it was an opportunist invader or merely part of the normal alimentary tract flora[190].

Minicoronavirus

A virus tentatively called a minicoronavirus and unrelated to coronavirus or BVD virus was found in Quebec, in the intestinal contents of one normal and seven diarrhoeal calves. Antibodies against the particles, 45–65 nm in size, were demonstrated in the serum of affected calves. Its role as an aetiological agent is still not known[186].

Calicivirus (Newbury agent)

The pathology is similar to that of rotavirus in that the virus attacks mature enterocytes, but it affects the cells towards the base rather than on the tip of the villi. Lactase activity is markedly reduced within 3 d of infection but activity recovers[274, 275].

Epithelial syncytial virus

Within 8 h of being given by mouth, this virus caused diarrhoea for only 24 h in gnotobiotic calves but resulted in mortality if an *E. coli* strain with K99 antigen was also present[354].

Bredavirus

This virus resembles coronavirus, but damages both the villi and crypts in the ileum and large intestines[276, 277].

Astrovirus

On its own, this virus[275], so called because of its stellate appearance, does not appear to cause disease in calves but in combination with bredavirus caused sloughing-off of the dome of cells above Peyer's patch.

Parvovirus

Bovine parvovirus appears to be fairly common in the UK, on the basis of the high prevalence of antibody in blood[263]. It has been isolated from clinically normal animals and from sick animals that showed no evidence of diarrhoea, so its role in calf enteric disease is not clear[262].

Attempts to produce diarrhoea in 4-day-old colostrum-deprived calves with a pool of five isolates of parvovirus were unsuccessful. Parvovirus has been isolated from calves in the first week of life together with other enteric viruses and *E. coli*, but the calves had probably received insufficient colostrum. Parvovirus has also been isolated from older calves receiving a liquid diet, at weaning and at 1–2 weeks after weaning when no other potential pathogens were detected. It was considered that defects in management were precipitating factors. Mortality rates were either nil or very low and morbidity was variable. Cryptosporidia were not present in association with parvovirus, but BVD virus and salmonella were present in two cases[262].

However, from Queensland it has been reported that calves developed intestinal parvovirus infection soon after weaning on three endemically infected properties, the infection being associated with diarrhoea. Maternally-derived antibody to parvovirus was found to have a half-life of 19 d[259]. Oral challenge of calves with parvovirus resulted in infected tonsils and lymph nodes, moderate small intestinal villous atrophy and fusion due to crypt damage together with lymphoid necrosis, predominantly associated with the intestinal tract and thymus[260].

It appeared that damage to the intestinal tract in experimentally induced enteritis was exacerbated by the existence of a subclinical coccidiosis infection and by the stress of weaning which was at 19 weeks of age[261].

Adenovirus

An adenovirus type 2[188] and type 5 has been associated with the so-called 'weak calf syndrome' which occurs in neonatal calves, particularly in E. Idaho and S.W. Montana in the USA, and is associated with polyarthritis, listlessness and diarrhoea[37]. This can be followed by dehydration and secondary infection[448]. Intravenous injection of adenovirus type 5 has produced the syndrome which occurs from birth to 10 d of age, has a morbidity of 6–15 per cent and a mortality of 60–80 per cent[37].

Calves born in cold and wet conditions in the spring are particularly affected[448]. The syndrome has also been associated with large variations in ambient temperature, i.e. between day and night[316]. Calves that survive have a reduced growth rate. The calves have very high levels of cortisol at birth which rapidly decrease during the next few days[448].

No difference has been found in the serum Ig levels of normal calves and those showing 'weak calf syndrome', but when calves were infected during the third trimester of pregnancy with microbial agents isolated from calves with 'weak calf syndrome', the IgG and IgM concentrations in precolostral serum were increased[45]. Haemorrhage in the subcutaneous tissue of the legs, especially of the hocks, may be due to a coagulation failure[315].

In young cattle of 9–12 months, adenovirus causes lesions in the blood vessels of the mucosa, resulting in a fatal disease.

It has been suggested that a deficiency of selenium might be associated with 'weak calf syndrome'. However, in a double blind control trial on four commercial dairy herds, injection of the dams with 50 mg selenium 10 d before parturition had no effect on the incidence of 'weak calf syndrome' although this treatment marginally increased the selenium status of the treated calves[264].

Chlamydia

The chlamydia, originally known as the psittacosis-lymphogranuloma venereum (PLV) group of organisms, were considered to be viruses, since they multiply in affected cells. They are now considered to be nearer to bacteria since they have a cell wall and contain both DNA and RNA.

Chlamydia have been associated with neonatal diarrhoea in calves. In Finland, 13 per cent of 458 blood serum samples showed group specific chlamydial antibodies, which it was suggested arose from previous clinical enteritis infections[253].

Histology of lesions in the small intestine

E. coli

When colostrum-deprived calves were protected against a septicaemia with IgM and were infected orally with O101:K2(A), stunting and fusing of the villi were seen in the distal half of the small intestine associated with adhesion of the challenge organism to the mucosa[382]. Within 36 h of infection adherence had spread from the ileum to 60 per cent of the small intestine, whilst stunting and thickening of the villi occurred within 6–12 h of challenge[344]. Figure 3.4 shows the damage that is done to the villi.

In colostrum-fed calves in Canada 10^{11} *E. coli* given orally produced colibacillosis, causing profuse diarrhoea accompanied by dehydration and depression. Stunting of the villi and adherence of *E. coli* occurred in the jejunum and ileum. Lesions in the duodenum were minimal or absent and there was no adherence of bacteria. In a calf examined up to 36 h after colibacillosis developed, there was focal degeneration and exfoliation of the absorptive epithelial cells at the tip of the jejunal and ileal villi and focal emigration of neutrophils, which were especially prominent in the dome area of aggregated lymphatic follicles (Peyer's patches)[343].

In further work in Canada, a comparison was made of the effect of giving 10^{11} *E. coli* of a strain without K99 antigen to colostrum-deprived calves compared to that of giving a strain with K99 antigen to colostrum-fed calves. The two strains were 210 (serotype $O9^+$:$K30^+$, $K99^-$, $F41^-$:H^-) and B44 (serotype $O9^+$:$K30^+$, $K99^+$,

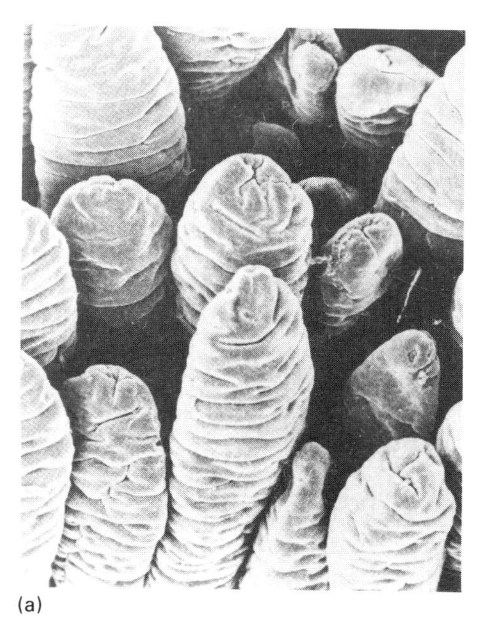
(a)

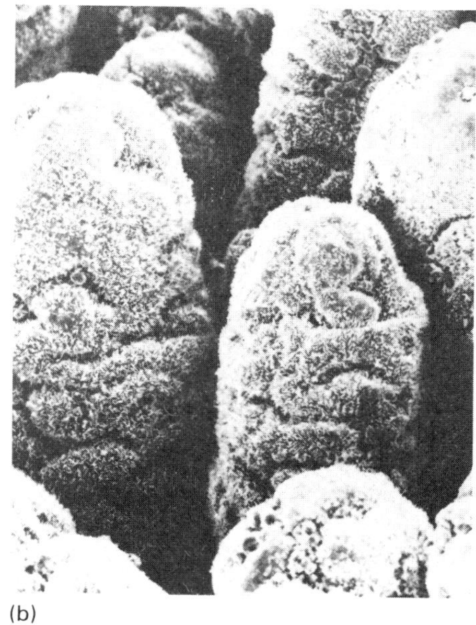
(b)

Figure 3.4 The year 1982. (*a*) Electron micrograph of the cranial portion of the small intestine of a healthy calf. There are long finger-shaped villi with prominent transverse furrows and depressions at the tip of some villi, and no indigenous flora is present. (*b*) Adhesion of ETEC strain 505 (serotype O101:K28, K99, Ent^+) to the intestinal villi of an infected calf. Layers of *E. coli* cover the entire surface of villi from the caudal portion of the small intestine[345]. (By permission of the American Veterinary Medical Association)

$F41^+:H^-$) and both gave similar histopathological changes, namely diarrhoea, atrophic villi with cuboidal epithelium and focal accumulations of a few neutrophils in the dome villi above Peyer's patches, but the changes with strain 210 were not as pronounced as with strain B44. Whereas B44 adhered as a continuous layer over most of the absorptive epithelial surface of both jejunum and ileum, adherence of strain 210 was restricted to the ileum and the bacteria often adhered focally in clumps rather than as a continuous layer[348]. Indeed, electron microscopy has shown that an interaction occurred between the glycocalyx of ETEC strain 210 in colostrum-deprived calves and the glycocalyx of the epithelial cells in the ileum[346].

In other work in the USA, colostrum-fed calves given 10^{11} enteropathogenic *E. coli* showed colonization of the middle and distal portions of the small intestine with 80 per cent of the organisms associated with the intestinal wall. Some organisms had dense fuzzy surface structures which resembled pili. Thick capsular material seemed to be in contact with the microvilli, which sometimes were elongated and projected into the intestinal lumen. The shortened and thickened villi were fused with the lamina propria, which was expanded by an inflammatory filtrate. Denudation of the tips of the villi and exposure of the lamina propria were only observed in samples taken after death and not in adjacent samples removed after general anaesthesia[345].

Of 10 strains of *E. coli* tested in E. Germany, five strains given 5–12 h after birth and 2–3 h before colostrum was given caused severe diarrhoea within 15–22 h and became attached to the epithelium of the small intestine and also of the colon, penetrating the lamina propria of the small intestine where they produced inflammatory changes[347].

Enteric viruses

All diarrhoeal viruses replicate within the epithelial cells of the small intestine resulting in variable degrees of villous atrophy[291].

Rotavirus

Whilst rotavirus may cause pathological lesions in the small intestine which may be accompanied by mild diarrhoea, it is not responsible for the high mortality that often occurs in neonatal diarrhoea. Calves quickly mount an immune response in the small intestine and antibodies begin to appear in the faeces within a few days[449].

Rotavirus causes the villi of the small intestine to be shortened and thickened with the tall columnar epithelium replaced by a short cuboid form[349]. Thus, stunting and fusion of the villi, exfoliation disarrangement and vacuolation of enterocytes and the presence of cuboidal enterocytes were observed in infected but not in rotavirus-free control calves[318]. In gnotobiotic calves these changes occurred particularly in the proximal and middle part of the small intestine where the virus is most abundant[311, 318, 350], the severity of the diarrhoea being related to the severity of the changes[311].

Thus, the normal villus length:crypt length ratio is 3.9, 5.4 and 3.8:1 for the anterior, middle and posterior small intestine respectively, whereas for calves infected with rotavirus, the ratios were 1.1, 1.1 and 1.4:1 respectively. This is a direct measurement of the loss of absorptive cells from the intestinal villi[351].

Whilst there is general agreement on the damage to the mucosa caused by rotavirus[349, 350], there is disagreement as to how far rotavirus spreads towards the ileum[350, 352].

The villous atrophy may persist after infection until the immature enterocytes have migrated from the villous crypts to replace the damaged mature enterocytes on the sides and tip of the villi; in normal neonatal calves this may take in excess of 48 h, but the period decreases with age[62].

Coronavirus

Coronavirus seems to affect the villi in a similar manner to rotavirus[349]. However, coronavirus produced more severe lesions than did rotavirus with lesions extending from the small intestine to the colon[356]. In addition a green mucoid faecal cast was commonly seen in the colon and was associated with a necrotizing enterocolitis[192].

Epithelial syncytial virus

This virus, unrelated to rotavirus and coronavirus, caused shedding of the infected epithelial cells within 2–3 h after the onset of diarrhoea, so that the villi had denuded tips or cuboidal to squamous epithelial cells. The infection started in the proximal portion of the small intestine and progressed distally[354].

Human infantile gastroenteritis virus

After passage through 3–4 calves, lesions from this virus were induced in gnotobiotic calves. Infection of the villous epithelial cells resulted in the replacement of tall columnar cells by cuboidal and squamous cells, shortening of the villi, enlargement of the reticular cells and lymphocytic infiltration of the villous lamina propria[341].

Mixed infections

When hypogammaglobulinaemic calves, naturally infected with rotavirus, were challenged with enteropathogenic *E. coli*, the main effect of rotavirus was stunting, thickening and fusion of villi, principally in the proximal and middle small intestines. In contrast, enteric colibacillosis lesions were in the distal ileum of the challenged calves and were associated with adhesion of *E. coli* and no denudation of the villi was observed. With the enteropathogenic strain and rotavirus, there were severe lesions in the distal ileum. There was no evidence of rotavirus infections spreading towards the ileum as suggested by other workers[353].

Similarly in colostrum-deprived calves, *E. coli* caused profuse liquid diarrhoea with no major lesions in the mucosa, whilst rotavirus produced a transitory diarrhoea with liquid or thick faeces and lesions in the gut which were never fatal. However, *E. coli* and rotavirus acted synergistically causing profuse liquid diarrhoea, total destruction of the digestive mucosa and death[355].

Diet

Soyabean protein

Some years ago, there was evidence that soya protein caused villous atrophy in the intestinal mucosa of calves[176]. More recently it has been shown that heated soyabean flour causes progressive changes in the mucosa, including villous atrophy, crypt hyperplasia, and lymphocyte infiltration. These abnormalities disappeared quickly after removal of soya from the diet but developed again within 24 h of its reintroduction[221]. Histological studies have shown that there are domes of pseudovilli over Peyer's patches caused by bulges in the lymphoid tissue. The pseudovilli are covered by specialized follicle-associated epithelium, which have poorly developed microvilli and possibly take up macromolecules from the lumen by pinocytosis and sensitize the lymphocytes[222].

Thus, heat-treated soya flour given at intervals from 6–8 weeks of age up to 12 weeks progressively produced larger increases in intestinal permeability to β-lactoglobulin and rising titres of serum antibodies to soyabean protein (IgG, IgA and IgM). The increase in permeability was transitory and after 24 h the permeability returned to normal and the β-lactoglobulin quickly disappeared from the circulation. The reaction, which was accompanied by diarrhoea, only occurred in three of six calves. The other three showed no sensitivity to soya antigens and no diarrhoea[220]. The increase in permeability began 1–2 h after ingestion, was greatly enhanced at 4–6 h and returned to normal by 24 h.

Malabsorption as a result of feeding soya protein concentrate or soya flour has been measured by the xylose absorption test and is probably associated with villous atrophy[224].

Soya protein has also been reported as causing flattening of the villi in the intestinal mucosa of the infant[225]. Abnormalities of the duodenal and jejunal mucosa in fattening calves were found to be similar to those in humans with malabsorption syndrome[226].

Although most reactions to dietary soya protein were associated with IgG antibodies, 3–5 per cent of calves given milk substitutes containing antigenic soya showed a transient IgE-mediated hypersensitivity immune reaction that initiated an inflammatory condition[269]. Other workers have found a transient IgE response, which coincided with physiological disturbances in the intestine[270]. The IgE has been successfully isolated and partially purified[271]. This type 3 hypersensitivity, causing a general inflammatory reaction, release of vasoactive amines and causing the villi to become shorter and broader, is similar to the changes that occur in gluten-sensitive enteropathy in man[342].

In infants and children with gastrointestinal disease, IgD and IgE antibodies specific for individual milk proteins and for soya protein are higher in basal and pancreozymin-stimulated duodenal fluid than in the duodenal fluid of healthy adults[268].

Since excessive proliferation of some microorganisms causes flattening of the villi[20] and is also associated with the death of calves given soya, the production of antibodies against soya presumably arises from the ingress of proteins through a damaged mucosa, the proteins having escaped digestion in the abomasum due to reduced acid and enzyme production and rapid outflow into the duodenum (see below).

Ovalbumin

Feeding calves with ovalbumin also led to high antibody levels and similar gastrointestinal reactions to those from soya protein[223]. Presentation to a young calf of ovalbumin antigen together with a bacterial lipopolysaccharide resulted in the generation of IgE antibody capable of sensitizing the skin of the recipient calf[273].

Effect of diarrhoea on metabolism

Food intake

In E. Germany, food intake was shown to decline with increasing severity of diarrhoea, particularly when diarrhoea reached its climax between 3 and 17 d[406]. In Scotland, rather controversially, it was considered that if fluid intake was maintained, there was no significant difference in fluid output of healthy calves, diarrhoeic surviving calves and diarrhoeic calves that died. Provided adequate plasma volume was maintained, the kidneys were able to compensate for excessive faecal fluid loss. Whilst plasma volume was maintained, losses from the extravascular pool manifested as a loss of live weight[378].

Electrolyte loss

Diarrhoea results in the loss, in the faeces, of large amounts of water, and the electrolytes bicarbonate, potassium, sodium and chloride, in this order of magnitude, because of continued gastrointestinal secretion and poor absorption of most nutrients, including vitamins. This results in the depletion of these electrolytes in the blood plasma and, consequently, there is movement of water and particularly of potassium from the cells of the body into the blood[31, 402].

Blood volumes have variously been reported as 58 ml in growing dairy cattle to 74 g to 106 ml/kg in calves[389, 390].

Although at first the kidneys of calves of <11 d of age are able to excrete an excess water load in the gut[405], to conserve body fluids the volume of urine produced declines and may virtually cease, and as a result urea[388] and potassium begin to accumulate in the blood. Blood urea may increase from normal values of 100–200 mg/l to values as high as 1200 mg/l, and plasma potassium, although declining within the cells of the body, may increase from normal values of 5 mM to 6–7 mM and even higher at the time of death[32, 33].

When ETEC O100:KA or O9:K30 were given by mouth to colostrum-fed calves at 24 h, acute diarrhoea occurred. The entire tract had one or more abnormalities in volume, digesta retention, pH, osmolality or electrolyte concentration. Large net losses of sodium, chloride and water occurred in the first 24 h but potassium loss was not affected. Net calcium and inorganic phosphorus losses were reduced in diarrhoea but the calves were still in positive balance. The diarrhoeal faeces contained 41 mmol HCO_3/l, and there was a loss before death of 181 mmol HCO_3. The total volatile fatty acid (VFA) concentration decreased in faeces of calves with diarrhoea but the total reducing sugar was increased due to malabsorption of carbohydrate. However, the carbohydrate was not fermented excessively because the concentration of total volatile fatty acids in the faeces was low. ETEC infection is thus fundamentally different from that of fermentative diarrhoea and enteritis in infants and is similar to Asiatic cholera[377].

Similarly when colostrum-deprived calves, protected against a septicaemia with IgM, were infected orally with O101:K?(A), severe diarrhoea occurred within 24 h and faecal dry matter (DM) concentration fell to less than 50 g/kg. By the fourth day, the calves were voiding 2.3 kg/d compared with 80–340 g/d for the control calves and PCV was high. Moreover, the calves were hyperkalaemic and uraemic. However, their haematology was unaffected and blood cultures were negative[382].

Acidosis

The loss of bicarbonate in the faeces and the accumulation of ketone bodies due to starvation results in a metabolic acidosis and the pH of the blood falls. If the blood pH falls much below 7.2, the calf will usually succumb.

The type of acidosis is affected by the age of the calf. Calves of 2–3 d of age showed severe dehydration and hypovolaemia together with hyperlactataemia, hyperalaninaemia and a considerable increase in other amino acids and an increase in plasma potassium and magnesium. In 15-day-old calves showing an acidosis, there were only minor changes in plasma amino acids, no hyperlactataemia and plasma magnesium was only slightly changed[370].

In viral-induced diarrhoea plasma lactate increased two-fold in seven calves and six- to seven-fold in two calves[375].

The degree of assistance at calving affects the acid-base balance of the newborn calf during the first 10 min of life. The duration of the passage through the birth canal and, within this period, the duration of the compression of the umbilical cord decided whether there would be a pronounced acidosis at birth when calves were removed by traction[441, 442]. Other work has shown that in the first 2 h of life calves showed a metabolic acidosis which subsided during 3–7 d[443].

Water loss

The body water in calves under 15 d of age has been estimated as total body water, 801 ml/kg live weight; extracellular fluid, 311 ml/kg; and plasma volume, 79 ml/kg. Diarrhoea caused a significant reduction in all three volumes, the greatest change being in plasma volume which was reduced by 19.4 per cent. The half-life of body water decreased from 1.43 d in normal calves to 0.74 d during diarrhoeic episodes[437].

The normal loss of water in faeces is about 0.88 litres per day. With faecal water losses of 2.3 litres per day in diarrhoeic calves, death occurred within 7 d[402]. Since there is a significant negative correlation between faecal water loss and urinary water loss and between blood PCV and liquid milk intake and a positive correlation between plasma volume and liquid milk intake, efforts should be made to maintain the fluid intake[378].

The extent of the fluid fluxes

A study of the sites of water and electrolyte loss in diarrhoea induced by *E. coli* showed that in the small intestine of normal calves there was a net secretion of sodium and chloride into the upper jejunum, but a net absorption from the distal ileum, whereas bicarbonate was absorbed from the proximal and secreted in the distal intestine, and potassium was secreted at both points. In the diarrhoeic calf, the net loss of sodium, potassium, chloride and water was increased at the distal small intestine, but there was partial compensation by the spiral colon, from which there was greater net absorption of water and sodium in diarrhoeic calves than in the control animals, although potassium losses in the colon were greater. Surprisingly there was no significant difference between the control and diarrhoeic calves for bicarbonate movement, and the fluxes in both directions of water and sodium (i.e. into and out of the small intestine) were reduced in diarrhoeic calves[39].

In conscious calves, the enterotoxin from *E. coli* resulted in increased intestinal secretion of water and a variable change in sodium[387], but in anaesthetized calves it resulted in a decrease in unidirectional fluxes of water and of sodium[39].

The fluxes of fluid into and out of the intestine are large, being 80 litres per day, although 5.5–6.0 litres per hour based on fluxes at the ileum has been suggested. In the normal healthy calf, the differences in the fluxes may be only 50 ml/d but in the diarrhoeic calf the difference may be 1 litre per day, and a difference of up to 2.5 litres per day will cause dehydration and death. The flux of fluid seems to occur into the abomasum as well, but the fluid is not gastric gland secretion since it is of high pH and hypertonic[377]. Although continuing to feed milk to calves that have diarrhoea can help maintain normal plasma volume, the practice does not appear to reduce mortality[378, 379].

The enterotoxin, at least in the rabbit, affects gut movement as well as secretion[380]. It is possibly similar to the vasoactive intestinal peptide that causes watery diarrhoea in man. In pigs this peptide caused diarrhoea associated with a hypokalaemia[381].

Although the colon seems at first to increase its absorption of fluid to make up for the loss into the small intestine[39], it in turn becomes overwhelmed. This situation could be aggravated by a secondary fermentation that may increase osmotic activity and stimulate fluid excretion into the colon. The calf may eventually succumb from dehydration and a metabolic acidosis.

Heart arrhythmias

Arrhythmias of the heart action occur in a large proportion of the calves that die, and it is probable that in many cases the diarrhoea produces primary cardiac failure as a result of depletion of the potassium in the heart muscle. Such depletion has been found to occur in calves dying as a result of diarrhoea, and also in calves which have had their blood pH reduced by acid infusion[34].

Hypoglycaemia

Hypoglycaemia may occur in calves with diarrhoea, with values of <2.2 mmol/l[214]. This may possibly result from *E. coli* endotoxin inhibiting glucogenesis in the liver or from depletion of glycogen in the liver. However, in viral-induced diarrhoea, hypoglycaemia (<400 mg/l plasma) occurred in only three of ten calves[375]. Similarly in acute diarrhoea in man, fasting blood glucose is low and insulin high[384]. In a study of diarrhoeic calves in W. Germany, hypoglycaemia was rarely observed[435].

Nutrient loss

In diarrhoeic calves there is a much greater volume of ileal digesta containing greater quantities of nitrogen, fat and reducing substances as well as of sodium and potassium[403]. Moreover increased transfer of serum protein to the gut lumen may occur in calves with diarrhoea[404].

Minerals

In calves with colibacillosis, intestinal iron concentration was increased, but sodium, molybdenum, copper, zinc and sulphate concentrations were decreased. In the jejunum magnesium concentration was decreased and inorganic phosphorus was increased. In the liver cobalt, aluminium, calcium and magnesium concentrations were decreased. Colibacillosis had no influence on potassium, total phosphorus and nickel concentrations[385].

In a comparison of calves surviving colienteritis and those that died, the latter had higher PCV (48 v. 42), unchanged (serum?) sodium (141 and 140 mg/l), much increased potassium (8 compared to 5 mg/l), reduced chloride (89 v. 100 mg/l) and bicarbonate (7 v. 17 mg/l), increased inorganic phosphorus (240 v. 80 mg/l) and residual nitrogen (1210 v. 490 mg/l)[361].

A study of the composition of blood of healthy calves and that of calves with diarrhoea in W. Germany showed that serum calcium, magnesium and inorganic phosphorus were only slightly affected. Serum sodium, pH and protein concentration were depressed, but hyperkalaemia was observed only rarely[435].

Comparison of faecal losses in normal and diarrhoeal calves

Data showing the faecal losses that result from diarrhoea and changes in the electrolyte pattern of the blood and heart muscle are given in Tables 3.1 and 3.2.

Enzymes

Since serum glutamic oxaloacetic acid transaminase (SGOT), now called aspartate amino transferase (AST) (EC 2.6.1.1)[459], serum glutamic pyruvic transaminase (SGPT), now called alanine amino transferase (ALT) (EC 2.6.1.2) and alkaline phosphatase (AP) (EC 3.1.3.1) were unaffected by viral-induced diarrhoea, it was considered that no damage occurred in the liver. However, aldolase (ALD) (EC 4.1.2.13), creatine phosphokinase (CPK) (EC 2.7.3.2) and hydroxybutyrate

Table 3.1 Amounts of various constituents excreted daily in the faeces of calves with diarrhoea compared with those in the faeces of normal calves[35]

	Normal	*Diarrhoeal*	*Diarrhoeal: normal ratio*
Water (g)	51.0	927	18.2
Dry matter (g)	12.5	93.5	7.5
Total fat (g)	4.1	37.4	9.1
Neutral fat (g)	1.5	10.6	7.0
Soaps (g)	1.9	8.3	4.4
Free fatty acids (g)	0.7	18.5	26.4
Volatile fatty acids (ml 0.1 M acid)	164.0	1056.0	6.4
N × 6.25 (g)	5.5	41.0	7.5
Ash (g)	1.5	10.6	7.1
Calcium (mmol)	10.8	49.4	3.7
Magnesium (mmol)	5.7	12.0	
Sodium (mmol)	5.0	41.6	11.3
Potassium (mmol)	2.2	39.9	
Phosphorus (mmol)	7.0	31.3	4.4
Coliform count × 10^8	118.0	2907.0	24.6
pH	6.8	6.0	

Table 3.2 Values for various characteristics of the blood and myocardial muscle of normal calves anad those with diarrhoea that survived or died[33, 34]

	Normal	*Diarrhoeal*	
		Survived	*Died*
Plasma Na (mM)	142	129	129
Plasma K (mM)	5.1	5.1	6.1
Plasma Cl (mM)	100	92	94
Plasma volume (ml/kg body weight)	66	59	57
Blood urea (mg/l)	160	410	910
Blood pH	7.4	7.3	6.9
HCO_3 (mM)	29	22	9
Myocardial K (g/kg muscle)	2.23	–	1.69

dehydrogenase (HBD) (EC 1.1.1.30) activities increased in two out of ten calves and CPK alone in two others, which indicated cardiac and possibly skeletal muscle damage[375]. Since body weight loss in calves that died was greater than was fluid loss, it is thought that muscle protein breakdown occurs to support metabolic functions[372]. In contrast in Egypt, an increase in SGOT occurred in most buffalo calves affected with enteric disease during the first 90 d of life. It reached 215 U/l serum in some calves with salmonella and 105 U/l in some calves with *Pseudomonas* infection. Large fluctuations occurred in SGPT concentration but it tended to increase with *Proteus*, *E. coli* and mixed infections[376].

Hormones

A study of plasma aldosterone in newborn Holstein-Friesian bull calves gave a value of 240 ng/l on the day of birth which declined to 55 ng/l on the first day and 10 ng/l by day 7. Neither intakes nor faecal and urinary excretions of sodium and potassium were correlated with plasma aldosterone concentrations[439].

However, plasma aldosterone concentrations vary inversely with those of plasma sodium. There is an initial increase from 1–3 d of age followed by a decrease to 20 d[373]. Moreover, a marked increase in plasma aldosterone concentration occurs during diarrhoea presumably in an attempt to maintain electrolyte homeostasis by increased absorption of sodium and water from the colon[372].

Calves that subsequently developed diarrhoea had a significantly lower aldosterone concentration at birth[373]. In contrast, greater plasma concentrations of corticosteroids were found in 4-day-old calves that subsequently developed diarrhoea between 5 and 20 d of age than in those of calves that were not affected[450].

Na:K ratios in plasma of diarrhoeic calves are considered consistent with adrenal insufficiency[374] but how far adrenal exhaustion and lack of aldosterone plays a part in causing death is not known[39].

However, other research is rather conflicting. Plasma progesterone and aldosterone were both found to decrease after 1 week of age. In calves that died after being given an enteropathogenic viral inoculum orally, whether or not they had been treated with electrolytes, corticosterone and aldosterone concentrations increased before death. In addition, hydrocortisone and progesterone concentrations increased in calves that were not given electrolytes[371,372]. In acute diarrhoea in man, basal concentration of pancreatic polypeptide and both basal and postprandial responses of motilin, enteroglucagon and vasoactive intestinal polypeptides (VIP) are higher, but there is no abnormality in plasma concentrations of gastrin, gastric inhibitory polypeptide or pancreatic glucagon[284]. Small amounts of VIP are released from the intestinal tract in response to vagal stimulation[383].

Fermentative diarrhoea

The infant

In infants fermentative diarrhoea, which is probably synonymous with many cases of so-called 'nutritional scours' in calves, is associated with excessive carbohydrate in the diet, or a deficiency of enzymes to degrade the carbohydrates.

The tendency to develop diarrhoea in response to meals containing unabsorbable carbohydrate is considered to depend more on the lack of colonic accommodation than on the rate of transit through the small intestine[165].

The young calf

The only sugars that can be utilized by the young calf are glucose and lactose, but too large a quantity of these in the diet may cause diarrhoea[35,56–59]. The preruminant calf cannot digest starch or the degradation products dextrin and maltose until at least 28 d of age, and no sucrase is present in the intestine for the hydrolysis of sucrose. Thus, the feeding of starch and its degradation products may also cause diarrhoea in the young calf.

Intakes greater than 200 g/d lactose and glucose caused the dry matter concentration in faeces (fat-free basis) to fall below 100 g/kg[35]. In general, more than 9 g/kg live weight of 'hexose equivalent' (i.e. the sum of the weight of glucose and the weight of lactose × 1.05)[59] tends to cause 'loose' faeces in calves. However, since a high content of carbohydrate in the diet is associated generally with a low content of fat, a 'hexose equivalent' of 12 g/kg may be fed to calves after the colostrum feeding period, provided that the fat intake is about 5.5 g/kg live weight. More recently an upper limit of 13.9 g hexose equivalent per kg live weight has been recommended for lambs[182].

In the Netherlands, the maximum limit for lactose was considered to be 10 g hexose equivalent per kg live weight, which was slightly higher than that for glucose and galactose. Above these limits, the changes in intestinal digesta and faecal characteristics were similar to those in infants and man suffering from hypolactasia. It was not known whether oligosaccharides originating in the ileum from conversion during lactose digestion had any beneficial effect, but there was none from the addition of other carbohydrates, starches or carbohydrate-splitting enzymes[166].

Increasing the intake of lactose to about 300 g/kg dry matter, i.e. a total intake of 200–480g/d, increased the frequency of diarrhoea during the first 10–12 d of life[167]. Similarly, all 24 calves aged 8–53 d had diarrhoea when given a high-lactose milk substitute of the following composition: 130 g fat, 400 g dried skim milk, 140 g whey and 300 g lactose per kg[168]. However, when the calves given a high-lactose milk substitute had access to fine hay or grass silage the frequency of diarrhoea was reduced, the rumen pH was higher and there were fewer Gram-positive cocci and rods in the rumen fluid. However some silage-fed calves had ulcers in their rumens. Moreover, the calves with access to dry food were also given 5 litres electrolyte solution daily[168].

The feeding of skim milk at a young age may cause diarrhoea, and this does not appear to be associated with disturbance in gastric function in that gastric acid production and gastric proteolysis are unimpaired. However, pancreatic protease secretion is reduced[60,61]. Since overall digestion of protein is unaffected, it would seem that the diarrhoea may be of the fermentative type.

Moreover, a diarrhoea associated with a mixed infection of rotavirus and *E. coli* resulted in reduced utilization of dietary lactose[26], and other work has shown decreased lactase synthesis in calves with diarrhoea[27], presumably as a result of damage to the epithelial cells of the small intestine known to be caused by rotavirus[15,311]. Reduced lactose digestion in the small intestine will result in more fermentation in the colon, with the production of a high concentration of short chain fatty acids, lowered pH and increased osmotic activity, which will stimulate solute and water secretion into the colon.

However, it has been suggested that in calves and lambs infected with rotavirus there is sufficient lactase remaining to digest the lactose in a milk diet[215]. On the other hand, decreased mucosal enzyme activity can decrease the absorption of amino acids and fats as well as of carbohydrates[310,460]

It is known that enzyme activity is greatest when an epithelial cell reaches the top of a villus in the small intestine; in the 2-week-old calf it takes 3 d for cells to divide in the crypts of Lieberkuhn, migrate along the villi and slough off[62].

The predominance of the saccharolytic flora in fermentative diarrhoea results in the production of organic acids of low molecular weight, especially lactic and acetic acids, which have an irritant effect on the intestine and produce hyperperistalsis. The irritant effect will be exacerbated if depletion of sodium bicarbonate is so great that there is insufficient to neutralize the acids in the large intestine. This usually results, in the case of infants, in faeces of low pH, but sometimes fermentative diarrhoea is accompanied by secondary putrefactive diarrhoea. In general, the consequences of fermentative diarrhoea are much less severe than those of putrefactive diarrhoea[14].

A condition in young calves is associated with the fermentation of starch in the stomach and the presence of large amounts of yeast cells of *Torulopsis glabrata* and a high blood alcohol concentration[169,170]. This condition, in which anorexia, intoxication, and death may occur, may also be brought about by the fermentation of glucose, when included in milk substitute diets. Dietary fat appears to inhibit the fermentation. A considerable loss of available energy also occurs, since the ethanol produced has only 0.30 of the energetic value of the sugar[170].

The older calf

Although fermentation of carbohydrates in the large intestine of the young calf is associated with an increased incidence of diarrhoea, in the older animal the resultant fatty acids probably supply available energy to the animal. However, from the results with other species it is very doubtful whether the microbial protein produced can be utilized[171].

Even at a live weight of 120–140 kg, the use of pregelatinized starch for calves resulted in a reduction in digestibility of energy, and faeces of a low pH and a dry matter concentration of less than 120 g/kg[172]. Similarly in 6-week-old lambs, increasing inclusion rates of partially hydrolysed starch resulted in larger amounts being digested in the large intestine[173]. With tropical starches 30–60 per cent of the high apparent digestible energy was in the form of volatile fatty acids and lactic acid produced in the large intestine[174]. Similarly, a large proportion of dehulled field beans, processed by various methods could not be digested in the small intestine at 33 d of age, but were fermented in the large intestine. However, the amount digested in the small intestine increased with age[175]. The detrimental effect of the oligosaccharides in soya flour may be the result of their fermentation in the large intestine. Supplementation with a growth promoter having antimicrobial activity resulted in a marked improvement in performance of calves given thermo-alkali treated soya flour, but a marked reduction in the disappearance of carbohydrate in the alimentary tract[176].

Whereas for milk protein diets most nitrogen at the ileum appears to be of endogenous origin, for soya, fish and single cell proteins, certain fractions such as the peptidoglycan from the cell wall of methanol-grown bacteria and the cell wall polysaccharides of yeast single cell protein were very resistant to digestion in the small intestine but were fermented in the large intestine. For cell wall polysaccharides the digestibility was only 40 per cent at the ileum but 77–90 per cent for the whole tract[177–179].

In a comparison of methanol-grown single cell protein and skim-milk based diets, ileal digesta always contained more threonine, serine, glycine, proline and cystine but less methionine, isoleucine, leucine, tyrosine, phenylalanine, lysine and arginine than did the faeces. It was suggested that the differences were probably related to the high proportion of endogenous protein in the ileal digesta, since threonine concentration is high in intestinal mucoproteins and in dietary protein, especially single cell protein. Bacterial protein produced in the intestines is known to have a high concentration of lysine, methionine, isoleucine, leucine, tyrosine and phenylalanine. The very high content of alanine in the ileal outflow from the single cell protein diet was probably related to the high amount of cell wall peptidoglycan[180]. The poor utilization of nitrogen in single cell protein is mainly due to the low digestibility of the bacterial mass but also to the high concentration of nucleic acids[181].

Predisposing factors in calf diarrhoea

The importance of predisposing causes in the aetiology of calf diarrhoea is illustrated by the findings that 30 per cent of healthy calves harbour one or more organisms capable of causing diarrhoea, and similarly no enteropathogenic organisms can be found in 30 per cent of calves with diarrhoea[149, 150].

The build-up of 'infection' in a calf house

The dominance of certain strains of *E. coli*, resulting from the continual passage of susceptible calves through a calf house with a resultant build-up of 'infection', has been demonstrated. In a calf house, maintained at 12.8–15.6°C and into which purchased newborn calves were introduced at intervals of every 1–3 d, each successive calf gained less weight than its predecessor, the incidence of diarrhoea (i.e. of faeces containing less than 100 g/kg dry matter) gradually increased and deaths eventually occurred from an *E. coli* localized intestinal infection, associated with the dominance of particular strains of *E. coli*[4, 66].

The rate of build-up is related to:

(a) the immune status of the calf;
(b) the post-colostral diet of the calf; and
(c) probably the air space in terms of cubic capacity per calf and the ventilation rate in the calf house.

Thus, in Czechoslovakia it was reported that 185 calves out of 189 born successively developed diarrhoea during the first 2 weeks of life and 35 showed a bacteraemia. When the diarrhoea occurred within the first 4 d of life, there was rapid dehydration[42].

Under farm conditions the use of an uncontaminated or disinfected house to break the cycle of infection and the rearing of calves on an all-in all-out basis appeared to have the greatest influence on the disease pattern[63, 129, 451]. Thus, the cycle of rotavirus infection in a calf house was broken by thorough cleaning and disinfection of the calf house with an iodophor disinfectant[313].

In E. Germany the incidence of virus diarrhoea in the first week of life was reduced to 15–25 per cent, the morbidity in the first 2–3 weeks was halved, mortality was reduced to 2–5 per cent and daily weight gains were increased if the chain of infection between older animals and young calves was broken. This involved strict isolation of a calf in the first week of life, feeding its dam's colostrum and instant elimination of suspected animals accompanied by improved standards of husbandry and work practice[314].

The purchase of calves for farms, which mainly rear home-bred calves, is another potent cause of increased mortality[452].

Diarrhoea and mortality is usually associated with an increase in the proportion of pathogenic to total *E. coli* in the faeces, a finding that often occurs following changes in diet[63]. The 'relative' immunological status of the calf will depend on the quantity and quality of colostrum ingested relative to the level of infection in the environment. Small amounts of colostrum will protect in one environment but not in another[64]. Moreover, calves in low-mortality herds in the USA (<10 per cent mortality from birth to 2 months) survived with blood Ig levels that would not protect calves in high-mortality herds (>15 per cent mortality from birth to 2 months)[65].

The post-colostral diet

Whole milk

As Theobald Smith pointed out in 1925[11], *E. coli* (putrefactive) diarrhoea is probably associated with abnormal amounts of protein and peptides in the chyme passing into the duodenum, and consequent disturbances in protein digestion. Overfeeding, which causes excessive curd formation in the abomasum, has been suggested as a predisposing factor to diarrhoea[76] but, in fact, very high levels of whole milk and good quality milk substitutes may be fed directly after the colostrum feeding period without necessarily causing diarrhoea[131] or excessive passage of undigested protein into the duodenum[77, 78]. Under good management conditions, even *ad libitum* feeding of whole milk or good quality milk substitute did not lead to diarrhoea[131].

Milk substitutes

Although differences in mortality between calves given whole milk and those given milk substitutes have not always been significant[132], the composition of the diet given after the colostrum feeding period can have a major effect on the incidence of enteric disease and of mortality[133].

The severity of the effect will depend on the immune status of the calf in relation to the burden of infection in the environment. The burden of infection, in turn, may be affected by the proportion of calves in a calf house that are receiving a detrimental diet.

Milk by-products

As milk substitutes are based to a large extent on skim milk powder, which usually comprises at least 50 per cent of the total dry matter, the quality of the skim milk powder is of considerable importance. If, in the processing of skim milk, denaturation of the whey proteins has occurred to a marked extent, then such milk powders will have a detrimental effect on the calf[35, 67–74, 134, 135]. Associated with the denaturation of the whey proteins is a reduction in ionizable calcium[88–96], the release of SH groups[97–102], poor clotting ability by rennet and reduced digestibility, but no effect on the biological value of the protein[72]. The effect of such diets may be shown in three ways:

1. In the absence of an active infective agent, weight gain may be reduced by as much as 30 per cent in the first 3 weeks of life[72].
2. The rate of build-up of infection[66] (see above) will be greater when large numbers of susceptible calves are passed through a calf house.
3. The incidence of diarrhoea and mortality will be higher once the 'infection' has been built up[70], as evidenced by the dominance of one or two strains of *E. coli*[4].

However, severe heat treatment of milk may have no ill- effects with calves reared out of doors[203].

Even if the whey proteins have not been denatured, too high a Na:Ca ratio in the diet will prevent clotting and cause diarrhoea[75]. Although it is known that the reduction in calcium content of a 'synthetic milk' from 1.26 to 0.73 g/l will induce diarrhoea[75], only a slight beneficial effect of calcium addition to a 'severely' heat-treated milk has been found, even though such addition improved the clotting of the milk *in vitro*[73].

Antibiotic supplementation does not appear to prevent the detrimental effect of a 'severely' heat-treated skim milk[35].

As might be expected from the finding that there is no change in biological value, supplementation with the amino acid methionine is also ineffective. Addition of undenatured whey proteins from liquid whey appears to be the only factor that will improve a severely heated skim milk powder[71]; this finding is in keeping with other results which showed that the presence of denatured whey proteins increases the time taken for the clotting of the casein of milk by rennet[103]. The presence of denatured whey proteins thus appears to have an important detrimental effect on the rate of clotting, and this is additional to the fact that heat treatment sufficient to cause denaturation also results in the binding of the calcium and a reduction in the concentration of ionizable calcium.

The feeding of milks that do not coagulate properly in the abomasum also result in reduced gastric acid secretion (an important protective barrier to the multiplication of *E. coli*). Thus, diarrhoea in calves has been associated with the rapid passage of milk from the abomasum without sufficient gastric digestion[35]. It is also known that, after feeding, there is a much more rapid fall in the pH in the abomasum and a much lower proportion of undigested protein passing into the duodenum of a calf given a diet containing a 'mildly' preheated spray-dried skim milk than when a diet containing a 'severely' preheated spray-dried skim milk is given[79]. Not only is less gastric acid secreted[60, 79, 80, 200] but also there is a tendency for a reduction in chymosin (rennin) and pepsin secretion[80] and reduced proteolysis[201, 202]. A 'severely' heat-treated milk also tends to cause a reduction in the rate of pancreatic secretion and protease outflow[60, 61].

It must be emphasized that the most marked detrimental effect of a poor quality skim milk powder occurs during the first 3 weeks of life under conditions of subclinical infection with *E. coli*, and that the breed of calf may also be important; calves of the Friesian breed may well be less susceptible than those of smaller breeds, possibly owing to their higher potential for digesting protein.

It has been suggested in Holland[104, 105] that, owing to poor rennet clotting of milk substitute diets, absorption in the small intestine of only partially broken down milk proteins may occur, with the resultant formation of antibodies, so that at a later stage anaphylactic shock symptoms may be observed. This suggestion has not been confirmed by others, who consider that some factor in the milk substitute, other than the milk protein, was involved[106].

With a good quality milk powder true digestibility of the protein will approach 100 per cent and the apparent digestibility will be about 94 per cent. However, at 1 week of age the digestibility will be slightly lower than at 4 weeks or at an older age.

Some of the common processing methods, together with their effect on the whey proteins, are listed in Table 3.3. Only a 'mild' preheating treatment before spray-drying gives a product closely similar to that of fresh milk. The content of non-casein nitrogen[416] (a measure of the whey proteins, together with a small amount of proteose-peptone nitrogen and non-protein nitrogen that are little affected by heat) as a percentage of the total nitrogen content indicates how much denaturation of the whey protein has occurred during processing. These analyses are meaningful if done on skim milk or skim milk with added fat but not if other products such as dried whey or vegetable protein are added. In raw milk the percentage is about 25, but in UHT sterilized milk it is only about 11. The inclusion of whey in the milk substitute will invalidate the results, because of its high

Table 3.3 Effect of heat treatment of milk on the calf[67, 87]

Treatment of milk	*Percentage total milk nitrogen as non-casein nitrogen*
No demonstrable effect on calf:	
Raw	25
Holder pasteurized (63°C for 30 min)	23
Spray-dried skim (preheating temperature 77°C for 15 s)	22
Detrimental to the calf, especially during the first 3 weeks of life:	
Spray-dried skim (preheating temperature 74°C for 30 min+)	15
Roller-dried skim (110°C)	13
UHT sterilized (135°C for 1–3 s)	11

concentration of proteose-peptone nitrogen and non-protein nitrogen; in dried whey, the majority of the proteins are usually denatured.

Adequate coagulation of a reconstituted milk powder should be obtained if the percentage is greater than 18; but a value of 22 is about the maximum that will be obtained in the most conservatively dried product. In the USA there is an American Dry Milk Institute (ADMI) grading system of milk powders – high-heat powder (<1.5 mg whey protein nitrogen (WPN)/g powder), medium-heat powder (>1.5–<6.0 mg WPN/g powder) and low-heat powder (>6.0 mg WPN/g powder[107]), but no such system exists in Europe[108]. These values of WPN, produced from the difference between the total nitrogen concentration and the values obtained by saturated sodium chloride precipitation of the casein together with the denatured whey, do not include proteose-peptone nitrogen and are thus 17 per cent less[109, 110] than the values obtained by the acid precipitation of casein used in the UK[111].

In Canada[112] it has been confirmed that a high-heat powder (0.2 mg WPN/g powder) caused an increased incidence of diarrhoea and lower weight gains than did a medium- or low-heat powder (4.0 or 8.1 mg WPN/g powder) when given to calves from 7 d of age, but there was no effect after 19 d on the experiment indicating that for older calves coagulation does not appear to be so important[117]. However, with the feeding of milk subjected to a very severe heat treatment sufficient to cause browning from the Maillard reaction, frequency of diarrhoea was increased in 6-week-old calves[231]. The Canadian workers suggested that gel formation should be used as a criterion of quality, since inclusion of ingredients other than fat and skim milk invalidates the assessment of quality, based on WPN determination.

Since the dividing line between suitable and detrimental skim milk powders is about 170 mg non-casein nitrogen/g total nitrogen[113], which is equivalent to 10.1 mg non-casein nitrogen/g dry matter, subtraction of 3.3 mg non-protein nitrogen/g dry matter (DM) gives a value of 6.8 mg WPN/g DM[111] equivalent to an ADMI value of 5.6 mg WPN/g DM (an allowance having been made for the proteose-peptone nitrogen not included in the ADMI grading) or 5.4 mg WPN/g powder (96 per cent dry matter). Thus, an ADMI low-heat powder in the USA would fulfil the necessary specification as a suitable skim powder for feeding to calves directly after the colostrum feeding period.

The results of a study of the amount of denaturation that was found in skim milk powders commercially available in the UK between 1965 and 1967 are given in Table 3.4 which gives some idea of the range of quality of products available for incorporation in milk substitute diets for calves.

Thus the firmness of the curd from the coagulation of milk protein increases with:

(a) a lower pH, over the range 5.6–6.6[115];
(b) a higher concentration of skim milk solids, over the range 50–200 g/l[115];
(c) a higher concentration of chymosin[79, 115];
(d) a lower preheating temperature before drying[79, 115];
(e) a higher temperature of the milk, over the range 30–37°C[115]; and
(f) a higher ionized calcium concentration[79].

Acid-precipitated casein and whey have a slower coagulation rate at pH 6.1 and 39°C and a weaker curd strength than does skim milk powder. They also markedly

Table 3.4 Ratios of non-casein nitrogen to total nitrogen in commercial skim milk powders produced at UK creameries (1965–67)

Creamery	*Type of dryer*	*Ratio of non-casein nitrogen to total nitrogen*
A	Spray	0.22
B	Spray	0.22
C	Spray	0.16
D	Spray	0.15
E	Spray	0.12
F	Spray	0.11
G	Spray	0.11
H	Roller	0.16
I	Roller	0.13

reduce the amount of fat retained in the clot[116]. Even reconstituting milk powder in water at 56°C for 2 min reduced the strength of the curd[115].

In *in vitro* experiments, low pressure dispersion of fat into dried skim milk powder caused higher curd firmness with chymosin than did homogenization of fat at all total solid concentrations (100–200 g/kg) and at all fat concentrations tested (100–400 g/kg DM). At higher total solid concentrations, curd firmness, clot weight, and the proportion of the fat included in the clot were all increased[114]. Low pressure dispersion of fat also promoted chymosin coagulation and a firmer coagulum than did homogenization when skim milk was partially replaced by insoluble fish protein concentrate-whey and isolated soya protein-whey mixtures[114].

A marked reduction in digestibility of milk organic matter during the period 14–20 d of age was brought about by causing partial or complete uncoaguability by adding sodium citrate or HCl to the diet before feeding[211].

Soyabean protein

Milk substitutes containing non-milk protein which do not coagulate in the abomasum[136] will similarly predispose calves to enteric infections.

Soyabean protein has a great potential for replacement of milk since it is readily available and has a reasonably well-balanced amino acid composition, although its methionine concentration is low. However, soyabeans contain trypsin inhibitor, haemagglutinins, and other detrimental factors which must be destroyed or removed by controlled toasting before they are suitable for inclusion in milk substitute diets. They also contain about 30 per cent of oligosaccharides, which are probably unavailable to the calf.

The inclusion of crude soyabean products in milk substitutes results in diarrhoea and a high mortality. Removal of soyabean oil[227], supplementation with methionine[228], and destruction of trypsin inhibitors and haemagglutinins by steaming or toasting[229, 230] or by treatment with acid or alkali[453] have produced soya flours that are still not satisfactory. Removal of the oligosaccharides produces a soya protein concentrate and even these may be poorly utilized if given to calves before 10 d of age and in the absence of access to additional dry food[230].

The increased size of pancreas that often occurs in calves given diets containing soya flour does not seem to be associated with the presence of trypsin inhibitor since the pancreas was found to be smaller for calves given raw soyabean or heated soyabean plus trypsin inhibitor than for those given heated soyabean alone or given heated plus raw soyabean plus trypsin inhibitor. However, all calves showed growth depression and reduced digestibility, except those given the heated soyabean alone[231].

Specific IgG antibodies against the soyabean globulins, glycinin and β-conglycinin, have been shown to appear in the blood serum of calves that have received several feeds of soyabean flour[232], causing a marked inhibition of abomasal emptying, followed by a reduction in transit time through the small intestine[233–235], reduced absorption of dietary protein, and a transitory rise in the permeability of the intestinal mucosa to protein molecules[223].

Feeds were given at intervals of 2–3 d and by the fifth feed digestive disturbances occurred[236]. The severity of the digestive disorders in calves given various soya products was related to the antibody titre[234], but only a proportion of calves appeared to become sensitized[220].

Heated or unheated soyabean flour caused inhibition of abomasal emptying for some hours, whereas soyabean concentrate prepared by alcohol extraction had little or no effect[234]. Since a suspension of heated soyabean flour put into the duodenum was found to slow the outflow of an abomasal feed containing casein[237], inhibition was clearly caused by a factor entering the duodenum. Although calves given the soyabean isolate showed lower antibody titres than did calves given soyabean flour, the holdback of nitrogen in the abomasum was more marked for the isolate than for the flour. After several hours of holdup, there was a rapid outflow of nitrogen from the isolate[234].

Under optimal *in vitro* conditions of pH for protease activity, β-conglycinin but not glycinin was unaffected by pepsin and both antigens were resistant to rennin and trypsin. The solubility of glycinin and β-conglycinin in saline extracts remained high over pH ranges of 2–3 and 6–10 but under conditions of intermediate pH, about 4.5, solubility was minimal. Thus preruminant calves suffer from gastrointestinal hypersensitive responses to certain soyabean products because the major proteases of the digestive tract fail to denature soluble antigenic constituents of the soyabean protein[272].

Treatment of soya with ethanol under specific conditions may overcome the problem[265]. Extracting fat-free raw soyabean with 75 per cent ethanol at 78–80°C gave somewhat better results than treating with alcohol and evaporating to dryness, a finding that suggests that the removal of the oligosaccharides as well as the beneficial effect of ethanol on the protein of the soya may be important. Although both treated soya products caused some changes in digesta movement, they were much less than that caused by heated soyabean[236, 266]. The concentration of ethanol in water and the temperature of extraction are critical if maximum destruction of glycinin and β-conglycinin is to be achieved. The recommended conditions are a concentration of between 55 and 76 per cent ethanol at a temperature between 70 and 80°C[454]. The haemagglutination inhibition assay has been found to be a suitable method of monitoring soyabean products to predict possible gastrointestinal allergic reactions[267].

Simpler methods based on steam rather than ethanol have now been developed for destruction of the glycinin and β-conglycinin.

Fat

An increased incidence of diarrhoea will occur in calves given diets containing high concentrations of fat, especially if the fat is of low inherent digestibility, but the diarrhoea is rarely watery. However, even with butterfat, concentrations of 490 and 550 g/kg dry matter resulted in a high incidence of diarrhoea[204]. Low-fat diets (skim milk containing only 10 g fat/kg dry matter) in comparison with high-fat diets (200 g/kg dry matter) also caused an increased incidence of diarrhoea; this possibly may be exacerbated if the diet contains only a low concentration of dry matter, i.e. 90–100 g dry matter/kg[205]. However, with once daily feeding high dry matter concentrations (200 g dry matter/kg) have been associated with an increased incidence of diarrhoea[206].

Although low-fat diets do not affect gastric proteolysis, pancreatic enzyme activity is reduced[60, 61]. The diarrhoea may be of a fermentative type similar to that produced by diets containing too much soluble carbohydrate.

However, it was found that calves given skim milk had 10^3 more coliforms in the proximal part of the small intestine than did calves given whole milk. The fatty acids responsible for the control of the coli are probably C8 and C10 acids[207]. The reason for the higher numbers of coli in the distal than in the proximal small intestine has been attributed to the fact that fat is mainly absorbed in the proximal 30 per cent of the intestines[208].

In Japan, the dry matter content of faeces and faecal pH were lower in calves given a milk substitute in which the fat was soyabean oil rather than tallow. The milk substitute contained 300 g milk protein and 100 g fat/kg dry matter. There was a tendency for increased viable counts of lactobacilli in faeces of calves given the diet containing soyabean oil[151].

Non-specific diarrhoea in children <3 years of age has also been associated with a low intake of dietary fat and may be controlled by increasing the fat content of the diet[152].

Effect of diet of the dam

Diarrhoea in suckled calves has been associated with changes in the composition of whole milk as a result of their dams grazing particular pastures[81] or consuming particular conserved grass products[302]. Moreover, suckler cows may be sufficiently deficient in calcium so that the coagulation of their milk is delayed from a normal time of 5–7 min up to more than 1 h or even 24 h. This has resulted in diarrhoea in their calves associated with a marked distension of the abomasum either due to gastric stasis or due to hypersecretion of fluid into the abomasum[130, 302]. Both these conditions of suckling calves could be alleviated by giving additional calcium, for instance in the form of calcium chloride, to the calves[81, 130].

In further studies of milk samples from 63 individual cows in two dairy herds in Caithness, it was found that 38 clotted in 0–7 min, but 11 took 16–60 min and seven took >60 min. Of 26 group samples submitted for brucellosis testing, each containing milk of up to five cows from five farms in Caithness and Orkney and involving 135 cows, only one sample clotted in 0–7 min, 14 took 16–60 min and seven took >60 min. All milk samples showed normal calcium concentration, but low ultrafiltrable calcium concentrations occurred in the milk with impaired clotting. Phosphate concentrations were also subnormal, especially the ultrafiltrable phosphate, in the milk with impaired clotting. Magnesium and citrate concentrations tended to be lower for the abnormal milk but the ultrafiltrable magnesium concentrations were normal.

The incidence of the condition in these dairy cows (40 per cent) was similar to that of the suckler cows in the earlier study (47 per cent). No seasonal effect on the incidence of the condition was detected[303].

The condition may also be associated with the secretion in the milk of excessive amounts of non-protein nitrogenous compounds. It is well known that the firmness of the curd in cheese making is associated with the concentration of casein in the milk as well as with the level of calcium. There is a marked variation in casein content of bulked milk between different areas of the country and in different months of the year[82]. In particular, milk produced between January and April, inclusive, has a casein content well below the annual average. Similarly, very high nitrogen fertilization has been shown to result in milk of poor coagulation characteristics[83].

Contamination of milk with E. coli

Contamination of milk with *E. coli* organisms has also been suggested as a predisposing cause[86], but whereas *E. coli* septicaemia can be produced in the colostrum-deprived calf by feeding specific strains of coli[6], the production of *E. coli* localized intestinal infection by feeding the organism isolated from calves that have died from this condition has often been unsuccessful[87]. However, experimental colibacillosis has been produced in colostrum-fed calves, for instance by infection with *E. coli* O101:K( ). In the first 24 h after onset of diarrhoea, the output and composition of the faeces were as shown in Table 3.5.

Table 3.5 Composition of faeces 24 h after onset of diarrhoea in experimental colibacillosis in colostrum-fed calves

	Control	*Infected*
Faeces(kg)	0.13	5.6
Supernatant of faeces		
Na (mM)	107	134
Cl (mM)	24	100
pH	5.5	7.1
HCO_3 (mM)	<10	41
Lactic acid (mM)	2.1	1.0
VFA (mM)	26.7	6.0

Vitamin A deficiency

Although supplementation of the newborn calf with vitamin A, or prepartum feeding of vitamin A to the dam, has not been shown to give any protection against *E. coli* infections[84, 85], a calf born of a vitamin A-deficient dam is likely to be more susceptible to damage to the intestinal epithelial cells and thus to diarrhoea. However, no relationship has been found between post-colostral serum carotene concentrations and health or growth of the calf[457].

Method of feeding

The rate of coagulation by chymosin is greater for calves sucking nurse cows or fed through a nipple rather than from a bucket, possibly due to the greater abomasal acidity. Moreover the milk may be exposed to more salivary lipase which may affect abomasal degradation of protein[455].

In calves fed without restriction from nipple feeders, abomasum pH remained constant at 3.5, whereas in bucket-fed calves given 1.5 or 4 litres milk per feed, pH increased steadily after feeding to 4.7 and 5.1 respectively and only returned to 2.7–2.9 after several hours[456].

Although it has been suggested that the raised abomasal pH after a meal may allow viable pathogens to pass into the small intestine, no difference in pH 2 h after feeding between diarrhoeic and normal calves was found[418]. Moreover, no relationship has been found between abomasal pH and isolation of ETEC, *Cryptosporidium,* rotavirus and coronavirus at the ileum[153].

In England, there was little difference between the costs of therapy for enteric disease between bucket feeding once or twice daily, but there was more diarrhoea with calves having milk *ad libitum* from a machine. The machine dispensed 35 g milk powder in 0.5 litres water, and dispensing was not possible within 2 min of the completion of a mixing cycle. The powder was formulated for minimal foaming and for good free-flow properties, but no mention is made of the quality in relation to its clotting ability[163]. In a further trial diarrhoea was more common when a machine dispensed reconstituted milk as fast as it was drunk than with once daily bucket feeding or with a machine dispensing in 2 min cycles[164].

In W. Germany, diarrhoea occurred in the first 3 d of life of 60 per cent of calves born of 53 cows that had dystokia or a caesarean and were given milk by buckets, whereas only 13 per cent of 30 similar calves showed diarrhoea when given unrestricted access to nipple feeders containing sour milk or cold whole milk. Only two calves in the first group and one in the second died. It was only considered necessary to feed warm milk when bucket feeding was used. No correlation was found between birth weight and the incidence of diarrhoea[162].

System of rearing

In a study of suckler cows at pasture with calves, home-reared or purchased early weaned calves and purchased veal calves in individual crates or in groups, 30 per cent of the records of treatment for infectious disease were for septicaemia or enteric disease. Treatment for enteric disease was almost entirely given to unweaned calves; the only groups requiring treatment after 6 weeks of age were veal calves. The proportion of group-housed veal calves treated for enteric disease in the first 2 weeks on the unit was higher than that of any other group. For early weaned calves, bucket feeding in individual pens or teat feeding in groups were equally successful[322].

The physical environment

Although the evidence is largely circumstantial, low air temperature, rapid changes in air temperature, increased air velocity and precipitation have all been considered as predisposing factors to diarrhoea[155–157]. However, no relationship was found between minimum temperature, rainfall and hours of sunshine and the incidence of diarrhoea in the first 3 weeks of life of calves kept outside from birth[159].

In Canada, it was reported that an increased incidence of diarrhoea occurred 2 d after snow storms, which were characterized by low and changeable temperatures[158]. In N. Dakota most diarrhoea occurred in March and April[400].

Environmental temperature

Environmental temperatures below the critical temperature of the calf may contribute to the pathogenesis of diarrhoea by increasing the rate of passage, and thus decreasing the apparent digestibility of milk substitutes and the inactivation of bacteria in the abomasum, so that increased amounts of undigested food reach the ileum[154]. Quite small decreases in digestibility will cause large increases in the quantity of digesta present at the ileum[161].

A survey of treatments for enteric disease in a climatic (mean air temperature 8.2°C, air movement 0.19 m/s) and an environmentally-controlled (mean air temperature 12.8°C, air movement 0.09 m/s) calf house, showed a trend for a higher proportion of calves being treated in the climatic house[322]. Not surprisingly, no difference in the incidence of diarrhoea in the first 2 weeks of life occurred between calves maintained at 14.8°C rather than at 20°C[40].

Relative humidity

An increased incidence of diarrhoea up to 6 weeks of age was associated with calves kept at a RH of 0.95 on wet wooden slats compared with those kept at a RH of 0.75 on dry wooden slats, irrespective of whether the environmental temperature was 7°C or 15°C[160]. It was suggested that this might have resulted from the increased heat loss involved in evaporating moisture from their coats, increased heat loss to the wet slats due to increased thermal conductivity and increased survival of enteric pathogens.

Breed of calf

It must be borne in mind that there are breed differences in the digestibility of protein and that, for this reason, the extreme dairy breeds may be more susceptible to the effects of excessive protein. It has also been reported that less diarrhoea occurred in calves from crossbred dams, but there was no difference in incidence for crossbred calves out of purebred parents and crossbred calves out of crossbred dams[400].

Age of dam

More calves from cows calving at 2 years of age had diarrhoea than had those from older dams[400].

Dystokia

A large proportion of calves from assisted deliveries in N. Dakota had diarrhoea[400].

Prophylaxis against neonatal diarrhoea

Vaccination against *E. coli* infection

The calf

On some farms where *E. coli* infections appear to occur regularly proprietary *E. coli* immune serum has been used as a preventative with good results. However, strains of coli against which the serum is ineffective may be introduced on the farm and other treatment will be necessary. (For immunization by mouth, see Chapter 2, p. 45).

The dam

Hyperimmunization of pregnant cows with *E. coli* antigen has been used to produce colostrum with augmented protection against *E. coli* infections[278–281].

Recently, by genetic engineering, a recombinant vaccine against the K99 antigen of *E. coli* has been produced. The possibility of producing a hyperimmune bovine colostral whey to cholera enterotoxin has also been mooted[283].

Europe

In Czechoslovakia, prevention of *E. coli* infections in calves is based on giving dried or liquid immune colostrum or serum obtained from pregnant cows injected with a polyvalent vaccine against enteropathogenic strains of *E. coli*[290]. Vaccination of pregnant dams against six strains of four 'O' groups, O9, O101, O8 and O141, reduced mortality in the calves from 16.3 to 2.3 per cent. Moreover, treatment of calves suffering from diarrhoea with a polyvalent serum against these strains (four doses of 50 ml) resulted in the recovery of calves suffering from diarrhoea[55].

In W. Germany, intramuscular vaccination of dams 5 and 2 weeks prepartum with a vaccine prepared from *E. coli* strains isolated from a calf gave unreliable results, but an oral vaccine given to calves from 1–10 d of age was more reliable[299]. Vaccination of pregnant cows in Belgium was found to increase the protective properties of colostrum against *E. coli* septicaemia[295].

In France, a hyperimmune colostral serum from cows hyperimmunized with *E. coli*, *Pasteurella multocida* and *Salmonella* spp. was given to calves at 8–12 d of age by subcutaneous injection. There were no deaths of the 64 calves treated and only 25 per cent needed further treatment. Of the 58 control calves, five died and 36 per cent needed treatment for diarrhoea[289]. In the same country vaccination of cows subcutaneously at 45 and 15 d before parturition with K99 antigen gave protection to calves, given 2 litres colostrum, against the homologous *E. coli* K99 strain, in that they were less prone to diarrhoea, but only half the calves were protected against a heterologous strain. The vaccine was most effective 20–40 d before parturition but it was not tolerated by all cows. It was suggested that protection was due to inhibition by the K99 antibodies of adherence to the small intestine epithelium[293, 297]. However, the use of a K99/F41 vaccine may prove to be ineffective in preventing diarrhoea caused by FY (Att 25)$^{+}$ strains[287].

America

In the USA, oral inoculation of approximately 1.2×10^7 viable *E. coli* to pregnant cows resulted in increased blood serum and colostral whey titres to the 'O' antigen. However, there was no correlation between the antibody titre of the colostrum ingested and either the serum antibody titre of the calf or its incidence of diarrhoea. The latter was greater in calves that did not ingest colostrum until after 12 h of age[292].

On the other hand, inoculation of cows at 4–6 and 2–3 weeks before calving with an inactivated vaccine produced from *E. coli* strain 1474 (K12, K99) protected all 24 calves against diarrhoea, dehydration and depression when challenged orally with *E. coli* strain O9:K30, K99 at 12 h of age, whereas of 20 control calves, 19 showed severe signs of infection within 12–24 h of challenge, 17 had watery diarrhoea and 10 died. However, the correlation between microagglutination antibody titres and protection against challenge was poor[294].

In Canada, a K99 vaccine was found to be effective against challenge in 1-day-old calves[301].

Asia

In Japan, hyperimmune bovine colostral whey from an *E. coli* serotype given in the final stages of gestation has been used for treatment of infants and as a prophylactic for premature infants in incubators[282].

Vaccination against rotavirus infection

The calf

Because most enteric viral infections occur in calves of <3 weeks of age, passive lactogenic immunity within the gut plays an important role in protection of susceptible intestinal cells[291].

Thus, newborn calves inoculated orally with attenuated bovine rotavirus responded with local IgA production; most of the rotavirus antibody-producing cells were located in the mucosa of the proximal small intestine[288]. Inoculation of gnotobiotic calves with an attenuated strain produced protection within 7–21 d but not within 3–5 d[296].

Of eight calves vaccinated with a live rotavirus vaccine in S. Africa, five showed mild diarrhoea after vaccination. However, calves remained healthy after challenge, although rotavirus particles were detected for a few days after vaccination and challenge[300].

Of six colostrum-deprived newborn calves in Belgium infected orally with a cell-culture bovine rotavirus after 4–6 passages in cells, none developed diarrhoea but all excreted virus for at least 2 weeks after inoculation and specific rotavirus antibodies appeared in their sera[312].

The dam

Hyperimmune colostral whey from pregnant cows vaccinated with rotavirus has been used to protect calves from rotavirus infection.

Europe

In N. Ireland, vaccination of cows with a combined inactivated, adjuvanted bovine rotavirus and *E. coli* K99 vaccine resulted in an increase in neutralizing antibody titre to rotavirus in serum and colostral whey. Evidence was obtained that vaccination resulted in a decreased incidence of rotavirus shedding and of abnormal faeces or diarrhoea in calves given colostrum and milk from the vaccinated dams. The *E. coli* component of the vaccine could not be tested as there was no evidence of natural challenge[285].

The difficulty of conducting field trials with vaccines against only two components of a multifactorial syndrome was illustrated by the trial in Scotland[286] where a vaccine against rotavirus and K99 from ETEC was given intramuscularly to pregnant cows 4–14 weeks before calving. Calves born and reared from ten vaccinated dams were protected from experimental rotavirus infection at 5 d compared with calves from six unvaccinated control cows.

However, in field trials on 40 farms, there was no significant diarrhoea problem in 31 herds (<10 per cent morbidity). Of the nine remaining farms, with >10 per cent morbidity from diarrhoea, four farms were associated with a cryptosporidiasis and one farm had no enteropathogen detected. All these 36 farms were excluded from the analysis. Of the remaining four farms, two beef suckler herds had concurrent rotavirus and cryptosporidial infection and vaccination was associated with a decreased excretion of rotavirus but not with a reduction in the incidence of diarrhoea. In the other two dairy herds (68 cows), which had a history of rotavirus infection, there was a significant decrease in the incidence of diarrhoea in the calves born to vaccinated cows. As in the N. Ireland trial, no natural challenge to ETEC occurred[286].

America

Rotavirus IgG_1 antibodies in colostrum and milk were elevated after intramammary vaccination of pregnant cows with an Ohio Agricultural Research and Development Centre rotavirus vaccine but not after intramammary immunization with a commercial rota-coronavirus vaccine. Colostrum from the former vaccination prevented diarrhoea and shedding of rotavirus by newborn calves challenged with rotavirus[291].

In Canada in two double blind trials in 18 herds neither modified live reo-like virus vaccine nor formalin-killed bacteria of six strains of enterotoxigenic *E. coli* used alone or in combination had any effect on the frequency and severity of calf diarrhoea. However, the results were complicated by the presence of a coronavirus in some herds[298].

Man

Since infection of calves *in utero* with rotavirus induced resistance to diarrhoeal disease caused by human as well as the homologous bovine virus, the latter might possibly be used in a human vaccine[198]. In fact, hyperimmune colostrum from cows vaccinated with human rotavirus at 10 d intervals from the eighth month of pregnancy has been used to prevent rotavirus infection in humans[284].

Probiotics

Lactobacilli

Milk soured with a culture of *Lactobacillus acidophilous* has been claimed as effective in controlling diarrhoea in calves in the USA[396], USSR[397, 398] and Czechoslovakia[399]. In India, all strains of *L. bulgaricus*, *L. acidophilous* and *L. lactis* and some strains of *L. plantarum* showed inhibitory action against most test organisms whether Gram-positive or negative, including pathogenic strains of *E. coli*. Inhibition was partly due to the presence of some antibacterial substances and partly to lactic acid, but hydrogen peroxide was not detected[391].

It has been claimed that, by giving 279 newborn calves regenerated lyophilized (freeze-dried) *Streptococcus faecalis* and *L. acidophilous* of calf origin at the rate of 1×10^{11} for 3 d and 1.25×10^{9} for 10 d, with colostrum being given 2 h after the first dose, mortality was reduced from 10.2 to 2.8 per cent and prevalence of diarrhoea from 82 to 34 per cent, but the comparison was with the results obtained in the previous year when the programme was not in operation[392].

However, in a study of the effects of acid and fat in controlling *E. coli* in the gastrointestinal tract, there was no confirmation that lactobacilli were antagonistic to *E. coli*[207].

Enterococci

A prophylactic system based on the biological competition between *E. coli* and enterococci plus lactobacilli, together with the disinfection of the udder before each suckling, has been advocated. Administration of the latter bacteria immediately after birth and before absorption of colostrum and continued during the first 3 d of life reduced mortality in a calf house from 7.2 to 3.1 per cent and incidence of diarrhoea from 62 to 20 per cent, but there were no contemporary untreated calves[393].

The use of *Enterococcus faecium* as a probiotic agent has been studied in purchased Friesian bull and Hereford × Friesian heifer calves. In the first trial the milk substitute, based either on skim milk powder or soya protein, contained virginiamycin and the probiotic was given daily. In the second trial the milk substitute was based on soya protein and only half the calves received virginiamycin. Virginiamycin had no apparent effect on the viability of *E. faecium* in the milk substitute. *E. faecium* had no beneficial effect on the health or performance of the calves up to 12 weeks of age[395].

Bacteriophage (phage)

A mixture of two phages protected calves against a potentially lethal oral infection with *E. coli* strain B44 (O9:K30, K99) when the phages were given before, but not after, the onset of diarrhoea. Calves that responded to phage treatment had fewer B44 and those that did not respond had higher numbers of mutants of B44 resistant to phage in their small intestines. Phage-treated calves that survived *E. coli* infection continued to excrete phage in their faeces, at least until numbers of B44 excreted were low. Calves that were inoculated orally with faecal samples from phage-treated calves, which contained sufficient B44 to cause a lethal infection, remained healthy[126].

Faeces from adult ruminants

When five colostrum-deprived calves were given orally 100 g faeces from an adult Jersey steer mixed in 500 ml milk substitute immediately on arrival at the Central Veterinary Laboratory at Weybridge, two exhibited a mild transient diarrhoea, one at 3 d and one at 6 d, and the others remained healthy. Of the six control calves, four died within 6 d after birth. There was no evidence of colostrum intake in the calves, the latex agglutination test giving values of <5 g Ig/l serum[394].

Acid

The tendency for the faeces of calves with putrefactive diarrhoea to have an alkaline pH has led to the suggestion that dilute acid (citric acid or a mixture of citric acid and hydrochloric acid) solutions should be given to calves to reduce the pH. For normal calves the modal pH was about 6.3, compared with 7.3 for calves with diarrhoea. Treatment with acid solution of the calves that had diarrhoea reduced the pH of the faeces to about the normal value and seemed to cause the replacement of the *E. coli* in the faeces by lactobacilli[36], although no objective evidence of this was presented.

It is claimed that acidified milk (pH 4.3–5.5) and possibly also fermented colostrum reduce the incidence of diarrhoea and that fermentation of mastitic milk (see Chapter 6) is a method of utilizing this milk for calf feeding without causing health problems[210].

Treatment of a milk substitute with hydrochloric acid to pH 4.25 and then dispersal of the casein clot by high-speed stirring reduced the incidence of diarrhoea compared to that with the untreated milk substitute, but appetite and digestibility during the first 3 weeks of life were reduced[211].

In Wales, the use of an acidified milk replacer up to a maximum intake of 5.5 litres per day was claimed to significantly reduce the faecal coliform count before 2 and also before 4 weeks of age from 10^9 to 10^8, but no serotyping of the *E. coli* was done[209].

However, calves given milk containing 32 mM propionic acid had 10^3 lower counts of lactobacilli in the entire gastrointestinal tract than did calves given milk without added propionic acid[207].

How far the acid supplied in an acidified milk alleviates a deficiency of gastric acid production because of the poor quality of the milk substitute used, e.g. of the skim milk powder (see above), is not known.

Measures that can be taken by the calf rearer

E. coli infections are undoubtedly associated with failures on the part of stockmen, farmers and dealers to exercise the rules of good animal husbandry, and on the part of manufacturers to produce milk substitutes that do not predispose to this condition.

By feeding adequate quantities of colostrum to the calf during the first 24 h of life, the incidence of *E. coli* septicaemia should be low. Antibodies against the strains of *E. coli* present in the cow's environment associated with the globulins of colostrum should protect the calf against death. However, new strains may be introduced with purchased calves, and the colostrum of down-calving cows or heifers purchased shortly before calving may not contain antibodies against the strains of *E. coli* on the buyer's farm. It is possible that strains present on a farm may increase in virulence as they pass through successive calves. Even though the feeding of colostrum should protect the majority of calves from an *E. coli* septicaemia, deaths from this cause still occur quite frequently under commercial conditions.

Deaths from *E. coli* infection often are associated with divergences from the finer aspects of good management; some of these faults are indicated below:

1. The interval between birth and the first feed of colostrum is delayed too long.
2. Colostrum from the first milking after calving is not given to the calf at the first feed.
3. Colostrum is diluted with water, so that the concentration of immune lactoglobulins is reduced, or heated sufficiently to denature the whey proteins, including the immune lactoglobulins.
4. Insufficient colostrum is offered to the calf.

5. Cows may be milked before parturition with the result that the postpartum secretion is of similar composition to that of milk[309].

Much can be done by the calf rearer to eliminate the predisposing causes of diarrhoea. The following points in particular should be borne in mind:

1. The younger the age of the calf when first given a milk substitute, the better must be the quality of the food. If given directly after the colostrum feeding period, the milk substitute must clot readily in the abomasum. This can be tested by adding 0.2 ml commercial rennet to 100 ml reconstituted powder at 37°C. A good-quality powder should form a firm clot within 3–4 min; a poor powder will take much longer and will produce only a flocculent curd.
2. A high-fat milk substitute is likely to give better results than a low-fat milk substitute, provided that the digestibility of the fat is high and fat globule size is small. This is especially so if the milk substitute is being fed at high levels.
3. Overfeeding of calves should be avoided, especially if their previous history when bought-in is unknown. If the calves are very thirsty on arrival, they may be offered glucose solution (30–50 g/l). If any calf has diarrhoea, it is probably wiser not to feed glucose, as it may aggravate the condition; water, isotonic saline or sodium bicarbonate is likely to be more suitable. When the faeces have returned to normal, glucose (30–50 g/l) may then be given, or the calf may be gradually introduced to a milk substitute. Thus, in calves given ETEC orally, there was a tendency for greater mortality in calves given a milk substitute diet rather than water[308].
4. Only freshly reconstituted milk, warmed to body temperature (36–38°C), should be given. If it is desired to feed at a lower temperature, the quality of the milk substitute is likely to be of even greater importance.
5. Feeding should be done at regular times and the utensils scrubbed and sterilized after each feed.
6. If there is a tendency for calves in a calf house to have diarrhoea, introduction of further susceptible animals should be avoided. During the winter months, the incidence of diarrhoea increases and reaches a peak in early spring. With bought-in calves, batch rearing on the 'all in – all out' method is likely to be more successful than the introduction of susceptible animals successively into the same building, a procedure that is liable to result in a 'build-up' of infection.
7. Calves that are purchased under 3 weeks of age should be isolated, or, alternatively, calves should not be bought-in until they have passed the susceptible age, i.e. 21 d[406]. If diarrhoea occurs in purchased calves, a simple blood test such as the ZST test (see Chapter 2) can be made to verify that the calves have received adequate colostrum.

In many cases of diarrhoea the duration of the attack is largely under the control of the calf rearer. During the sixth to twelfth day of life, the calf appears to be particularly susceptible. By careful observation it is possible to anticipate the onset of diarrhoea on the day before it occurs, and milk can then be withheld or reduced in quantity, with the result that the calf recovers quickly.

The following signs of impending diarrhoea should be looked for: dry muzzle, thick mucus appearing from the nostrils, very firm dung, refusal of milk, a tendency to be prostrated and a high rectal temperature (over 39.3°C).

Sudden changes in diet, the use of sour milk, poor hygiene and exposure of the calf to stress in transit and to cold and damp may all play their part in increasing the susceptibility to infection.

Treatment of neonatal diarrhoea

Rehydration

General principles

The intestinal loss of fluid and electrolytes may involve hypersecretion, malabsorption or both. Oral rehydration is particularly useful under practical conditions and is effective whether diarrhoea is due to bacteria or viruses. A calf of 50 kg live weight with 10 per cent dehydration will have a water deficit of 5 litres[358].

Intravenous electrolyte replacement therapy, together with transfusion of plasma, is often advocated but, in general, it has not been very successful under farm conditions. Moreover, in a comparison of whole blood transfusions, balanced fluid replacement therapy and fluids together with antibiotics given to calves having spontaneous diarrhoea and ZST values of 5–15, none of the treatments had any effect on the survival or the incidence of diarrhoea[368].

The principle of rehydration is that if salt and water are presented slowly enough to the intestine, then absorption should be complete. Secondly, the reabsorption is linked to and enhanced by glucose and amino acid absorption. Indeed, it was in Oxford in 1949 that glucose was shown to promote the absorption of water and salt from the gut that eventually led to oral rehydration therapy in man[369]. This is the basis of all salt and sugar solutions used in the treatment of cholera and infant diarrhoea[359].

Tonicity of solution

There is some disagreement as to whether solutions should be isotonic, i.e. 300 mosm/kg or hypotonic. It is suggested that the decision must depend on the severity of the condition. In calves that developed diarrhoea a few days after being bought-in at 4.5 d of age and were given 2–4 litres milk twice daily, the fall in sodium and chloride and rise in potassium concentration in plasma was much greater in calves that died than in those that survived. Of calves that died, plasma osmolality was lowest 4 d before death, was subsequently isotonic for a short period and then hypertonic on the day before death in spite of considerable reduction in sodium and chloride concentrations[367].

Oral rehydration

Until proprietary solutions containing an amino acid were available, isotonic saline (pH 6.2, 9 g NaCl/l), isotonic sodium bicarbonate (pH 8.0, 12 g $NaHCO_3$/l) or Darrow's repair solution (4 g sodium chloride/l, 52 mM sodium lactate, 2.7 g potassium chloride/l)[41] were the usual electrolyte solutions suggested[31]. In Czechoslovakia a solution for oral use containing Na, 90 mM; K, 20 mM; Cl, 75 mM; HCO_3, 33 mM; glucose, 120 mM has been found suitable[42]; the best results for oral rehydration were given by a solution containing (mM/l) NaCl 45, KCl 20, $NaHCO_3$ 45 and glucose 140[42].

In Hungary, a solution containing 0.15 g KCl, 0.25 g $CaCl_2$, 4.3 g NaCl, 30 g glucose, 5 g $NaHCO_3$ and 5 g Na citrate dissolved in camomile tea (/l?) was advocated. The highest rate of recovery from coli-enteritis of 92 per cent occurred in calves treated with this salt mixture + sulphotrim (5 parts sodium sulfachlorpyridazine:1 part trimethoprim) together with a single injection of adrenocortical hormone[361].

It has been recommended that at the onset of diarrhoea, the amount of milk should be reduced in proportion to the severity of the diarrhoea with complete withdrawal for up to three feeds in severe cases and 9 g NaCl/l solution given[362]. In the USA, a comparison between giving Jersey calves suffering from diarrhoea a solution of sodium chloride and sodium bicarbonate, or sodium chloride, sodium bicarbonate and potassium chloride and withholding milk, or reducing milk intake and giving an antibiotic disclosed no difference between treatments[434].

Recommendations for the use of one proprietary solution for diarrhoeic calves is to withdraw milk entirely and feed 2 litres electrolyte solution twice daily for 2 d. In the subsequent 2 d period, 1 litre solution and 1 litre milk at each feed is recommended and then a return to milk alone. For bought-in calves, it is suggested that 2 litres of electrolyte solution is given in place of milk for the first feed and 1 litre electrolyte and 1 litre milk at the next feed and then a return to milk alone.

Glucose and glycine

Although there is a net loss of water and electrolyte into the intestine as a result of the multiplication of enterotoxaemic strains, glucose absorption can still actively occur[43]. Orally administered xylose (200 mg/kg body weight) was also found to be absorbed in diarrhoeic calves although the serum concentrations were lower and time to reach maximum concentration was delayed in diarrhoeic compared with healthy calves[360].

However, glucose solutions should be isotonic (i.e. 50 g/l), since hypertonic solutions have been shown to aggravate dehydration in infants, probably as a result of their osmotic effects[44]. In W. Germany, it was recommended that glucose should always be included in an electrolyte solution if diarrhoea persists for longer than 1 d[363].

The glucose-glycine-electrolyte (GGE) powder of the following composition: 675.3 g glucose, 143.4 g NaCl, 103 g glycine, 8.1 g citric acid, 2.1 g potassium citrate and 68 g potassium dihydrogen phosphate/kg has been evaluated. The powder was stored in twin sachets, containing a total of 64 g, to prevent a 'browning reaction' between glucose and glycine which would otherwise occur in the presence of moisture. Before use, this amount of powder was dissolved in 2 litres warm water and was given in one feed. It is claimed that citrate improves absorption of the solution in the small intestine.

To test the solution (GGES), 150 calves were given an oral inoculation of enteropathogenic *E. coli* (O9:K30,K99). Of these calves, 100 became diarrhoeic and were shown to absorb glucose and glycine at a similar rate to that of healthy calves. The GGES yielded water for absorption more rapidly than did isotonic saline solution or than did a GGES without citrate ions. A comparison was made of GGES, given instead of milk for 2 d followed by a mixture of GGES and milk (1:1) for 2 d, with the same treatment together with 400 mg amoxycillin twice daily for 4 d and with a milk diet plus 400 mg amoxycillin. Mortality was highest in a control group given milk alone, and all three treatments reduced the duration of diarrhoea in the survivors. The GGES plus the antibiotic treatment was the most effective[436]. After infection, PCV was raised and mean blood pH lowered on all treatments, but the blood pH, but not the PCV, returned more rapidly to normal in the treated group.

In the Republic of Ireland an oral GGES alone was as effective as the solution given with an oral triple sulphonamide and streptomycin, but this finding was attributed to the widespread antibiotic resistance[444].

The beneficial effect of inclusion of glucose and glycine in oral rehydration solutions has been shown in E. Germany, where 1080 calves with diarrhoea associated with *E. coli* and rotavirus were given either a solution containing 0.35 g KCl, 0.35 g $CaCl_2$, 0.2 g $MgCl_2$, 3.1 g CH_3COONa and 5.0 g NaCl/l or a new solution containing 4.8 g NaCl, 4.8 g $NaHCO_3$, 20.2 g glucose and 10.1 g glycine/l. The calves were treated by partial or total withdrawal of milk for not more than three meals and an equal amount of electrolyte solution substituted, followed by a gradual return to milk. The new solution was more acceptable to the calves and their illness was milder and of shorter duration[364].

It has been suggested in France that sodium acetate should be included in GGES to replace part of the sodium chloride. Sodium acetate was found to cause much greater absorption of water and sodium in the jejunum and there was still substantial absorption of sodium in the ileum although this was not as great as with sodium chloride. Similarly, acetate was readily absorbed in the jejunum whilst chloride was mainly absorbed in the ileum. In practice, sodium acetate cannot entirely replace sodium chloride due to the large faecal losses of chloride. Moreover, acetate formulations caused a concomitant secretion of chloride and bicarbonate[438].

From further work it was reported that sodium propionate, as well as sodium acetate, stimulated water and sodium absorption especially in the jejunum. In addition, the propionate had a glucogenic effect after absorption. It was also claimed that acetate metabolism gave rise to a progressive alkalinizing effect whilst providing energy. They suggested the following range of concentrations for oral glucose-sodium acetate formulations (mmol/l), Na^+, 70–80; K^+, 20–30; Mg^{++}, 3–5; Ca^{++}, 0–5; Cl^-, 50–60; CH_3COO^-, 30–50; $C_2H_5COO^-$, 0–20; $H_2PO_4^-$, 5–10; glucose, 70–120 with a pH of 6.0–6.5 and a tonicity of 290–350 mosm/l.

The formulation they were testing in the field contained (mmol/l): glucose, 80; Na^+, 80; K^+, 25; Mg^{++}, 4; Cl^-, 54; CH_3COO^-, 40; $C_2H_5COO^-$, 10; $H_2PO_4^-$, 5 with a pH of 6.2[440]. This formulation had been used to treat more than 100 000 calves with a success rate of >95 per cent[445].

For rotavirus or E. coli infections

It is claimed in the Netherlands that it is possible to differentiate between the diarrhoea caused by rotavirus and ETEC.

For all recumbent diarrhoeic calves, it is advocated that 500 ml $NaHCO_3$ (42 g/l) and 100 ml NaCl (9 g/l) should be given intravenously with trimethoprim/sulpha (20 mg/kg body weight) followed by 1 litre glucose-electrolyte solution (per litre, glucose 14.4 g; NaCl 3.8 g; KCl 0.44 g and $NaHCO_3$ 2.94 g) given by oesophageal feeder with 2.5 mg colistin sulphate (a polymyxin antibiotic)/kg body weight. During the first 24 h, 2 litres glucose-electrolyte solution and on the second day 1 litre milk + 1 litre electrolyte solution should be given three times daily.

For the successful treatment of enteric colibacillosis, the hypovolaemia must be corrected and it is suggested that 1 litre of a solution containing 68 g glucose, 6.8 g $NaHCO_3$, 3.1 g NaCl, 1.7 g KCl and 0.4 g $MgSO_4$, and 500 ml $NaHCO_3$ (42 g/l) should be given intravenously together with an oral electrolyte (per litre, glucose 14.4 g; CH_3COONa, 6.6 g; KCl 1 g and NaCl, 1.2 g). The addition of acetate to the glucose seemed to improve its effect on rehydration[366].

Rotavirus-infected calves often have a severe metabolic acidosis, which persists even after repeated intravenous infusion of sodium bicarbonate solutions. Thus, for calves thought to be infected with rotavirus, the same intravenous solution is recommended, but since the digestive capacity of the small intestine is disturbed for at least 2 weeks, small quantities of colostrum (500 ml twice daily) and good quality milk (4–6 times per day) supplemented with 500 ml electrolyte solution containing 0.45 g KCl, 3 g $NaHCO_3$ and 6.5 g NaCl per litre should be given. It is considered that the prolonged replacement of milk by glucose-electrolyte solution is dangerous because of the risk of cachexia[366].

Effect on absorption of antibiotics
The absorption of oxytetracycline, and to a lesser extent that of amoxycillin, was impaired by giving it in milk or water, whereas it was increased when the antibiotics were given in a glucose-glycine-electrolyte solution[365].

Intravenous rehydration

Mild dehydration can be overcome by oral administration, but advanced dehydration will require parenteral administration. Isotonic saline is considered by some to be unsuitable for intravenous administration, as it will tend to aggravate the metabolic acidosis. The benefit of plasma in addition to electrolytes seems to lie in its protein content and thus in its buffering effect on the blood pH. For intravenous infusion 400–800 ml plasma followed alternately by 500 ml Darrow's solution and 500 ml glucose solution (50 g/l) until a total of about 4.5 litres has been infused has been suggested as a suitable procedure. The initial infusion should be given at a slow rate (6 ml/min) and then, after 200 ml has been infused, the rate can be increased to the maximum under gravity flow until 400–600 ml has been infused, with the remainder infused at the rate of 2–3 ml/min[31]. The absence of a sucking reflex indicates the necessity of intravenous therapy with a solution containing Na, 94; K, 15; Cl, 64; lactate or acetate, 45 and glucose, 110 mmol/l[42].

Use of antibiotics

Diarrhoea in calves has not been reliably controlled by the use of antibiotics or by vaccination. Certain sulphur drugs and antibiotics have been used to good effect but, in general, there is no specific cure for all outbreaks of *E. coli* infection.

In Saskatchewan in a clinical trial with 254 calves with acute undifferentiated diarrhoea during a period of 60 d, in which all calves were given fluid therapy to maintain hydration, half the calves were starved for 24 h after admission and half were given milk. Each group was further subdivided into three groups. The first received ampicillin orally, the second a combination of chloramphenicol intravenously and nifuraldezone orally and the third no antibiotic treatment. No effect of the dietary or antibiotic treatment was apparent, the mortality varying between 16 and 40 per cent for the different groups[331].

Antibiotic-resistant strains of E. coli

To enhance the value of medication with antibiotics, bacterial sensitivity tests of the organisms isolated against various antibiotics may be necessary, to ascertain the most efficacious antibiotic to use.

The knowledge that *E. coli* develops resistance to antibiotics both when used therapeutically and when used prophylactically at low levels in proprietary feedstuffs, as has been the custom in the USA[46–49], and the knowledge that resistance may be of a multiple and infective nature[50–54] and may be passed to other organisms such as salmonella by contact, has led to the use of alternative methods of treatment.

Incidence

Europe. Veterinary Investigation Service surveys in the UK indicate little change in prevalence of drug resistance of *E. coli* from 1971 to 1977. Whilst dosing animals with antibiotics leads to immediate increase in resistant organisms in the gut, the numbers level off after treatment[334].

In the Limousin department of France, all *E. coli* strains were reported to be sensitive to gentamicin and amikacin and almost all to flumequine[147], but by 1985 in France[335], as in the UK[336], resistance to gentamicin and apramycin in *E. coli* from calves had been reported. These are the most recently introduced deoxystreptamine aminoglycosides used in veterinary therapy in France.

In Czechoslovakia, a study of 200 strains of *E. coli* from healthy calves and 60 strains from diarrhoeic calves showed that 3–5 d of therapy with chloramphenicol increased the frequency of occurrence of resistant strains[328].

America. In Canada, the majority of ETEC isolated in 1973 were resistant to commonly used antibiotics, except gentamicin, nifuraldezone and polymyxin B[139]. In 1976, it was reported that the majority of 300 isolates were resistant to tetracycline and streptomycin, but all were susceptible to nitrofurantoin and sulfachlorpyridazine[219].

Now that antibiotic-resistant organisms exist, they cannot be eradicated even by avoiding the use of antibiotics. Thus in the USA in pigs, in which no antibiotics were used for 10 years, there was at first a marked fall in the number of antibiotic-resistant organisms, but the proportion then remained constant for several years[334].

Asia. In Iran it was considered that antibiotic resistance was not widespread in animals whose diets did not contain antibiotics. Of 619 isolates, 16.5 per cent were resistant to one or more drugs and 26 per cent of these resistant strains were capable of transferring either part or their entire resistance pattern to sensitive recipient strains[327].

Relationship with antigens of E. coli

A close association has been found between K99 inheritance and resistance to streptomycin, oxytetracycline and sulphonamides[326]. Similarly of 337 strains showing drug resistance, 29 per cent had colicinogenic characters[329].

It has been reported from Denmark that subinhibitory concentrations of tiamulin, a pleuromulin derivative, have a significant effect on the haemagglutinating actions of *E. coli* strains possessing the fimbrial antigens K88 and K99. It thus may influence the pathogenesis of subclinical and possibly clinical *E. coli*-associated enteric syndromes by a direct inhibitory effect on *E. coli* localization in the small intestine. If this process can be prevented, 'flushing' of organisms out of the small intestine is likely to occur[414].

Source of strains

The faecal *E. coli* flora is very complex and is continually changing with time.

In one study it was shown that new *E. coli* 'O' groups associated with increased antibiotic resistance occurred in the faecal flora of calves even if they had received no antibiotics and even though they were separated from other calves by at least 1.5 miles. Moreover, a 5 d prophylactic course of furazolidone did not lead to any additional increase in the proportion of strains resistant to ampicillin, chloramphenicol, kanamycin, streptomycin, sulphafurazole and tetracycline. It was suggested that this increase in antibiotic resistance that occurred relatively rapidly and independently of treatment was due either to the proliferation of a minority flora in the gastrointestinal tract or more likely the tract became colonized by strains from the farm environment that had a high antibiotic resistance as a consequence of selection arising from the therapeutic and prophylactic use of antibiotics over a number of years[325].

Further work indicated that the farm environment was likely to be the source. Thus in intensively-reared purchased calves, the increase in resistance to antibiotics during the first 21 d of life occurred independently of oral drug therapy and resulted from the substitution of the original flora by multiple resistant strains rather than with the dominance of a few resistant clones[323].

Similarly, in dairy heifer replacement calves the incidence of antibiotic resistance was low at 1–2 d of age when they were housed with their dams, but rose rapidly when they were moved into pens with strains being resistant to four or five of the six drugs tested. However, the incidence of drug resistance fell from the third week to low levels by the time the calves were 5 months old due to the increasing proportion of sensitive *E. coli* strains. However, these sensitive strains differed from those that colonized the calves in the first few days of life[324].

Earlier work in Norway had shown that a high proportion of calves at 1–2 weeks of age harboured resistant strains, but these were replaced by sensitive ones as the calves grew older[330].

Antibiotic-resistant *E. coli* have been reported as commonly appearing in the faeces of calves being given milk from cows undergoing intramammary antibiotic therapy for clinical mastitis[332, 333] (see Chapter 6).

The effect of permitted feed-additive antibiotics

The use of zinc bacitracin, flavomycin and nitrovin resulted in greater *E. coli* concentrations in the faeces by $10–10^2$/g. It is considered that at nutritional levels, the flora-stabilizing effect that these antibiotics produce does not exclude the existence of multiple drug-resistant faecal *E. coli*, but it is only therapeutic levels of standard antibiotics and sulphonamides that produce a definite selection of drug-resistant strains. When the use of such antibiotics is stopped, there remains a high proportion of organisms resistant to them, but the number of compounds against which they are insensitive diminishes[329].

Milk powders may be imported, which often contain antibiotics that may not have a product licence in the UK or may be included at a level outside the UK product licence. The responsibility for control of these imports must rest with the Customs authorities[340].

The effect of penicillinases

Broad-spectrum penicillins, such as ampicillin, may be inactivated by penicillinases produced by *E. coli*. Until recently amoxycillin potentiated by clavulanic acid was the only available compound to overcome this problem but it was restricted to oral use and should only be used for preruminant calves[338].

Now, sulbactam-ampicillin combines ampicillin, a β-lactam antibiotic, with sulbactam, an irreversible β-lactamase inhibitor. The sulbactam prevents degradation of ampicillin by several major classes of bacterial β-lactamases and restores the activity of ampicillin against most strains of bacteria, in which resistance is mediated by β-lactamase production. When given subcutaneously rather than intramuscularly, higher peak plasma concentrations of both drugs were obtained in calves of 98–119 kg body weight. Sulbactam has no significant antibacterial activity against animal pathogens *per se* but is highly effective in combination with ampicillin against ampicillin-resistant strains[337].

In a comparison of sulbactam-ampicillin or ampicillin alone with a control group involving 300 Friesian and Ayrshire calves 3–10 d of age and of known immune (ZST) status, the mortality rates were 26.4, 14 and 9.5 per cent for the control, ampicillin and sulbactam-ampicillin groups respectively. Similarly, the probability of diarrhoea subsequent to the initiation of treatment was 0.5, 0.44 and 0.35 for the three treatments respectively. The differences in mortality and incidence of diarrhoea were statistically significant[339].

Measures that can be taken by the calf rearer

Various methods have been suggested for the treatment of diarrhoea in calves by the farmer and two of them are given below:

1. At the first onset of diarrhoea warm water only or warm water containing 0.9 per cent salt (9 g sodium chloride/l) should be fed for 24 h at the rate of 1.1 kg per feed in three feeds per day. During this period no other dry or liquid feed should be given. For the next 24 h 1.7 kg milk with 2.6 kg water should be given, and for the following 24 h feed 2.6 kg milk with 1.7 kg water in each case in three feeds per day. If the calf is now normal, undiluted milk may then be offered.
2. This method is simpler than the preceding one. It consists of starving the calf for one feed and then reducing the daily ration of milk to that required for maintenance of body weight under normal conditions; for a 25 kg calf, this is about 1.8 kg/d, for a 35 kg calf about 2.7 kg/d and for a 45 kg calf 3.4 kg/d. The calf should be maintained on this ration, given in two or three feeds, until the faeces show signs of becoming firmer. The quantity of milk may then be gradually increased.

Nutmeg

Nutmeg has been used therapeutically and prophylactically for diarrhoea in calves. Although there were no control calves, all calves given 3–4 g freshly grated nutmeg immediately after diarrhoea was observed, recovered. Similarly, 2 g (one tea-spoonful) of nutmeg given daily in milk for 3–4 d after birth prevented all diarrhoea during a 2 year trial period[401].

Economic loss from diarrhoea

It was estimated in 1974 that the national loss from inadequate colostrum consumption was £8 million/year, mainly due to losses from mortality but also to the costs of treatment[38].

The economic loss from neonatal diarrhoea is not only from calf mortality and cost of treatments, but from poor growth of calves that recover[142].

References

1. SOJKA, W. J. *Escherichia coli* in domestic animals and poultry *Rev. Ser. Commonw. Bur. Anim. Hlth* No. 7 (1965)
2. LOVELL, R. *Vet. Revs Annot.* **1**, 1 (1955)
3. VETERINARY INVESTIGATION SERVICE. *Vet. Rec.* **76**, 1139 (1964)
4. WOOD, P. C. *J. Path. Bact.* **70**, 179 (1955)
5. BRIGGS, C., LOVELL, R., ASCHAFFENBURG, R., *et al. Br. J. Nutr.* **5**, 356 (1951)
6. FEY, H. *Schweiz. Arch. Tierheilk.* **104**, 1 (1962)
7. FEY, H. and MARGADANT, A. *Zentbl. VetMed.* **9**, 767 (1962)
8. FEY, H., LANZ, E., MARGADANT, A. and NICOLET, J. *Dt. tierärztl. Wschr.* **69**, 581 (1962)
9. INGRAM, P. L. *Symp. Soc. gen. Microbiol.* **14**, 122 (1964)
10. INGRAM, P. L., LOVELL, R., WOOD, P. C. *et al. J. Path. Bact.* **72**, 561 (1956)
11. SMITH, T. and ORCUTT, M. L. *J. exp. Med.* **41**, 80 (1925)
12. SMITH, H. W. *J. Path. Bact.* **90**, 495 (1965)
13. INGRAM, P. L. Observations on the pathology and pathogenesis of experimental colibaccilllosis in calves. *PhD Thesis,* University of London (1962)
14. WEIJERS, H. A. and VAN DE KAMER, J. H. *Nutr. Abstr. Rev.* **35**, 591 (1965)
15. WOODE, G. N. and BRIDGER, J. C. *Vet. Rec.* **96**, 85 (1975)
16. McNULTY, M. S., McFERREN, J. B., BRYSON, D. G., LOGAN, E. F. and CURRAN, W. L. *Vet. Rec.* **99**, 229 (1976)
17. SOJKA, W. J., WRAY, C., HUDSON, E. B. and BENSON, J. A. *Vet. Rec.* **96**, 280 (1975)
18. ACRES, S. D., LAING, C. J., SAUNDERS, J. R. and RADOSTITS, O. M. *Can. J. comp. Med.* **39**, 116 (1975)
19. FISHER, E. W. and MARTINEZ, A. A. *Br. vet. J.* **132**, 127 (1976)
20. MEBUS, C. A., STAIR, E. L., RHODES, M. B., UNDERDAHL, N. R. and TWIEHAUS, M. J. *Annls Rech. vét.* **4**, 71 (1973)
21. MEBUS, C. A., WHITE, R. G., BASS, E. P. and TWIEHAUS, M. J. *J. Am. vet. med. Ass.* **163**, 880 (1973)
22. MARIN, M., LARIVIÈRE, S. and LALLIER, R. *Can. J. comp. Med.* **40**, 228 (1976)
23. SALAJKA, E., SĂRMANOVÁ, Z., HORNICH, M. and ULMANN, L. *Vet. Med., Praha* **20**, 645 (1975)
24. CORLEY, L. D., STALEY, T. E., BUSH, L. J. and JONES, E. W. *J. Dairy Sci.* **60**, 1416 (1977)
25. SMITH, H. W. and HUGGINS, M. G. *J. gen. Microbiol.* **92**, 335 (1976)
26. HALPIN, C. G. and CAPLE, I. W. *Aust. vet. J.* **52**, 438 (1976)
27. BYWATER, R. J. and PENHALE, W. J. *Res. vet. Sci.* **10**, 591 (1969)
28. WARD, D. E., BALDWIN, B. H., STEVENS, C. E. and TENNANT, B. *Gastroenterology* **70**(5), Pt 2, A-118/976 (1976)
29. ØRSKOV, L. F., ØRSKOV, H., SMITH, H. W. and SOJKA, W. J. *Acta path. microbiol. scand.* **B83**, 31 (1975)
30. SIVASWAMY, G. and GYLES, C. L. *Can. J. comp. Med.* **40**, 241 (1976)
31. WATT, J. G. *Vet. Rec.* **77**, 1474 (1965)
32. ROY, J. H. B., SHILLAM, K. W. G., HAWKINS, G. M., LANG, J. M. and INGRAM, P. L. *Br. J. Nutr.* **13**, 219 (1959)
33. FISHER, E. W. *Br. vet. J.* **121**, 132 (1965)
34. FISHER, E. W. and McEWAN, A. D. *Br. vet. J.* **123**, 4 (1967)
35. BLAXTER, K. L. and WOOD, W. A. *Vet. Rec.* **65**, 889 (1953)
36. COWIE, R. S. *Vet. Rec.* **76**, 1516 (1964)
37. CUTLIP, R. C. and McCLURKIN, A. W. *Am. J. vet. Res.* **36**, 1095 (1975)
38. IRWIN, V. C. R. *Vet. Rec.* **94**, 406 (1974)
39. BYWATER, R. J. and LOGAN, E. F. *J. comp. Path.* **84**, 599 (1974)
40. ROY, J. H. B., STOBO, I. J. F., GASTON, H. J., GANDERTON, P., SHOTTON, S. M. and OSTLER, D. C. *Br. J. Nutr.* **26**, 363 (1971)
41. MEDICAL RESEARCH COUNCIL. The treatment of acute dehydration in infants. *Med. Res. Coun. Memo.* No. 26 (HMSO, London, 1952)
42. RÁSKOVÁ, H., SECHSER, T., VANĚČEK, J. *et al. Zentbl. VetMed.,* **B23**, 131 (1976)
43. BYWATER, R. J. *J. comp. Path.* **80**, 565 (1970)
44. IRONSIDE, A. G., TUXFORD, F. and HEYWORTH, B. *Br. med. J.* **iii**, 20 (1970)
45. IVANOFF, M. R. and RENSHAW, H. W. *Am. J. vet. Res.* 36, 1129 (1975)
46. SMITH, H. W. and CRABB, W. E. *Vet. Rec.* **68**, 274 (1956)

47. SMITH, H. W. and CRABB, W. E. *J. gen. Microbiol.* **15**, 556 (1956)
48. INGRAM, P. L., SHILLAM, K. W. G., HAWKINS, G. M. and ROY, J. H. B. *Br. J. Nutr.* **12**, 203 (1958)
49. GLANTZ, P. J. *Cornell Vet.* **52**, 552 (1962)
50. OCHIAI, K., YAMANAKA, T., KIMURA, K. and SAWADA, O. *Nippon Iji Shimpo* No. 1861, 34 (1959)
51. AKIBA, T., KOYAMA, K., ISHIKI, Y., KIMURA, S. and FUKUSHIMA, T. *Jap. J. Microbiol.* **4**, 219 (1960)
52. ANDERSON, E. S. *Br. med. J.* **ii**, 1289 (1965)
53. SMITH, H. W. *Vet. Rec.* **80**, 464 (1967)
54. WALTON, J. R. *Lancet* **ii**, 1300 (1966)
55. SALAJKA, E., ULMANN, L., SĂRMANOVÁ, Z. and HORNICH, M. *Vet. Med., Praha* **20**, 659 (1975)
56. ROJAS, J., SCHWEIGERT, B. S. and RUPEL, I. W. *J. Dairy Sci.* **31**, 81 (1948)
57. FLIPSE, R. J., HUFFMAN, C. F., WEBSTER, H. D. and DUNCAN, C. W. *J. Dairy Sci.* **33**, 548 (1950)
58. CUNNINGHAM, H. M. and BRISSON, G. J. *Can. J. Anim. Sci.* **37**, 152 (1957)
59. WALKER, D. M. and FAICHNEY, G. J. *Br. J. Nutr.* **18**, 209 (1964)
60. TERNOUTH, J. H., ROY, J. H. B. and SIDDONS, R. C. *Br. J. Nutr.* **31**, 13 (1974)
61. TERNOUTH, J. H., ROY, J. H. B., THOMPSON, S. Y., TOOTHILL, J., GILLIES, C. M. and EDWARDS-WEBB, J. D. *Br. J. Nutr.* **33**, 181 (1975)
62. MOON, H. W. and JOEL, D.D. *Am. J. vet. Res.* **36**, 187 (1975)
63. RAY, C. and THOMLINSON, J. R. *Vet. Rec.* **96**, 52 (1975)
64. ROY, J. H. B. In *Perinatal Ill-health in Calves* (ed. J. M. Rutter), Commission of the European Communities, p. 125 (1975)
65. FERRIS, T. A. and THOMAS, J. W. *J. Dairy Sci.* **57**, 641 (1974)
66. ROY, J. H. B., PALMER, J., SHILLAM, K. W. G., INGRAM, P. L. and WOOD, P. C. *Br. J. Nutr.* **9**, 11 (1955)
67. SHILLAM, K. W. G. and ROY, J. H. B. *Proc. VIIth Int. Congr. Anim. Prod., Hamburg* **3**, 276 (1961)
68. SHILLAM, K. W. G., DAWSON, D. A. and ROY, J. H. B. *Br. J. Nutr.* **14**, 403 (1960)
69. SHILLAM, K. W. G., ROY, J. H. B. and INGRAM, P. L. *Br. J. Nutr.* **16**, 267 (1962)
70. SHILLAM, K. W. G., ROY, J. H. B. and INGRAM, P. L. *Br. J. Nutr.* **16**, 585 (1962)
71. SHILLAM, K. W. G., ROY, J. H. B. and INGRAM, P. L. *Br. J. Nutr.* **16**, 593 (1962)
72. SHILLAM, K. W. G. and ROY, J. H. B. *Br. J. Nutr.* **17**, 171 (1963)
73. SHILLAM, K. W. G. and ROY, J. H. B. *Br. J. Nutr.* **17**, 183 (1963)
74. SHILLAM, K. W. G. and ROY, J. H. B. *Br. J. Nutr.* **17**, 193 (1963)
75. KASTELIC, J., BENTLEY, O. G. and PHILLIPS, P. H. *J. Dairy Sci.* **33**, 725 (1950)
76. SHEEHY, E. J. *Scient. Proc. R. Dubl. Soc.* **21**, 73 (1934)
77. MYLREA, P. J. *Res. vet. Sci.* **7**, 407 (1966)
78. ROY, J. H. B., GASTON, H. J., SHILLAM, K. W. G., THOMPSON, S. Y., STOBO, I. J. F. and GREATOREX, J. C. *Br. J. Nutr.* **18**, 467 (1964)
79. TAGARI, H. and ROY, J. H. B. *Br. J. Nutr.* **23**, 763 (1969)
80. WILLIAMS, V. J., ROY, J. H. B. and GILLIES, C. M. *Br. J. Nutr.* **36**, 317 (1976)
81. SHANKS, P. L. *Vet. Rec.* **62**, 315 (1950)
82. HARDING, F. and ROYAL, L. *Dairy Inds* **39**, 372 (1974)
83. CHAPMAN, H. R., GILLESPIE, F. T., KNIGHT, D. J. and CHEESEMAN, G. C. *Rep. natn. Inst. Res. Dairy, 1971–72*, p. 121 (1973)
84. ASCHAFFENBURG, R., BARTLETT, S., KON, S. K., *et al. Br. J. Nutr.* **7**, 275 (1953)
85. MURPHY, T. *J. Dep. Agric. Repub. Ire.* **50**, 5 (1953–54)
86. THOMPSON, S. *J. Hyg., Camb.* **54**, 311 (1956)
87. ROY, J. H. B. *Vet. Rec.* **76**, 511 (1964)
88. LAMPITT, L. H. and BUSHILL, J. H. *Biochem. J.* **28**, 1305 (1934)
89. VERMA, I. S. and SOMMER, H. H. *J. Dairy Sci.* **33**, 397 (1950)
90. BERNARDONI, E. A. and TUCKEY, S. L. *J. Dairy Sci.* **33**, 409 (1950)
91. HARMAN, T. D. and SLATTER, W. L. *J. Dairy Sci.* **33**, 409 (1950)
92. HILGEMAN, M. and JENNESS, R. *J. Dairy Sci.* **34**, 483 (1951)
93. BAKER, J. M., GEHRKE, C. W. and AFFSPRUNG, H. E. *J. Dairy Sci.* **37**, 643 (1954)
94. CHRISTIANSON, G., JENNESS, R. and COULTER, S. T. *Analyt. Chem.* **26**, 1923 (1954)
95. HOSTETTLER, H. and STEIN, J. *Landw. Jb. Schweiz* **7**, 163 (1958)
96. DAVIS, D. T. and WHITE, J. C. D. *Proc. XVth Int. Dairy Congr., London* **3**, 1677 (1959)
97. GOULD, I. A. Jr. and SOMMER, H. H. *Tech. Bull. Mich. agric. Exp. Stn* No. 164 (1939)

98. JOSEPHSON, D. V. and DOAN, F. J. *Milk Dir* **29**(2), 35 (1939)
99. TOWNLEY, R. C. and GOULD, I. A. *J. Dairy Sci.* **26**, 843 (1943)
100. LARSON, B. L. and JENNESS, R. *J. Dairy Sci.* **33**, 896 (1950)
101. HUTTON, J. T. and PATTON, S. *J. Dairy Sci.* **35**, 699 (1952)
102. ZWEIG, G. and BLOCK, R. J. *J. Dairy Sci.* **36**, 427 (1953)
103. KANNAN, A. and JENNESS, R. *J. Dairy Sci.* **44**, 808 (1961)
104. FRENS, A. M., VAN DER GRIFT, J. and DAMMERS, J. *Tijdschr. Diergeneesk.* **86**, 255 (1961)
105. FRENS, A. M. *Tijdschr. Diergeneesk.* **86**, 1636 (1961)
106. BOOGAERDT, J. and KOETSVELD, E. E. *Tijdschr. Diergeneesk.* **86**, 1287 (1961)
107. AMERICAN DRY MILK INSTITUTE INC., CHICAGO *Bulletin* No. 916 (1971)
108. KNIPSCHILDT, M. E. *J. Soc. Dairy Technol.* **22**, 201 (1969)
109. HARLAND, H. A. and ASHWORTH, U. S. *J. Dairy Sci.* **28**, 879 (1945)
110. O'SULLIVAN, A. C. *J. Soc. Dairy Technol.* **24**, 45 (1971)
111. ROWLAND, S. J. *J. Dairy Res.* **9**, 42 (1938)
112. LISTER, E. E. and EMMONS, D. B. *Proc. XIXth Int. Dairy Congr., New Delhi, India* **1E**, 120 (1974)
113. ROY, J. H. B. *Proc. Nutr. Soc.* **28**, 160 (1969)
114. JENKINS, K. J., EMMONS, D. B. and LASSARD, J. R. *Can. J. Anim. Sci.* **61**, 393 (1981)
115. EMMENS, D. B. and LISTER, E. E. *Can. J. Anim. Sci.* **56**, 317 (1976)
116. JENKINS, K. J. *J. Dairy Sci.* **65**, 1652 (1982)
117. LISTER, E. E. and EMMENS, D. B. *Can. J. Anim. Sci.* **56**, 327 (1976)
118. VASIU, C. and VASIU, A. *Proc. IXth Int. Congr. Dis. Cattle Paris* **1**, 267 (1976)
119. LOGAN, E. F., PEARSON, G. R. and McNULTY, M. S. *Vet. Rec.* **101**, 443 (1977)
120. CONTRRPOIS, M., DUBOURGUIER, H. C., BORDAS, C. and GOUET, P. *Recl Méd. vét.* **155**, 553 (1979)
121. SHERWOOD, D., SNODGRASS, D. R. and LAWSON, G. H. K. *Vet. Rec.* **113**, 208 (1983)
122. ABDUL-RUDHA, G. S., HASSAN, F. K. and SHARMA, V. K. *Vet. Rec.* **114**, 39 (1984)
123. DE GRAAF, F. K. and ROORDA, I. *Infect. Immun.* **36**, 751 (1982)
124. MORRIS, J. A., THORNS, C. J., SCOTT, A. C., SOJKA, W. J. and WELLS, G. A. W. *Infect. Immun.* **36**, 1146 (1982)
125. MORRIS, J. A., THORNS, C. J. and SOJKA, W. J. *J. gen. Microbiol.* **118**, 107 (1980)
126. SMITH, H. W. and HUGGINS, M. B. *J. gen. Microbiol.* **129**, 2659 (1983)
127. CHANTER, N., MORGAN, J. H., BRIDGER, J. C., HALL, G. A. and REYNOLDS, D. J. *Vet. Rec.* **114**, 71 (1984)
128. HADAD, J. J. and GYLES, C. L. *Am. J. vet. Res.* **39**, 1651 (1978)
129. MENSIK, J., DRESSLER, J. and FRANZ, J. *Vet. Med., Praha* **22**, 463 (1977)
130. JOHNSTON, W. S. and MacLACHLAN, G. K. *Vet. Rec.* **101**, 325 (1977)
131. LEIBHOLZ, J. *N.S.W. vet. Proc.* **11**, 19 (1975)
132. HARTMAN, D. A., EVERETT, R. W., SLACK, S. Y. and WARNER, R. G. *J. Dairy Sci.* **57**, 576 (1974)
133. STAPLES, G. E. and HAYGSE, C. N. *Br. vet. J.* **130**, 374 (1974)
134. LAKSESVELA, B., SLAGSVOLD, P. and LANDSVERK, T. *Acta vet. scand.* **19**, 159 (1978)
135. LISTER, E. E. and EMMONS, D. B. *J. Dairy Sci.* **57**, 650 (1974)
136. TOULLEC, R., MATHIEU, C. M. and PION, R. *Annls Zootech.* **23**, 75 (1974)
137. ACRES, S. D. *J. Dairy Sci.* **68**, 229 (1985)
138. AGRICULTURAL AND FOOD RESEARCH COUNCIL. *Calf Scour: Know Your Microbial Enemy.* At Royal Show (1985)
139. ACRES, S. D., SAUNDERS, J. R. and RADOSTITS, O. M. *Can. vet. J.* **18**, 113 (1977)
140. SNODGRASS, D. R., SHERWOOD, D., TERZOLO, H. G. and SYNGE, B. A. *Proc. XIIth Wld Congr. Dis. Cattle, Amsterdam, Netherlands,* p. 380 (1982)
141. SMITH, T. *J. exp. Med.* **41**, 81 (1925)
142. MEBUS, C. A. *Am. J. clin. Nutr.* **30**, 1851 (1977)
143. MARSOLAIS, G., ASSAF, R., MONTPETIT, C. and MAROIS, P. *Can. J. comp. Med.* **42**, 168 (1978)
144. MADR, V., PSIKAL, J., MENSIK, J., FRANZ, J. and VALICEK, L. *Arch. exp. VetMed.* **37**, 277 (1983)
145. RHODES, M. B., STAIR, E. L., McCULLOUGH, R. A. McGILL, L. D. and MEBUS, C. A. *Can. J. comp. Med.* **43**, 84 (1979)
146. LARIVIÈRE, S., LALLIER, R. and MORIN, M. *Am. J. vet. Res.* **40**, 130 (1979)
147. NICOLAS, J. A., GAYAUD, C. and NOEL, F. *Recl Méd. vét.* **160**, 107 (1984)
148. VANNIER, P., TILLON, J. P., MADEC, F. and MORISSE, J. P. *Annls Rech. vét.* **14**, 450 (1983)

149. SNODGRASS, D. R., TERZOLO, H. R., SHERWOOD, D., CAMPBELL, I., MENZIES, J. D. and SYNGE, B. A. *Vet. Rec.* **119**, 31 (1986)
150. REYNOLDS, D. J., MORGAN, J. H., CHANTER, N., *et al. Vet. Rec.* **119**, 34 (1986)
151. ABE, M., TAKASE, O., SHIBUJ, H. and IRIKI, T. *Br. J. Nutr.* **46**, 543 (1981)
152. ANONYMOUS *Nutr. Rev.* **38**, 240 (1980)
153. MOON, H. W., McCLURKIN, A. W., ISAACSON, R. E., *et al. J. Am. vet. med. Ass.* **173**, 577 (1978)
154. WEBSTER, A. J. F. In *Environmental Aspects of Housing for Animal Production* (ed. J. A. Clark), Butterworths, London, p. 217 (1981)
155. REISINGER, R. C. *J. Am. vet. med. Ass.* **147**, 1377 (1965)
156. EDGSON, F. A. *Vet. Rec.* **76**, 1351 (1964)
157. MOLL, T. *J. Am. vet. med. Ass.* **147**, 1364 (1965)
158. RADOSTITS, O. M. and ACRES, S. D. *Can. vet. J.* **21**, 243 (1980)
159. ROY, J. H. B., SHILLAM, K. W. G. and PALMER, J. *J. Dairy Res.* **22**, 252 (1955)
160. KELLY, T. G., DODD, V. A., RUANE, D. J., TUITE, P. J., FALLON, R. J. and DEMPSTER, J. F. *Proc. II Int. Livestk Envir. Symp. Iowa State Univ., Ames, Iowa,* p. 392 (1982)
161. WILLIAMS, P. E. V. and INNES, G. M. *Res. vet. Sci.* **32**, 383 (1982)
162. ZAREMBA, W. and GRUNERT, E. *Dt. tierärtzl. Wschr.* **88**, 130 (1981)
163. ANDREWS, A. H. and READ, D. J. *Br. vet. J.* **139**, 423 (1983)
164. ANDREWS, A. H. and READ, D. J. *Br. vet. J.* **139**, 431 (1983)
165. READ, N. W., MILES, C. A., FISHER, D., *et al. Gastroenterology* **79**, 1276 (1980)
166. HOF, G. *Meded. LandbHogesch., Wageningen* 80-10 (1980)
167. SLAGSVOLD, P., LAKSESVELA, B., FLATLANDSMO, K., KROGH, N., ULSTEIN, T. L., EK, N. and LANDSVERK, T. *Acta vet. scand.* 18, 194 (1977)
168. LAKSESVELA, B., OMMUNDSEN, A. and LANDSVERK, T. *Acta vet. scand.* **19**, 543 (1978)
169. ABE, R. K., MORRILL, J. L., BASSETTE, R., MUSSMAN, H. C. and OEHME, F. W. *J. Dairy Sci.* **49**, 727 (1966)
170. WHITE, R. W. *J. agric. Sci., Camb.* **83**, 185 (1974)
171. UDEN, P. and VAN SOEST, P. J. *Br. J. Nutr.* **47**, 267 (1982)
172. HONING, Y. VAN DER, SMITS, B., LENIS, N. and BOEVE, J. *Z. Tierphysiol. Tierernähr. Futtermittelk.* **33**, 141 (1974)
173. THIVEND, P., CLARK, C. F. S. ORSKOV, E. R. and KAY, R. N. B. *Annls Rech. vét.* **10**, 422 (1979)
174. ASSAN, B. E. and THIVEND, P. *Proc. Nutr. Soc.* **35**, 104A (1976)
175. BESLE, J. M., LASSALAS, B. and THIVEND, P. *Reprod. Nutr. Dev.* **21**, 629 (1981)
176. ROY, J. H. B., STOBO, I. J. F., SHOTTON, S. M., GANDERTON, P. and GILLIES, C. M. *Br. J. Nutr.* **38**, 167 (1977)
177. GAILLARD, B. D. E. and VAN WEERDEN, E. J. *Misc. Pap. LandbHogesch., Wageningen* No. 11, p. 67 (1975)
178. GAILLARD, B. D. E. and VAN WEERDEN, E. J. *Br. J. Nutr.* **36**, 471 (1976)
179. GUILLOTEAU, P., PATUREAU-MIRAND, P., SAUVANT, D. and TOULLEC, R. *3rd EAAP Symp. Protein Metabolism and Nutrition* **1**, 227 (1980)
180. GUILLOTEAU, P., TOULLEC, R. and PATUREAU-MIRAND, P. *Annls Biol. anim. Biochim. Biophys.* **19**, 949 (1979)
181. ROTH, F. X., KIRCHGESSNER, M. and MULLER, H. L. *Z. Tierphysiol. Tierernähr. Futtermittelk.* **41**, 313 (1979)
182. GIBNEY, M. J. and WALKER, D. M. *Aust. J. agric. Res.* **29**, 133 (1978)
183. WELLEMANN, G., ANTOINE, H., BOTTON, Y. and VAN OPDENBOSCH, E. *Annls Méd. vét.* **121**, 411 (1977)
184. DE LEEUW, P. W., ELLENS, D. J., STRAVER, P. J., VAN BALKEN, J. A. M., MOERMAN, A. and BAANVINGER, T. *Res. vet. Sci.* **29**, 135 (1980)
185. DEA, S., ROY, R. S. and ELAZHARY, M. A. S. Y. *Annls Rech. vét.* **13**, 351 (1982)
186. DEA, S., ROY, R. S. and ELAZHARY, M. A. S. Y. *Can. J. comp. Med.* 47, 88 (1983)
187. PASTORET, P. P., BURTONBOY, G., JOSSE, M., KAECKENBEECK, A. and SCHOENEERS, F. *Annls Méd. vét.* **122**, 679 (1978)
188. STAUBER, E., RENSHAW, H. W., BORO, C., MATTSON, D. and FRANK, F. W. *Can. J. comp. Med.* **40**, 98 (1976)
189. BRIDGER, J. C. and WOODE, G. N. *Br. vet. J.* **131**, 528 (1975)
190. HORNER, G. W., HUNTER, R. and KIRKBRIDGE, C. A. *N.Z. vet. J.* **23**, 98 (1975)

191. MCNULTY, M. S., ALLAN, G. M., CURRAN, W. L. and MCFERRAN, J. B. *Vet. Rec.* **98**, 463 (1976)
192. LANGPAP, T. J. and BERGELAND, M. E. *Am. J. vet. Res.* **40**, 1476 (1979)
193. MEBUS, C. A., WYATT, R. G., SHARPEE, R. L., *et al. Infect. Immun.* **14**, 471 (1976)
194. TAKAHASHI, E., INABA, Y., KUROGI, H., SATO, K., SATODA, K. and OMORI, T. *Nat. Inst. Anim. Hlth Q., Tokyo* **17**, 32 (1977)
195. DURHAM, P. J. K. and BURGESS, G. W. *N.Z. vet. J.* **25**, 233 (1977)
196. MOHAMMED, K. A. *Diss. Abstr. Int., Sect.* **B39**, 2143 (1978)
197. ALMEIDA, J. D., CRAIG, C. R. and HALL, T. E. *Vet. Rec.* **102**, 170 (1978)
198. WYATT, R. G., MEBUS, C. A., YOLKEN, R. H., *et al. Science N.Y.* **203**, 548 (1979)
199. SCHROEDER, B. A., KALMAKOFF, J., HOLDAWAY, D. and TODD, B. A. *N.Z. vet. J.* **31**, 114 (1983)
200. GARNOT, P., TOULLEC, R., THAPON, J. L. *et al. J. Dairy Res.* **44**, 9 (1977)
201. LEIBHOLZ, J. *Aust. J. agric. Res.* **26**, 623 (1975)
202. JOHNSON, R. J. and LEIBHOLZ, J. *Aust. J. agric. Res.* **27**, 903 (1976)
203. DONNELLY, P. E., DEAN, R. J. and KEVEY, C. *Proc. N.Z. Soc. Anim. Prod.* **36**, 87 (1976)
204. LEIBENBERG, L. H. P. and VAN DER MERWE, F. J. *S. Afr. J. Anim. Sci.* **4**, 21 (1974)
205. ROY, J. H. B., STOBO, I. J. F. and GASTON, H. J. *Br. J. Nutr.* **24**, 459 (1970)
206. JENNY, B. F., MILLS, S. E., JOHNSTON, W. E. and O'DELL, G. D. *J. Dairy Sci.* **61**, 765 (1978)
207. WARD, G. E. and NELSON, D. I. *Am. J. vet. Res.* **43**, 1165 (1982)
208. LANDSVERK, T. *Acta vet. scand.* **20**, 572 (1979)
209. SIMM, G., CHAMBERLAIN, A. G. and DAVIES, A. B. *Vet. Rec.* **107**, 64 (1980)
210. KEYS, J. E. *Hoard's Dairym.* **122**, 1219 (1977)
211. TOULLEC, R., FRANTZEN, J. F. and MATHIEU, C. M. *Annls Zootech.* **23**, 359 (1974)
212. WEBSTER, A. J. F. In *The Control of Infectious Diseases in Farm Animals,* BVA Trust, London, p. 28 (1982)
213. MINISTRY OF AGRICULTURE, FISHERIES AND FOOD. *Manual of Veterinary Investigation Laboratory Techniques Part 2 Bacteriology* (1978)
214. TENNANT, B., HARROLD, D. and REINA-GUERRA, M. *Cornell Vet.* **58**, 136 (1968)
215. FERGUSON, A., PAUL, G. and SNODGRASS, D. R. *Gut* **22**, 114 (1981)
216. SMITH, H. W. and LINGGOOD, M. A. *J. med. Microbiol.* **5**, 243 (1972)
217. GRAAF, F. K., KLAASEN-BOOR, P. and VAN HEES, K. E. *Infect. Immun.* **30**, 125 (1980)
218. ANDREWS, A. H. In *Calf Management and Disease Notes,* A. H. Andrews, Harpenden (1983)
219. SIVASWAMY, G. and GYLES, C. L. *Can. J. comp. Med.* **40**, 247 (1976)
220. KILSHAW, P. J. and SLADE, H. *Clin. exp. Immunol.* **41**, 575 (1980)
221. KILSHAW, P. J., HEPPEL, L. M. J., EDWARDS-WEBB, J. D. and SLADE, H. M. *Res. natn Inst. Res. Dairy,* p. 114 (1981)
222. LANDSVERK, T. *Acta vet. scand.* **22**, 198 (1981)
223. KILSHAW, P. J. *Cur. Top. Vet. Med. Anim. Sci.* **12**, 203 (1981)
224. SEEGRABER, F. J. and MORRILL, J. L. *J. Dairy Sci.* **62**, 972 (1979)
225. AMENT, M. E. and RUBIN, C. E. *Gastroenterology* **62**, 227 (1972)
226. SANCHEZ-GARNICA Y MONTE, C., VINAS BORRELL, L. and ESPINOSA, F. L. *IXth Int. Congr. Dis. Cattle, Paris. Rep. and Abstr.* **1**, 341 (1976)
227. RAVEN, A. M. *J. Sci. Fd Agric.* **21**, 352 (1970)
228. GORRILL, A. D. L. and NICHOLSON, J. W. G. *Can. J. Anim. Sci.* **49**, 315 (1969)
229. GORRILL, A. D. L. and THOMAS, J. W. *J. Nutr.* **92**, 215 (1967)
230. NITZAN, Z., VOLCANI, R., HASDAI, A. and GORDIN, S. *J. Dairy Sci.* **55**, 811 (1972)
231. KAKADA, M. L., THOMPSON, R. M., ENGLESTAD, W. E., BEHRENS, G. C. and YODER, R. D. *J. Dairy Sci.* **57**, 650 (1974)
232. KILSHAW, P. J. and SISSONS, J. W. *Res. vet. Sci.* **27**, 366 (1979)
233. SMITH, R. H., HILL, W. B. and SISSONS, J. W. *Proc. Nutr. Soc.* **29**, 6A (1970)
234. SMITH, R. H. and SISSONS, J. W. *Br. J. Nutr.* **33**, 329 (1975)
235. SMITH, R. H. and WYNN, C. F. *Proc. Nutr. Soc.* **30**, 75A (1971)
236. SISSONS, J. W., SMITH, R. H. and HEWITT, D. *Br. J. Nutr.* **42**, 477 (1979)
237. SISSONS, J. W. and SMITH, R. H. *J. Physiol., Lond.* **283**, 307 (1978)
238. BARBER, D. M. L., MacLENNAN, W., LAWSON, G. H. K. and ROWLAND, A. C. *Vet. Rec.* **97**, 386 (1975)
239. SMITH, H. W. *J. Path. Bact.* **93**, 499 (1967)
240. BARBER, D. M. L., DOXEY, D. L. and MacLENNAN, W. *Vet. Rec.* **97**, 424 (1975)

241. BARBER, D. M. L. and MacLENNAN, W. *Vet. Rec.* **97**, 362 (1975)
242. LEWIS, C. J. *Vet. Rec.* **97**, 506 (1975)
243. SKIRROW, M. B. *Br. med. J.* **ii**, 9 (1977)
244. AL-MASHAT, R. R. and TAYLOR, D. J. *Vet. Rec.* **107**, 459 (1980)
245. ALLSUP, T. N. and HUNTER, D. *Vet. Rec.* **93**, 389 (1973)
246. AL-MASHAT, R. R. and TAYLOR, D. J. *Vet. Rec.* **112**, 54 (1983)
247. TERZOLO, H. R., LAWSON, G. H. K., ANGUS, K. W. and SNODGRASS, D. R. *Res. vet. Sci.* **43**, 72 (1987)
248. JEFFREY, M. and HOGG, R. A. *Vet. Rec.* **122**, 89 (1988)
249. EAGLESOME, M. D., GARCIA, M. M., HAWKINS, C. F. and ALEXANDER, F. C. M. *Vet. Rec.* **119**, 299 (1986)
250. MORGAN, G., CHADWICK, P., LANDER, K. P. and GILL, K. P. W. *Vet. Rec.* **116**, 111 (1985)
251. AL MASHAT, R. R. and TAYLOR, D. J. *Vet. Rec.* **112**, 141 (1983)
252. ANTSBERG, G., BISPING, W., KRABISCH, P. and MATTHIESEN, I. *Zentlb. VetMed.* **B24**, 104 (1977)
253. NEUVONEN, E. and ESTOLA, T. *Acta vet. scand.* **15**, 256 (1974)
254. POPOFF, M. R. and LECOANET, J. *Vet. Rec.* **121**, 591 (1987)
255. APPLEYARD, W. T. and MOLLISON, A. *Vet. Rec.* **116**, 522 (1985)
256. CLARKSON, M. J., FAULL, W. B. and KERRY, J. B. *Vet. Rec.* **116**, 467 (1985)
257. LANDSVERK, T. *Res. vet. Sci.* **42**, 299 (1987)
258. WILLIAMS, R. O. and BURDEN, D. J. *Res. vet. Sci.* **43**, 264 (1987)
259. DURHAM, P. J. K., JOHNSON, R. H., ISLES, H., PARKER, R. J., HOLROYD, R. G. and GOODCHILD, I. *Res. vet. Sci.* **38**, 234 (1985)
260. DURHAM, P. J. K., LAX, A. and JOHNSON, R. H. *Res. vet. Sci.* **38**, 209 (1985)
261. DURHAM, P. J. K., JOHNSON, R. H. and PARKER, R. J. *Res. vet. Sci.* **39**, 16 (1985)
262. LUCAS, M. H. and WESTCOTT, D. G. F. *Vet. Rec.* **116**, 698 (1985)
263. HUCK, R. A., WOODS, D. W. and ORR, J. P. *Vet. Rec.* **96**, 155 (1975)
264. RICE, D. A., McMURRAY, C. H., KENNEDY, S. and ELLIS, W. A. *Vet. Rec.* **119**, 571 (1986)
265. SISSONS, J. W. and SMITH, R. H. *Br. J. Nutr.* **36**, 421 (1976)
266. SISSONS, J. W. and SMITH, R. H. *Rep. natn. Inst. Res. Dairy*, p. 62 (1975–76)
267. SISSONS, J. W., SMITH, R. H., HEWITT, D. and NYRUP, A. *Br. J. Nutr.* **47**, 311 (1982)
268. FREIER, S., LEBENTHAL, E., FREIER, M., SHAH, P. C., PARK, B. H. and LEE, P. C. *Immunology* **49**, 69 (1983)
269. BARRATT, M. E. J. and PORTER, P. *J. Immun.* **123**, 676 (1979)
270. KILSHAW, P. J. and SISSONS, J. W. *Res. vet. Sci.* **27**, 361 (1979)
271. BARRATT, M. E. J., SENIOR, S. J., MAY, K., HALL, H. and PORTER, P. *Res. vet. Sci.* **37**, 93 (1984)
272. SISSONS, J. W. and THURSTON, S. M. *Res. vet. Sci.* **37**, 242 (1984)
273. BARRATT, M. E. J., TWOHIG, B. M. A., HALL, H. and PORTER, P. *Res. vet. Sci.* **39**, 62 (1985)
274. HALL, G. A., BRIDGER, J. C. and BROOKER, B. E. *Proc. XIIth Wld Congr. Dis. Cattle, Amsterdam, Netherlands,* p. 217 (1982)
275. WOODE, G. N. and BRIDGER, J. C. *J. med. Microbiol.* **11**, 441 (1978)
276. POHLENZ, J. F., WOODE, G. N., CHEVILLE, N. F., MOKRESH, A. H. and MOHAMMED, K. *Proc. XIIth Wld Congr. Dis. Cattle, Amsterdam, Netherlands,* p. 252 (1982)
277. WOODE, G. N., REED, D. E., RUNNELS, P. A., HERRIG, M. A. and HILL, H. T. *Vet. Microbiol.* **7**, 221 (1982)
278. SCHIPPER, I. A., POMMER, J., LANDBLOM, D., DANIELSON, R. and SLANGER, W. *N. Dak. Fm Res.* **41**, 23 (1984)
279. RICHTER, H., PERLBERG, K. W. and URBANECK, D. *Tierzücht* **37**, 232 (1983)
280. RENAULT, L., ESPINASSE, J. and COURTAY, B. *Bull. mens. Soc. vet. prat. Fr.* **68**, 141 (1984)
281. CASTRUCCI, G., FRIGERI, F., FERRARI, M. *et al. Comp. Immun. Microbiol. Infect. Dis.* 7, 11 (1984)
282. BALLABRIGA, A. *Acta paediat. jap. (Overseas Edn)* **24**, 235 (1982)
283. McCLEAD, R. E. and GREGORY, S. A. *Infect. Immun.* **44**, 474 (1984)
284. EBUIA, T., SATO, A., UMEZU, K. *et al. Lancet* **ii**, 1029 (1983)
285. McNULTY, M. S. and LOGAN, E. F. *Vet. Rec.* **120**, 250 (1987)
286. SNODGRASS, D. R. *Vet. Rec.* **119**, 39 (1986)
287. CONTREPOIS, M. G. and GIRARDEAU, J. P. *Infect. Immun.* **50**, 947 (1985)
288. VONDERFECHT, S. L. and OSBURN, B. I. *J. clin. Microbiol.* **16**, 935 (1982)
289. ALLONCLE, F. Prevention and treatment of diseases of newborn calves with hyperimmune colostral serum. *Ecole Nat. Vet. Alfort Thesis* (1980)
290. MENSIK, J., SALAJKA, E., STEPANEK, J., ULMANN, L., PROCHAZKA, Z. and DRESSLER, J. *Annls Rech. vét.* **9**, 255 (1978)

291. SAIF, L. J. and SMITH, K. L. *J. Dairy Sci.* **68**, 206 (1985)
292. WARD, A. C. S., WALDHEIM, D. G., FRANK, F. W., MEINERSHAGEN, W. A. and DUBOSE, D. A. *J. Am. vet. med. Ass.* **170**, 340 (1977)
293. NAVETAT, H., CONTREPOIS, M., DUBOURGUIER, H. C., GIRARDEAU, J. P., GOBY, J. F. and GOUET, P. In *Gastro-enteritis Néonatales du Veau* (ed. J. Espinasse), Maisons-Alfort, France, Soc. Fr. Buiatrie, p. 191 (1979)
294. LENTSCH, R. H. and BORDT, D. E. *Mod. vet. Pract.* **64**, 729 (1983)
295. KAECKENBEECK, A. *Proc. IXth Int. Congr. Dis. Cattle, Paris* 193 (1976)
296. WOODE, G. N., BEW, M. E. and DENNIS, M. J. *Vet. Rec.* **103**, 32 (1978)
297. CONTREPOIS, M., GIRARDEAU, J. P., DUBOURGUIER, H. C., GOUET, P. and LEVIEUX, D. *Annls Rech. vét.* **9**, 385 (1978)
298. ACRES, S. D. and RADOSTITS, O. M. *Can. vet. J.* **17**, 197 (1976)
299. HOFFMANN, W. and WEBER, A. *Dt. tierärztl. Wschr.* **86**, 47 (1979)
300. THEODORIDIS, A., PROZESKY, L. and ELS, H. J. *Ondersterpoort J. vet. Res.* **47**, 31 (1980)
301. ACRES, S. D., ISAACSON, R. E., BABIUK, L. A. and KAPITANY, R. A. *Infect. Immun.* **25**, 121 (1979)
302. JOHNSTON, W. S., MacLACHLAN, G. K. and HOPKINS, G. F. *Vet. Rec.* **106**, 174 (1980)
303. JOHNSTON, W. S., HOPKINS, G. F. and MacLACHLAN, G. K. *Vet. Rec.* **118**, 637 (1986)
304. LINTON, A. H. reported in *Vet. Rec.* **117**, 373 (1985)
305. SHERWOOD, D., SNODGRASS, D. R. and O'BRIEN, A. D. *Vet. Rec.* **116**, 217 (1985)
306. FORMAL, S. B., GEMSKI, P., GIANELLA, R. A. and TAKEUCHI, A. Ciba Foundation Symp. 42, Elsevier, Amsterdam, p. 27 (1976)
307. MOON, H. W. *Adv. vet. Sci. comp. Med.* **18**, 179 (1974)
308. BYWATER, R. J. *Vet. Rec.* **107**, 549 (1980)
309. ASCHAFFENBURG, R., BARTLETT, S., KON, S. K. *et al. Br. J. Nutr.* **5**, 343 (1951)
310. LANDSVERK, T. *Acta vet. scand.* **22**, 449 (1981)
311. LOGAN, E. F., PEARSON, G. R. and McNULTY, M. S. *Vet. Rec.* **104**, 206 (1979)
312. SCHWERS, A., PASTORET, P. P., BROECKE, C. V. *et al. Annls Méd. vét.* **126**, 59 (1982)
313. McNULTY, M. S. and LOGAN, E.F. *Vet. Rec.* **113**, 333 (1983)
314. SENF, W. *Mh. VetMed.* **37**, 761 (1982)
315. BORDER, M. *Can. J. comp. Med.* **40**, 265 (1976)
316. STAUBER, E. H. *J. Am. vet. med. Ass.* **168**, 223 (1976)
317. MEYLING, A. *Acta vet. scand.* **15**, 457 (1974)
318. REYNOLDS, D. J., HALL, G. A., DEBNEY, T. G. and PARSONS, K. R. *Res. vet. Sci.* **38**, 264 (1985)
319. BELLINZONI, R. C., MATTION, N., LA TORRE, J. L. and SCODELLER, E. A. *Res. vet. Sci.* **42**, 257 (1987)
320. MUNIAPPA, L., GEORGIEV, K. G., DIMITROV, D., MITOV, K. B. and HARALAMBIEV, E. H. *Vet. Rec.* **120**, 23 (1987)
321. HALL, G. A., PARSONS, K. R., BRIDGER, J. C., GHATEI, M. A., YING, Y. C. and BLOOM, S. R. *Res. vet. Sci.* **38**, 99 (1985)
322. WEBSTER, A. J. F., SAVILLE, C., CHURCH, B. M., GNANASAKTHY, A. and MOSS, R. *Br. vet. J.* **141**, 472 (1985)
323. HINTON, M., HEDGES, A. J. and LINTON, A. H. *J. appl. Bact.* **58**, 27 (1985)
324. HINTON, M., LINTON, A. H. and HEDGES, A. J. *J. appl. Bact.* **58**, 131 (1985)
325. HINTON, M. and LINTON, A. H. *Vet. Rec.* **112**, 567 (1983)
326. DUBOURGUIER, H. C., CONTREPOIS, M. and GOUET, P. In *Gastroenteritis Néo-natales du Veau* (ed. J. Espinasse), Maisons-Alfort, France, Soc. Fr. Buiatrie, p. 161 (1979)
327. NAZER, A. H. K. *Vet. Rec.* **103**, 587 (1978)
328. SOKOL, A., PAULIK, S., VARGOVCIKOVA, A. and BANYAI, S. *VetMed., Praha* **22**, 333 (1977)
329. GADEK, B. and SCHAL, E. *Zentbl. VetMed.,* **B23**, 89 (1976)
330. LARSEN, H. E. and LARSEN, J. L. *Nord. VetMed.* **27**, 65 (1975)
331. RADOSTITS, O. M., RHODES, C. S., MITCHELL, M. E., SPOTSWOOD, T. P. and WENKOFF, M. S. *Can. vet. J.* **16**, 219 (1975)
332. JONES, T. O. *Vet. Rec.* **120**, 399 (1987)
333. JONES, T. O. *Preventive Medicine in Bovine Practice.* Brit. Cattle Vet. Assoc., p. 2 (1979)
334. HINTON, M. reported in *Vet. Rec.* **117**, 374 (1985)
335. CHASLUS-DANCLA, E. and LAFONT, J. P. *Vet. Rec.* **117**, 90 (1985)
336. THRELFALL, E. J., ROWE, B., FERGUSON, J. L. and WARD, L. R. *Vet. Rec.* **113**, 627 (1983)

337. GRIMSHAW, W. T. R. and COLMAN, P. J. *Br. vet. J.* **143**, 361 (1987)
338. MARSHALL, A. B., BURGESS, M., HEWETT, G. R., PALMER, G. H. and WEST, B. *Proc. XIIth Wld Congr. Dis. Cattle. Amsterdam, Netherlands,* p. 1167 (1982)
339. GRIMSHAW, W. T. R., COLMAN, P. J. and PETRIE, L. *Vet. Rec.* **121**, 162 (1987)
340. COOKE, B. C. *Vet. Rec.* **116**, 301 (1985)
341. MEBUS, C. A., WYATT, R. G. and KAPIKIAN, A. Z. *Vet. Pathol.* **14**, 273 (1977)
342. BARRATT, M. E. J., STRACHAN, P. J. and PORTER, P. *Proc. Nutr. Soc.* **38**, 143 (1979)
343. BELLAMY, J. E. C. and ACRES, S. D. *Am. J. vet. Res.* **40**, 1391 (1979)
344. PEARSON, G. and LOGAN, E. F. *Vet. Rec.* **105**, 159 (1979)
345. HADAD, J. J. and GYLES, C. L. *Am. J. vet. Res.* **43**, 41 (1982)
346. CHAN, R., LIAN, C. J., COSTERTON, J. W. and ACRES, S. D. *Can. J. comp. Med.* **47**, 150 (1983)
347. GUNTHER, H., SCHULZE, F. and HEILMANN, P. *Mh. VetMed.* **38**, 96 (1983)
348. BELLAMY, J. E. C. and ACRES, S. D. *Can. J. comp. Med.* **47**, 143 (1983)
349. MEBUS, C. A. *J. Dairy Sci.* **59**, 1175 (1976)
350. MEBUS, C. A. and NEWMAN, L. E. *Am. J. vet. Res.* **38**, 553 (1977)
351. WOODE, G. N., SMITH, C. and DENNIS, M. J. *Vet. Rec.* **102**, 340 (1978)
352. PEARSON, C. R., McNULTY, M. S. and LOGAN, E. F. *Vet. Rec.* **102**, 454 (1978)
353. MEBUS, C. A., STAIR, E. L., UNDERDAHL, N. R. and TWIEHAUS, M. J. *Vet. Path.* **8**, 490 (1971)
354. MEBUS, C. A., RHODES, M. B. and UNDERDAHL, N. R. *Am. J. vet. Res.* **39**, 1223 (1978)
355. MANDARD, O., DUBOURGUIER, H. C., CONTREPOIS, M. and GOUET, P. In *Gastroenteritis Néo-natales du Veau* (ed. J. Espinasse) Maisons-Alfort, France, Soc. Fr. Buiatrie, p. 81 (1979)
356. MEBUS, C. A., STAIR, E. L., RHODES, M. B. and TWIEHAUS, M. J. *Vet. Path.* **10**, 45 (1973)
357. PETZINGER, E. *Berl. Munch. tierärtzl. Wschr.* **97**, 83 (1984)
358. BYWATER, R. J. *Proc. XIIth Wld Congr. Dis. Cattle, Amsterdam, Netherlands,* p. 291 (1982)
359. ALLISON, S. P. *Proc. Nutr. Soc.* **45**, 163 (1986)
360. ČELEDA, L., ČERNÝ, J., FENDRICH, Z., *et al. Zentbl. VetMed.* **B30**, 189 (1983)
361. HORVAY, MAGDOLNA S., ROMVARY, A. and MISLEY, A. *Magy. Allatorv. Lap.* **32**, 381 (1976)
362. STEINBACH, G., HARTMANN, H., BARNAU, F., MEYER, H. and LUSTERMANN, S. *Mh. VetMed.* **36**, 377 (1981)
363. HARTMANN, H., MEYER, H. and STEINBACH, G. *Mh. VetMed.* **36**, 371 (1981)
364. HARTMANN, H., GENTSCH, S., ENGELMANN, H., MEYER, H. and STEINBACH, G. *Mh. VetMed.* **37**, 691 (1982)
365. BYWATER, R. J., PALMER, C. H. and STANTON, A. *Proc. XIIth Wld Congr. Dis. Cattle Amsterdam, Netherlands,* p. 301 (1982)
366. VAN BRUINESSEN KAPSENBERG, E. G. and BREUKINK, H. J. *Proc. XIIth Wld Congr. Dis. Cattle Amsterdam, Netherlands,* p. 1225 (1982)
367. HARTMANN, H., MEYER, H., STEINBACH, G., ROSSOW, N. and LESCHE, R. *Mh. VetMed.* **38**, 292 (1983)
368. BUNTAIN, B. J. and SELMAN, I. E. *Vet. Rec.* **107**, 245 (1980)
369. FISHER, R. B. and PARSON, D. S. (quoted in *The Times* Nov 18 1986, no. 62618, p. 23)
370. DEMIGNE, C., CHARTIER, F. and REMESY, C. *Annls Rech. vét.* **11**, 267 (1980)
371. LOPEZ, G. A. *Diss. Abstr. Int., Sect. B.* **35**, 1007B (1974)
372. LOPEZ, G. A., PHILLIPS, R. W. and LEWIS, L. D. *Am. J. vet. Res.* **36**, 1245 (1975)
373. CABELLO, G. *Br. vet. J.* **136**, 160 (1980)
374. OPLISTIL, M. *Acta vet., Beogr.* **39**, 17 (1970)
375. LEWIS, L. D., PHILLIPS, R. W. and ELLIOTT, C. D. *Am. J. vet. Res.* **36**, 413 (1975)
376. NADIM, M. A. and MOTTELIB, A. A. *Zentbl. VetMed.* **A21**, 539 (1974)
377. WARD, D. E. *Diss. Abstr. Int., Sect. B.* **37**, 1133 (1976)
378. FISHER, E. W. and MARTINEZ, A. A. *Res. vet. Sci.* **20**, 302 (1976)
379. FISHER, E. W. and MARTINEZ, A. A. *Br. vet. J.* **134**, 234 (1978)
380. BURNS, T. W., MATHIAS, J. R., CARLSON, G. M. and MARTIN, J. L. *Am. J. Physiol.* **235**, E311 (1978)
381. MODLIN, I. M., BLOOM, S. R. and MITCHELL, S. J. *Gastroenterology* **75**, 1051 (1978)
382. PEARSON, G. R., McNULTY, M. S. and LOGAN, E. F. *Vet. Path.* **15**, 92 (1978)
383. BLOOM, S. R. and EDWARDS, A. V. *J. Physiol., Lond.* **299**, 437 (1980)
384. BESTERMAN, H. S., CHRISTOFIDES, N. D., WELSBY, P. D., ADRIAN, T. E., SARSON, D. L. and BLOOM, S. R. *Gut* **24**, 665 (1983)
385. GRYS, S. and MALINOWSKA, A. *Acta microbiol. pol., Ser. B***6**, 185 (1974)

386. FIELD, M., GRAF, L. H., LAIRD, W. J. and SMITH, P. L. *Proc. natn Acad. Sci. USA* **75**, 2800 (1978)
387. BYWATER, R. J. *Res. vet. Sci.* **14**, 35 (1973)
388. TENNANT, B., HARROLD, D. and REINA-GUERRA, M. *J. Am. vet. med. Ass.* **161**, 993 (1972)
389. TURNER, C. W. and HERMAN, H. A. *Mo. Agric. Expt Stn Res. Bull.* **159** (1931)
390. STAHL, P. R. and DALE, H. E. *J. Dairy Sci.* **40**, 617 (1957)
391. SINGH, J. and LAXMINARAYANA, H. *Indian J. Dairy Sci.* **26**, 135 (1973)
392. FLICOTEAUX, J. R., OSTRE, L., VAAST, R. and CANCE, M. *Bull. mens. Soc. vet. prat. Fr.* **62**, 563 (1978)
393. TOURNUT, J., BEZILLE, P. and REDON, P. *Proc. IXth Int. Congr. Dis. Cattle, Paris* **1**, 277 (1976)
394. ROBERTS, D. H. and LUCAS, M. H. *Vet. Rec.* **119**, 459 (1986)
395. KAY, R. M. and POOLE, P. *Anim. Prod.* **46**, 499 (1988)
396. SHAW, J. N. and MUTH, O. H. *J. Am. vet. med. Ass.* **90**, 171 (1937)
397. KONOVALOV, V. *Moloch. Prom.* **10**, No. 5, 30 (1949)
398. DILANYAN, Z. *Moloch. Prom.* **24**, No. 2, 37 (1963)
399. MARKOVIC, P. *Sborn. csl Akad. zemed. Ved* **23**, 168 (1950)
400. HAUGSE, C. N. and STRUM, G. *N. Dak. Res. Rep.* **69**, (1978)
401. STAMFORD, I. F., BENNETT, A., and GREEN-HALF, J. *Vet. Rec.* **103**, 14 (1978)
402. FISHER, E. W. and DE LA FUENTE, G. H. *Res. vet. Sci.* **13**, 315 (1972)
403. MYLREA, P. J. *Res. vet. Sci.* **9**, 14 (1968)
404. MARSH, C. L., MEBUS, C. A. and UNDERDAHL, N. R. *Am. J. vet. Res.* **30**, 163 (1969)
405. DALTON, R. G. *Br. vet. J.* **124**, 371 (1968)
406. JENTSCH, D., BUNGER, U. and KLEINER, W. *Arch. Tierz.* **24**, 237 (1981)
407. MORRIS, J. A., THORNS, C. J., WELLS, G. A. H., SCOTT, A. C. and SOJKA, W. A. *J. gen. Microbiol.* **129**, 2753 (1983)
408. TO, S. C. M. *Infect. Immun.* **43**, 549 (1984)
409. GIRARDEAU, J. P., DOUBOURGUIER, H. C. and CONTREPOIS, M. *Bull. Grpe tech. Vet.* **80-4-B-190**, 49 (1980)
410. POHL, P., LINTERMANS, P., VAN MUYLEM, K. and SCHOTTE, M. *Annls Méd. vét.* **126**, 569 (1982)
411. MORRIS, J. A., SOJKA, W. J. and READY, R. A. *Res. vet. Sci.* **38**, 246 (1985)
412. MORRIS, J. A., THORNS, C. J., BOARER, C. and WILSON, R. A. *Res. vet. Sci.* **39**, 75 (1985)
413. MORRIS, J. A., CHANTER, N. and SHERWOOD, D. *Vet. Rec.* **121**, 189 (1987)
414. LARSEN, J. L. *Vet. Rec.* **121**, 431 (1987)
415. MOORTHY, A. R. S. *Vet. Rec.* **116**, 159 (1985)
416. ASCHAFFENBURG, R. and DREWRY, R. *Proc. XVth Int. Dairy Congr.* p. 1631 (1959)
417. BERGEY'S *Manual of Determinative Bacteriology*, 8th edn, Williams and Wilkins, Baltimore (1974)
418. SMITH, H. W. *J. Path. Bact.* **84**, 147 (1962)
419. POHL, P. *Annls Méd. vét.* **19**, 435 (1975)
420. JOHNSTON, N. E., ESTRELLA, R. A. and OXENDER, W. D. *Am. J. vet. Res.* **38**, 1323 (1977)
421. DE RYCKE, J., BOIVIN, R. and LE ROUX, P. *Annls Rech. vét.* **13**, 385 (1982)
422. FARID, A., IBRAHIM, M. S. and REFAI, M. *Zentbl. VetMed.* **B23**, 38 (1976)
423. BYWATER, R. J. *Res. vet. Sci.* **18**, 107 (1975)
424. BYWATER, R. J. *J. S. Afr. vet. Ass.* **47**, 193 (1978)
425. MOON, H., WHIPP, S. and DONTA, S. *Proc. IXth Int. Congr. Dis. Cattle, Paris* **1**, 273 (1976)
426. ISAACSON, R. E. and SCHNEIDER, R. A. *Am. J. vet. Res.* **39**, 1750 (1978)
427. BURGESS, M. N., BYWATER, R. J., COWLEY, C. M., MULLAN, N. A. and NEWSOME, P. M. *Infect. Immun.* **21**, 526 (1978)
428. KAECKENBECK, A., SCHOENAERS, F., PASTORET, P. P. and JOSSE, M. *Annls Méd. vét.* **121**, 55 (1977)
429. MYERS, L. L. and GUINEE, P. A. M. *Infect. Immun.* **13**, 1117 (1976)
430. POHL, P., LINTERMANS, P., KAECKENBEECK, A. DE MOL, P., VAN MUYLEM, K. and SCHOTTE, M. *Annls Méd. vét.* **127**, 37 (1983)
431. WOODE, G. N. *Rep. Inst. Res. Anim. Dis.* p. 33 (1977)
432. DE GRAFF, F. K., WIENTJES, F. B. and KLAASEN-BOOR, P. *Infect. Immun.* **27**, 216 (1980)
433. TZIPORI, S. *Vet. Rec.* **108**, 510 (1981)
434. MOORE, E. D., HOLLON, B. F., DAVLEN, H. H., HALL, R. F. and OWEN, J. R. *J. Dairy Sci.* **58**, 147 (1975)
435. FISCHER, W. and BUTTE, R. *Dt. tierärztl. Wschr.* **81**, 567 (1974)
436. BYWATER, R. J. *Am. J. vet. Res.* **38**, 1983 (1977)

437. THORNTON, J. R. and ENGLISH, P. B. *Br. vet. J.* **134**, 445 (1978)
438. DEMIGNE, C., REMESY, C., CHARTIER, F. and LEFAIVE, J. *Am. J. vet. Res.* **42**, 1356 (1980)
439. SAFWATE, A., DAVICCO, M. J., BARLET, J. P. and DELOST, P. *Reprod. Nutr. Dev.* **22**, 689 (1982)
440. DEMIGNE, C., REMESY, C. and CHARTIER, F. *Proc. XIIth Wld Congr. Dis. Cattle, Amsterdam, Netherlands,* p. 305 (1982)
441. SZENCI, O., KUTAS, F. and HARASZTI, J. *Magy. Allatorv. Lap.* **37**, 601 (1982)
442. SZENCI, O. *Magy. Allatorv. Lap.* **38**, 273 (1983)
443. FIGUEREDO, J. M. and CAPOTE, J. M. *Mh. VetMed.* **39**, 118 (1984)
444. GREENE, H. J. *Annls Rech. vét.* **14**, 548 (1983)
445. DEMIGNE, C., REMESY, C., CHARTIER, F. and KALIGIS, D. *Annls Rech. vét.* **14**, 541 (1983)
446. WOODE, G. N. *Vet. Rec.* **103**, 44 (1978)
447. MISCIATTELI, M. E., BELLETTI, G. L., GUARDA, F., MANSTRETTA, G. and BIANCARDI, G. *Schweiz. Arch. Tierheilk.* **122**, 403 (1980)
448. FARWELL, S. O., KAGEL, R. A. and GUTENBERGER, S. K. *Analyt. Chem.* **55**, 985A (1983)
449. ANONYMOUS. *Dep. Agric. North. Irel. Ann. Rep. Tech. Work.,* p. 176 (1978)
450. HUDSON, S., MULLORD, M., WHITTLESTONE, W. G. and PAYNE, E. *Br. vet. J.* **132**, 551 (1976)
451. MENSIK, J., POSPISIL, Z., CEPICA, A., and DRESSLER, J. *Vet. Med., Praha* **22**, 475 (1977)
452. BOYD, J. W., BAKER, J. R. and LEYLAND, A. *Vet. Rec.* **95**, 310 (1974)
453. RAMSEY, H. A. and WILLARD, T. R. *J. Dairy Sci.* **58**, 436 (1967)
454. SISSONS, J. W., NYRUP, A., KILSHAW, P. J. and SMITH, R. H. *J. Sci. Fd Agric.* **33**, 706 (1982)
455. WISE, G. H., MILLER. P. G., ANDERSON, G. W. and LINNERUD, A. C. *J. Dairy Sci.* **59**, 97 (1976)
456. ZAREMBA, W. *Prakt. Tierartz* **64**, 977 (1983)
457. FORD, E. J. H., BOYD, J. W. and HOGG, R. A. *Index Res. Meat Livestock Comm.,* p. 4 (1978)
458. DARDILLAT, C. In *Perinatal Ill-health in Calves* (ed J. M. Rutter), Commission of the European Communities p. 111 (1975)
459. INTERNATIONAL UNION OF BIOCHEMISTRY. *Enzyme Nomenclature,* Academic Press, New York and London (1984)
460. MOON, H. W. *J. Am. vet. med. Ass.* **172**, 443 (1978)

Chapter 4

Salmonellosis

Incidence

Infection in calves is usually by the organisms *Salmonella dublin or S. typhimurium*, with infection by the former organism being, until 1969, the more common[2,3]. Thus, between 1958 and 1960, 64 per cent of cases of salmonellosis diagnosed at Veterinary Investigation Centres were associated with *S. dublin* and 32 per cent with *S. typhimurium*. Of the post-mortem examinations made on calves up to 6 months of age during the period 1959–61, 16 per cent were associated with a salmonella septicaemia and 7 per cent with a salmonella gastroenteritis[1]. For the years 1961–64 the ratio of *S. typhimurium* to *S. dublin* infections in cattle (mainly calves) was 1:3, with a marked increase in the incidence of *S. typhimurium* in 1965, so that the ratio decreased to 1:1.3[4].

The number of isolations of salmonellae in cattle reached a peak in 1969, due to a dramatic rise in *S. dublin* outbreaks. Since then there has been a decline in the number of *S. dublin* incidents with a concomitant increase in *S. typhimurium* incidents, which reached a peak in 1973. At the same time there was an increase in the importance of other serotypes which accounted for 15 per cent of the incidents[8]. By 1976 the continued decline in bovine salmonellosis, particularly in *S. dublin* incidents, had resulted in a ratio of *S. typhimurium* to *S. dublin* of 1.2:1 whilst the other serotypes had increased to 17 per cent of the incidents.

In 1981, the number of incidents in calves (these may be an individual or a group of calves) revealed a ratio of *S. typhimurium* to *S. dublin* of 2.6:1. In that year, other serotypes were associated with only 9 per cent of the incidents in calves[32]. Between 1981 and 1985, the peak incidence of outbreaks in England and Wales occurred in 1983, and in Scotland, a very marked increase occurred in 1985. By 1985, the ratio of *S. typhimurium* to *S. dublin* incidents in calves was 3.2:1 with only 4.6 per cent of incidents associated with other serotypes[58].

Most salmonellosis in calves in 1980 and 1981 was caused by three phage types of *S. typhimurium*, DT204, DT204a and DT204c. A study of calves purchased in the market in 1979 and 1982 showed that DT193 was endemic in 1979 and DT204c in 1982[35]. By 1985, the epidemic spread of phage type DT204c increased so that it represented 77 per cent of isolations of *S. typhimurium* from calves, whereas in 1982 it represented only 15 per cent. This strain is highly pathogenic to calves, resistant to a wide range of antibiotics and capable of acquiring new resistant patterns rapidly. It is also important as a cause of human food poisoning (see below)[58,61]. Salmonellosis in cattle has recently been reviewed[73].

Of other serotypes, *S. enteritidis* has been responsible for an outbreak in which 33 per cent of the calves died[33]. An unusual lactose-fermenting *S. typhimurium* has also been isolated from calves[34].

In 1985, a survey in the UK of calf diarrhoea showed that salmonella was associated with 12 per cent of the cases[36]. In large calf rearing units, the incidence of faecal excretion of *S. dublin* on at least one occasion was between 3.6 and 5.5 per cent[37, 38].

Geographical distribution

S. dublin is endemic in certain areas of the UK, such as S. Wales and Somerset, so that 0.5–1 per cent of adult cattle are excreting the organism in their faeces. These areas are often of rough grazing where cattle are drinking from streams or dykes, and where liver fluke infection is also common. On the other hand, *S. typhimurium* does not appear to be endemic in particular areas. Its relative absence in S. Wales and Somerset suggests that endemic *S. dublin* may protect calves against *S. typhimurium* infection.

Geographical variations in the ratio of *S. typhimurium* to *S. dublin* in both 1980 and 1981 showed the highest ratio of 5.6:1 in Scotland and the lowest 1.3:1 in Wales with an intermediate value for England of 2.0:1[32]. By 1985, the ratio in Scotland was 10.6:1, in Wales 1.6:1 and in England 2.6:1[58]. Possibly the high ratio in Scotland reflects the movement of large numbers of calves from S.W. England for fattening.

Clinical symptoms

The two infections produce similar clinical symptoms, which are variable. In acute cases a septicaemia may occur, but in milder forms diarrhoea, often bloodstained, may be the only symptom and this may be transient. Calves affected more severely become emaciated, although appetite may not be completely lost. Some of these calves make a slow recovery, but others show a short period of weakness and prostration before death. Sometimes the disease is accompanied by lameness, associated with arthritis, and often the calves are also infected with 'virus pneumonia'. In purchased calves that were infected with *S. typhimurium*, 40 per cent showed no clinical symptoms.

It is claimed that there is a more profound disturbance of body fluid balance in salmonellosis than in other neonatal disease syndromes, since there is a greater insensible water loss associated with pyrexia, a greater reduction in food intake and a larger faecal dry matter excretion, resulting in a weight loss of 187 g/kg[39].

Histological and haematological findings

The intestinal epithelial cells are relatively undamaged after the salmonellae pass through, since they leave the epithelium comparatively undisturbed and the damaged brush borders normally regenerate. The salmonellae often concentrate in the Peyer's patches. However, there is a progressive inflammatory reaction in the lamina propria and this leads to necrotic enteritis in the ileum and large intestine[40].

Lesions are uncommon in other organs[41]. In a study of *S. dublin* infection, the organism spread to the lymph nodes but not to the liver and spleen[12].

In the acute phase of infection, calves become hypoferraemic. This appears to be part of the calf's defence mechanism, since the ability of a salmonella to compete for iron may be considered as a virulence factor[42].

Age at infection

The calf may become infected as early as the fourth day of life, although the incidence is highest during the period 1–4 weeks of age. Calves that recover do not usually continue to excrete the organism in their faeces. In one study, excretion of mainly *S. dublin* ceased within 6 weeks of a calf's arrival at the unit[37], and in another peak excretion occurred 2 weeks after arrival with excretion declining to 6 weeks by which time few calves were excreting. Only on one occasion was excretion detected at 12 weeks, and there was no evidence of any animal being a carrier when slaughtered after 1 year of age[38].

When *S. typhimurium* infection was induced in calves, pyrexia and diarrhoea lasted about 11 d[41]. With calves purchased from the market, isolation of *S. typhimurium* from rectal swabs on arrival was very low, rose to a peak at 2–3 weeks and was self-limiting within 5 weeks[35]. The highest counts occurred in faeces of calves that died. With those that survived, 10^2–10^5 organisms/g faeces occurred continuously for up to a maximum of 20 d, followed by intermittent excretion[41].

The apparently lower incidence of *S. typhimurium* in the older ruminant calf may be associated with the finding that rumen fluid that contains high concentrations of VFA has an inhibitory effect *in vitro* on *S. typhimurium*. The greatest bactericidal effect was with a VFA (C2–C5 acids) concentration of 91.7 mM/l and pH 6.1[43].

Aetiology

Prevention depends on ensuring that the calf has adequate passive immunity and removal of the source of infection.

Passive immunity from colostrum

The Ig requirement for protection against salmonellosis appears to be about twice that required for protection against *E. coli*[44]. Since in calves infected with salmonellosis there was a negative correlation between serum IgM concentration and faecal output, it was suggested that in dying calves, hypercatabolism of Ig occurred leading to depletion of serum Ig. High serum Ig concentrations, particularly of IgM, should thus protect calves from death. For salmonellosis, this protection appeared to be non-specific[45].

Purchased calves

In endemic areas, which are mainly breeding areas, the outbreaks in calves are often quite mild and the mortality sporadic, but in purchased calves, including those bought-in from endemic areas, which have been exposed to further infection in markets and transport, to chilling and to changes in diet, the disease may be serious with a high mortality rate. In a survey in 1964, salmonellosis was responsible for 6 per cent of the total deaths in home-bred calves and 30 per cent in purchased calves[1].

Outbreaks of *S. typhimurium* are usually associated with purchased calves. Thus the current epidemic is considered to be due to the movement of calves and the constant mixing of infected with non-infected but susceptible calves[58]. However, in contrast, in Scotland it was reported in 1988 that most outbreaks of *S. typhimurium* were not associated with purchased calves[63].

To counteract the spread of multiresistant *S. typhimurium* 204c, it has been suggested that the State Veterinary Service should introduce a register of calf dealers and ensure adequate supervision of their hygienic procedures[72].

Purchased calves should be segregated from older batches, and should be carefully examined on arrival so that any infection may be quickly diagnosed and recovery from the stress of the journey expedited. In particular, if there is any sign of looseness of faeces, they should forego a feed, but have access to water. There is considerable diversity of opinion as to whether glucose solution (30–50 g/l) should be given to the calves on arrival. If the journey has been short and the calves recently fed, additional carbohydrate may predispose to fermentative diarrhoea, especially if the milk substitute that had been used contained more carbohydrate than is present in whole milk. On the other hand, if the calves have been without food for a considerable length of time, then the feeding of glucose, as a readily available source of energy, can be recommended.

An attempt has been made to produce a skin test for the identification of latent *S. dublin* infection in calves[31].

Housing

With purchased calves, an all-in all-out system of rearing is recommended[46]. Although it is generally recommended that sick calves should be kept warm, fluctuations in environmental temperature did not affect the symptoms or duration of excretion in four calves artificially infected with *S. dublin*[47].

The lowest incidence of salmonellosis was found in calves penned singly and the highest where calves were grouped[37]. However, to obtain further information on the effect of penning, 25 batches of calves, purchased from markets, comprising 589 calves on 11 farms had their faeces examined for 28 d after arrival. Salmonellae were found in 51 per cent of 423 calves from 18 of 21 batches on nine farms. The calves were housed in single pens on eight farms and in groups on the others.

Thus, type of penning and farm were confounded and the results must therefore be treated with caution. The trend in new excretors and the proportion of *S. typhimurium* excretors was similar for both types of penning. The peak of excretion was slightly higher in single-penned calves, being reached 18–19 d after arrival. Also, the peak of new excretors appeared earlier and declined sooner in single-penned calves. On average, single-penned calves excreted salmonellae for a longer period.

S. dublin was only found in single-penned calves and the trend in excretion was similar to that for *S. typhimurium* in single-penned calves, except that the peak was at about 28 d[66].

Breed

Friesian calves appear to be more resistant to *S. dublin* than are Ayrshires[48], and also more resistant to *S. typhimurium* than are Jerseys[41].

Adult cows

An adult cow may be a carrier of the disease, having recovered from an infection which may have been so mild that no symptoms were present. In self-contained dairy herds in E. Anglia, *S. dublin* was isolated from 0.7 per cent of the 1486 calvings that were monitored[49]. Pregnant cows that either develop clinical salmonellosis or become active carriers may abort their calves, or if the calves are carried to full term they may be infected *in utero*. Such calves may die shortly after birth. If not infected *in utero*, the calves are likely to acquire the infection from the carrier dam soon after birth. In a study of 111 cases of abortion associated with *S. dublin* infection, excretion of the organism in milk and vaginal mucus did not persist beyond 4–5 weeks and faecal excretion was usually transient. If the cows were not faecal excretors, there was no evidence of congenital infection at the next calving. However, if the cows were constant faecal excretors, congenital infection of the calf and/or transient vaginal excretion could be demonstrated[50]. Samples from abortions may give a negative complement fixation test, although the titres from a serum agglutination test are indicative of an infection[51].

In contrast to these findings, calves obtained by caesarean section from cows with a history of excreting *S. dublin* following enteritis or abortion showed no evidence of transplacental infection, in spite of the fact that many of the cows were excreting the organism in faeces, vaginal discharge or milk after parturition. It was thus considered that early calfhood infection arose after birth[49, 52].

Calves born of infected dams do not usually die, presumably because of the protection afforded by the dam's colostrum, but they may infect other calves whose illness may be much more severe. With *S. dublin* infections a carrier cow may continue to pass large numbers of organisms in its faeces for many years, but with *S. typhimurium* the carrier state is usually of shorter duration. Moreover, organisms may be present in an internal organ, such as a lymph gland, and may not be excreted in the faeces until the animal is exposed to stress. Milk from a clinically infected cow may be the cause of infection, but milk from carrier cows is seldom infected unless contaminated by faeces. Identification of a carrier cow is possible by laboratory examination of the sera and faeces of suspected cows, although sometimes the organisms are only excreted intermittently.

Concurrent disease

Although in N. Ireland it was thought that salmonella infection might be dependent on a concurrent liver fluke infection, statistical analysis has shown that the two conditions are independent of one another, although both organisms are killed or inhibited by drought and survive better on wet pastures[53].

A concurrent infection with infectious bovine rhinotracheitis (IBR) resulted in many deaths[46]. In addition, BVD virus was found to exacerbate the effects of salmonella infection. With *S. dublin*, faecal excretion was similar for the dual as for the single infection, but a protracted bacteraemia occurred only in two dual-infected calves, one of which died from a suppurative meningitis. With *S. typhimurium*, body temperature, passage of abnormal faeces and isolation rate of salmonella from the faeces was significantly increased[59].

Feed

Pasture

The possibility of contamination of pastures by salmonellae must be borne in mind. Whereas in dungheaps the salmonella organism is normally killed by the temperature rise in the heap, the spreading on pasture of faeces and urine in the form of fresh slurry may increase the risk of spread of salmonellosis. However, a period of storage of 1 month will significantly reduce the number of salmonellae in the slurry, and the risk can be further reduced by avoiding grazing pastures on which slurry has been spread until 1 month has elapsed.

Since survival in slurry is influenced by the normal microbial flora producing acids that are toxic to *S. dublin*, the possibility exists of increasing the rate of decline of the organisms by addition of small concentrations of organic acids[54].

Silage

Survival of salmonella in maize silage that was contaminated with faeces inoculated with salmonella was greatest when the manure had an initial pH of 6.0–6.5, whereas none survived when the pH was 4.0–4.5. The ensiling temperature had a pronounced effect on salmonella survival after 4 d with none recovered when the temperature was 35°C, one strain recovered at 25°C and 25 of 27 strains recovered at 15°C. At 25°C and 35°C, the pH of the feed ensiled was lower than when ensiled at 5°C or 15°C[55].

Concentrates

Concentrate feeds have often been implicated as the source of exotic serotypes. In particular, feeds of animal origin have frequently been shown to be contaminated with salmonella[8]. In 1975, a survey showed that 19 per cent of meat and bone meals used in poultry feeds were infected at the rate of 60–390 salmonellae/kg, but none of the fish meals were infected[56]. Pelleting of a meal resulted in a 1000-fold reduction in numbers of Enterobacteriaceae[56]. However, in Belgium a study over the years 1975–1982 showed that the ingredients used in calf feeds were not an important source of infection of *S. typhimurium*, *S. typhimurium* var. Copenhagen or *S. dublin*[57].

To reduce infection from feeds, two new orders have been introduced and came into effect in 1982. They are the Diseases of Animals (Protein Processing) Order (SI 676) 1981 and the Importation of Processed Animal Protein Order (SI 677) 1981. Since 1982, about 30 per cent of batches of imported animal protein have been contaminated, except that for white fish meal and herring meal the value has been about 10 per cent[61].

Early weaning

The development of the early-weaning system, in which the energy intake is low during the most susceptible period of life, may also be responsible for the increased incidence of the disease.

Other species

Whereas *S. dublin* is mainly a pathogen of cattle but may infect man, *S. typhimurium* is able to infect practically all species, including man. A human carrier may therefore be a source of infection. Mice rather than rats are the most important rodent vector of *S. typhimurium*[9]. There was no report of isolation from rats in the UK between 1958 and 1967, nor in 1980 or 1981, nor in 1985[61]. Although *S. typhimurium* has been isolated from pigeons in the UK, it was not considered a prime source of *S. typhimurium* var. Copenhagen in Belgium[57].

Number of organisms

Calves

To produce *S. typhimurium* infection by mouth, 10^8–10^9 organisms were required with 10^4–10^7 giving less consistent results[41]. For *S. dublin* given by mouth to calves up to 3 weeks of age, 5×10^{10} organisms caused a high death rate with peak mortality at 6–7 d[10].

Pregnant heifers

When 10^{10}–10^{11} *S. dublin* were given to nine heifers at 180 d of pregnancy four were unaffected, and for the remainder the severity of clinical symptoms varied from pyrexia and anorexia to mild diarrhoea and dysentery. Those given 10^{10} organisms did not appear to be infected at calving but those given 10^{11} were infected and one calf was born infected or became infected shortly after birth[11].

To produce an active infection of *S. dublin* in cows a single large inoculation of the rumen was required rather than the same total number given intermittently[12].

Prevention and treatment

Disinfection

If an outbreak on a farm is diagnosed, sick animals should be segregated. On no account should children and dogs have access to these animals, owing to the risk of infection, and the attendant should take precautions, by using protective clothing, to prevent the spread of disease to other stock or into his home.

After an outbreak, the building must be thoroughly cleaned, disinfected and fumigated with formalin or a proprietary aerosol, as the organism may survive for up to 6–12 months in dust as well as in faeces. However, the author has experience of a *S. typhimurium* infection with the same phage type in a calf house in a number of successive years in spite of vigorous disinfection and fumigation. Similarly, in another report, salmonella was isolated from the environment in six of nine farms studied, even after cleaning and disinfection. It was considered that the persistence of salmonellae in the environment deserved more attention in the formulation of control programmes[66].

Nevertheless to control *S. typhimurium* infection, calf houses should be thoroughly cleaned between batches of calves[65].

Vaccination

A live vaccine (*S. dublin* strain 51) is available, which gives a good degree of immunity against *S. dublin* and should be injected subcutaneously at least 1 week before the calves leave the premises on which they were born[5]. This vaccine also gives a degree of cross-immunity to *S. typhimurium*, as two of the seven 'O' antigens which occur are common to both organisms[6]. Buyers of calves should ensure that the calves have been vaccinated and have come from herds free from adult carriers. Even if vaccination is delayed until calves arrive on the premises of the rearer, with the disadvantage that the calves by this time may already be infected, vaccination still has some protective value[7]. As antibiotics or other antibacterial drugs, if given after vaccination, may reduce the immune response, only such drugs given orally, which are not readily absorbed from the intestine, should be used.

A live oral vaccine of a streptomycin-dependent mutant of *S. dublin* has been developed in E. Germany. When given daily for 10 d, it gave complete immunity and the development of 'H' agglutinin antibodies within 2 weeks after the last antigen was administered which persisted up to the age of 6 months[13]. This live vaccine gave better results than a heat-inactivated dead antigen when given orally, but not when given as a single subcutaneous injection[14].

Passive immunity from colostrum affects the production of antibodies by the calf; when a live *S. dublin* vaccine was given at 3 d of age, complement-fixing antibodies were not detected in colostrum-fed calves but were detected in colostrum-deprived calves. In contrast, with vaccination at 3 months, complement-fixing and serum agglutinin antibodies persisted for at least 3–6 months[51].

A study of the protection against salmonella afforded to five groups of calves immunized twice at intervals of 2 weeks with 2 ml of a combined salmonella, *E. coli* and pasteurella vaccine could not give meaningful results since the calves did not meet a significant challenge with salmonella. However, calves that were heavier and therefore possibly older gave a better serological response to *S. typhimurium*[67].

Antibiotics

Prophylaxis

Ampicillin has been used as a prophylaxis against salmonellosis, as it is rapidly absorbed after oral administration and high concentrations are found in the bile and urine. At a level of 50–200 mg/d for 2 weeks[16], ampicillin prevented the development of diarrhoea and increased the weight gains of calves exposed to *S. dublin* infection. It has been suggested that the particular beneficial effect of ampicillin is associated with its high concentration in bile, and the fact that, in chronic infections, salmonella organisms are likely to be localized in the gall bladder[15]. However, prophylactic feeding of antibiotics practised on nine farms appeared to have no influence on excretion of salmonella[66].

Treatment

Amoxycillin, similar to ampicillin, when given parenterally gave a rapid response in the treatment of experimental *S. dublin*. However, when given orally results were moderate if the calves were receiving milk substitute alone and were even less effective if given to calves consuming hay and concentrates[48]. Treatment of calves within 48 h of infection with *S. dublin* with five daily doses of trimethoprim and sulphadiazine (1:5 ratio) was very effective although individually the drugs were ineffective[10].

Antibiotic-resistant strains

In 1974, a study of four large calf rearing units showed that *S. dublin* was the dominant serotype and that multiple drug resistance was not a problem[37]. Moreover, in the following year, it was reported that the continuous feeding of

Table 4.1 Proportion of salmonellae isolated from cattle that were resistant to concentrations of drugs (%) (*S. typhimurium* in parentheses)[18, 19, 58]

	1972	*1973*	*1974*	*1982*	*1983*
No. of isolates	1538	2643	1993	1962	2728
Compound sulphonamides (50 µg)	35 (60)	67 (76)	65 (78)	63 (83)	67 (84)
Sulphonamides (500 µg)				36 (58)	43 (64)
Streptomycin (10 µg)	96 (97)	76 (88)	50 (56)	45 (38)	46 (51)
Streptomycin (25 µg)				15 (21)	18 (25)
Chlortetracycline (10 µg)	2.5 (3.0)	5.2 (5.8)	4.2 (8.9)	27 (44)	33 (50)
Trimethoprim/sulphamethoxazole (25 µg)	0 (0)	0 (0)	0 (0)	9 (15)	27 (40)
Ampicillin (10 µg)	0.3 (0.8)	0.6 (1.4)	0.8 (1.8)	6.9 (11)	23 (35)
Neomycin (10 µg)	5.7 (11.2)	1.5 (3.7)	1.5 (3.1)	7.0 (12)	23 (36)
Chloramphenicol (10 µg)	1.6 (13.2)	0.3 (0.7)	0.1 (0.1)	6.1 (10)	23 (35)
Furazolidone (15 µg)	0.3 (0.4)	1.1 (2.1)	0.5 (1.0)	<0.1 (0)	<0.1 (<0.1)
Apramycin (15 µg)					20 (19)
Sensitive to all drugs	3.8 (2.6)	7.0 (3.8)	20.0 (13.4)	18.0 (14.2)	23.8 (14.1)

350 mg chlortetracycline daily with or without 350 mg Sulphamezathine (sulphadimidine) to 8–9-week-old calves beginning 5 d before inoculation with *S. typhimurium* did not increase the incidence or persistence of salmonella and did not select for chlortetracycline- or sulphonamide-resistant strains. By the end of 2 weeks after inoculation only half the calves were shedding salmonella. However, the age of the calves was beyond the normal range for most natural infections[17]. Table 4.1 shows the proportion of salmonellae from cattle that in England and Wales were resistant to antibiotics between 1972 and 1974[18, 19], and 1982 and 1983[58].

In the USA, it was claimed that the feeding of 10 mg flavomycin (an antibiotic permitted in the UK for routine feeding to calves without a veterinary prescription) daily to 8-week-old calves infected with *S. typhimurium*, in an experiment in which precautions were taken to eliminate cross-infection, reduced the duration and prevalence of salmonella excretion in the faeces. It was also found that flavomycin reduced the number of salmonella serotypes that were resistant to streptomycin, ampicillin and oxytetracycline[20].

Since 1977 a strain of *S. typhimurium* type 204 resistant to chloramphenicol, streptomycin, sulphonamides and tetracyclines has been widespread in cattle. A related strain (phage type 193), resistant to ampicillin, kanamycin, streptomycin, sulphonamides and tetracyclines and derived from the above type 204, comprised 28 per cent of the total *S. typhimurium*. In 1979 it had been isolated from 700 cattle, on at least 300 farms, from streams and rivers and in food intended both for human and animal consumption[21]. In 1987, it was reported that four different antibiotic-resistant patterns were detected among 111 isolates from 73 outbreaks[64].

Similarly in France, it was claimed in 1979 that resistance to antibiotics was much higher for *S. typhimurium* and *S. dublin* isolated from veal calves than for those of human or porcine origin. The multiresistance of *S. typhimurium* during the previous 4 years increased concomitantly with a considerable increase in the incidence of *S. typhimurium* compared with *S. dublin* and other serotypes. In 38 strains of *S. typhimurium* and 20 strains of *S. dublin* tested against 14 antibiotics, a large proportion were resistant to tetracyclines, streptomycin, erythromycin, sulphonamides, chloramphenicol, kanamycin, ampicillin and neomycin[22].

Resistance to cephalosporins (cephalonium, cephaloridine and cefoperazone) has been found in *S. typhimurium* 204c isolates but not in five other phage types examined. This suggested that intramammary antibiotic preparations are involved in selection of populations of antibiotic-resistant Enterobacteriaceae, including *S. typhimurium* 204c. Since the recommendation for dry cow therapy with a preparation containing cephalonium has a period of activity in the udder of up to 10 weeks, some consumption by the calf of cephalonium, at least from colostrum, must occur when the gut flora is being established[68].

Public health aspects

Adult cows and parturition

A post-calving salmonellosis infection with *S. typhimurium* (phage type 49) was associated with disease in a farmer's family and in the attending veterinary surgeon[62]. In Denmark, *S. typhimurium* infection involving cows and calves on four farms and humans on two of the farms was reported. Up to 75 per cent of cows on one farm developed clinical salmonellosis and two died. On another farm, 25 per cent of cows were infected and there were two human cases. On three farms a number of the animals that recovered continued to excrete salmonella for 1–3 months after treatment. The same biotype occurred on two farms and tetracycline resistance was demonstrated on one farm. The source of infection was not known, except that on one farm the feed was the source; it was suspected that the owner spread the infection through his herd by walking on the feed in contaminated boots[60].

Milk

Infection of man by milk-borne salmonellosis has been a particular problem in Scotland. Up to 1974, infection by *S. typhimurium,* although of several phage types, was the most common cause. Since then other serotypes have appeared[23].

Between 1980 and 1982, 1090 persons, of whom eight died, were affected by milk-borne salmonellosis in Scotland in 21 outbreaks involving serotypes, including 12 phage types, of *S. typhimurium*[24]. The problem in Scotland was exacerbated by the large size of dairy farms and the greater proportion of raw milk consumed than in the rest of the UK. This was hopefully corrected in August 1983, when the Statutory Instrument No. 1866 (S 168) of 1980 the Milk (Special Designations) (Scotland) Order for pasteurization of virtually all cows' milk came into effect. However, 55 persons were farm workers and families taking raw milk from dairy farms, although the milk for sale was normally pasteurized[24].

Although excretion of salmonella in milk is relatively rare[70], it has been reported that salmonella persisted in the milk of one herd during a period of 2 years; eventually *S. typhimurium* (phage type 49a) was recovered from the mammary tissue of an individual cow[69].

Carcasses

After the stress of transport of calves excreting *S. dublin* in their faeces to a slaughterhouse, the organism was isolated from several organs at slaughter and previously unaffected calves became cross-infected.

During slaughter, carcasses may become contaminated on their surfaces. From one contaminated carcass, salmonella was recovered after chilled storage at 0°C for 1 week and also after freezing at −20°C for 1 month[47]. Of 720 veal calves slaughtered at a commercial abattoir, salmonellae were isolated from 4.3 per cent of the carcasses. The highest incidence was in the hepatic lymph nodes, followed by the liver, mesenteric lymph nodes and carcass surface. Since the calves had very low serum titres to the 'O' and 'H' antigens of salmonella, it was suggested that *pre mortem* or *post mortem* infection or contamination had occurred[25]. For ruminant cattle, it is suggested that regular feeding during transport and lairage and thus a low rumen pH would help to reduce *S. typhimurium* infection[43].

Sewerage

A study of the incidence of salmonella in the sewerage system and abattoir effluent between 1982 and 1984 of a town in N. Scotland revealed 20 different *S. typhimurium* phage types and four different *S. enteritidis* phage types. Ten of the phage types were isolated from livestock in the district. There were seven recorded incidences of human infection, involving four salmonella serotypes, only three of which were isolated concurrently from sewerage[71].

General

The Central Public Health Laboratories in England have always viewed farm animals as a source of antibiotic-resistant strains of enteric organisms in man and there is circumstantial evidence that multiresistant strains of *S. typhimurium*, probably arising from the use of particular antibiotic therapy, have been associated with a large number of cases of food poisoning in man[26,7]. Thus, phage type DT204c of *S. typhimurium* has become of increasing importance as a cause of human food poisoning, the number of human cases in Great Britain increasing from 26 in 1982 to 214 in 1985[58].

However, there was no evidence in the UK or the Netherlands that the use of subtherapeutic levels of antibiotics had selected for multiple resistance in any salmonella serotype and it was considered that multiple resistance of *S. typhimurium* arose as a consequence of the use of antibiotics in therapy rather than in growth promotion. It was concluded that the penetration of multiple resistant strains through the calf population was due to the stress involved in the movement of calves at a tender age and that further controls additional to those of the Medicines Act 1968 were unwarranted[28].

This Act made it illegal to feed antibiotics to calves without a veterinary prescription with the exception of certain antibiotics that are not used for human medicine. The legislation arose from a fear that resistance might be transferred in 'infectious resistance' from *E. coli* to *S. typhimurium* in both cattle and humans, and thence to *S. typhi* in humans. In particular, it was hoped that the use of chloramphenicol could be controlled since this is the most effective antibiotic for this latter disease.

Although the veterinary profession has been blamed for their frequent lax and empiric use of antibiotics, such as in the treatment of sick calves with products that had no useful effect against multiple-resistant salmonellae[28], the medical profession must also accept reproach for the uncritical prescribing of antibiotics[29]. For instance, it has been claimed that the medical rather than the veterinary use of tetracyclines has created the main selection pressure for the high incidence of tetracycline-resistant organisms in the human population[30].

The Zoonoses Order

The Zoonoses Order 1975 made a number of regulations dealing with diseases which man can contact from animals or birds; this order applied to organisms of the genera *Brucella* and *Salmonella*. In December 1988, the Zoonoses Order 1988 included bovine spongiform encephalopathy (see Chapter 7) under the regulations.

If an animal or a bird is known or suspected of carrying these organisms or to be infected with the disease, a veterinary inspector can declare the place where it is, or has been kept, as an infected place under the Diseases of Animals Act 1950 and its successor the Animal Health Act of 1981, and may impose restrictions on movement and insist on isolation and disinfection of the premises or vehicles involved.

Where either of the organisms has been isolated from an animal or bird, their carcass or surroundings, anyone who knows or suspects that an animal or bird is, or was, carrying the organism must report the matter to a veterinary inspector or other officer of the Ministry of Agriculture, Fisheries and Food or the Department of Agriculture and Fisheries for Scotland.

References

1. VETERINARY INVESTIGATION SERVICE. *Vet. Rec.* **76**, 1139 (1964)
2. GIBSON, E. A. *Vet. Rec.* **73**, 1284 (1961)
3. GIBSON, E. A. *Agriculture, Lond.* **73**, 213 (1966)
4. STEVENS, A. J., GIBSON, E. A. and HUGHES, L. E. *Vet. Rec.* **80**, 154 (1967)
5. SMITH, H. W. *J. Hyg., Camb.* **63**, 117 (1965)
6. RANKIN, J. D., NEWMAN, G. and TAYLOR, R. J. *Vet. Rec.* **78**, 765 (1966)
7. RANKIN, J. D., TAYLOR, R. J. and NEWMAN, G. *Vet. Rec.* **80**, 720 (1967)
8. SOJKA, W. J., WRAY, C., HUDSON, E. B. and BENSON, J. A. *Vet. Rec.* **96**, 280 (1975)
9. HUNTER, A. G., LINKLATER, K. A. and SCOTT, J. A. *Vet. Rec.* **99**, 145 (1976)
10. WHITE, G. and PIERCY, D. W. T. *Proc. IXth Int. Congr. Dis. Cattle, Paris* **1**, 299 (1976)
11. HALL, G. A. and JONES, P. W. *Br. vet. J.* **135**, 75 (1979)
12. HALL, G. A., JONES, P. W. and AITKEN, M. M. *J. comp. Path.* **88**, 409 (1978)
13. MEYER, H., HARTMANN, H., STEINBACH, G. *et al. Arch. exp. VetMed.* **31**, 71 (1977)
14. STEINBACH, G., MEYER, H., HARTMANN, H. *et al. Arch. exp. VetMed.* **31**, 95 (1977)
15. ACRED, P., LARKIN, P. J. and MIZEN, L. *Vet. Rec.* **79**, 103 (1966)
16. KERR, W. J. and BRANDER, G. C. *Vet. Rec.* **76**, 1105 (1964)
17. LAYTON, H. W., LANGWORTH, B. F., JAROLMEN, H. and SIMPKINS, K. L. *Zentbl. VetMed.*, **B22**, 461 (1975)
18. SOJKA, W. J. and HUDSON, E. B. *Br. vet. J.* **132**, 95 (1976)
19. SOKJA, W. J., WRAY, C. and HUDSON, E. B. *Br. vet. J.* **133**, 292 (1977)
20. DEALY, J. and MOELLER, M. W. *J. Anim. Sci.* **44**, 734 (1977)
21. THRELFALL, E. J., WARD, L. R. and ROWE, B. *Vet. Rec.* **104**, 60 (1979)
22. MARTEL, J. L. and FLEURY, C. *Bull. Soc. Sci. vet. Med. comp. Lyon* **81**, 203 (1979)
23. SHARP J. C. M., PATERSON, G. M. and FORBES, G. I. *J. Infect.* **2**, 333 (1980)
24. REILLY, W. J., SHARP, J. C. M., FORBES, G. I. and PATERSON, G. M. *Vet. Rec.* **112**, 578 (1983)
25. NAZER, A. H. K. and OSBORNE, A. D. *Br. vet. J.* **132**, 192 (1976)
26. THRELFALL, E. J., WARD, L. R. and ROWE, B. *Br. med. J.* **ii**, 997 (1978)
27. THRELFALL, E. J., WARD, L. R., ASHLEY, A. S. and ROWE, B. *Br. med. J.* **280**, 1210 (1980)
28. LINTON, A. H. *Vet. Rec.* **108**, 328 (1981)
29. ANONYMOUS. *Vet. Rec.* **111**, 287 (1982)
30. RICHMOND, M. H. and LINTON, K. B. *J. antimicrob. Chemother.* **6**, 33 (1980)
31. RICHARDSON, A. and PARKE, J. A. C. *Vet. Rec.* **97**, 15 (1975)
32. MINISTRY OF AGRICULTURE, FISHERIES AND FOOD, WELSH OFFICE AGRICULTURE DEPARTMENT, DEPARTMENT OF AGRICULTURE AND FISHERIES FOR SCOTLAND. Animal Salmonellosis Annual Summaries (1980/1981)
33. PETRIE, L., SELMAN, I. E., GRINDLAY, M. and THOMPSON, H. *Vet. Rec.* **101**, 398 (1977)
34. JOHNSTON, K. G. and JONES, R. T. *Vet. Rec.* **98**, 276 (1976)
35. HINTON, M., ALI, E. A., ALLEN, V. and LINTON, A. H. *J. Hyg., Camb.* **91**, 33 (1983)
36. AGRICULTURAL AND FOOD RESEARCH COUNCIL. *Calf Scour: Know Your Microbial Enemy.* At the Royal Show (1985)
37. LINTON, A. H., HOWE, K., PETHIYAGODA, S. and OSBORNE, A. D. *Vet. Rec.* **94**, 581 (1974)
38. OSBORNE, A. D., LINTON, A. H. and PETHIYAGODA, S. *Vet. Rec.* **94**, 604 (1974)
39. FISHER, E. W. and MARTINEZ, A. A. *Br. vet. J.* **131**, 643 (1975)
40. MORRIS, J. A. In *Function and Dysfunction of the Small Intestine* (eds R. M. Batt and T. L. J. Lawrence), Liverpool University Press, p.247 (1984)
41. WRAY, C. and SOJKA, W. J. *Res. vet. Sci.* **25**, 139 (1978)
42. PIERCY, D. W. T. *J. comp. Path.* **89**, 309 (1979)
43. CHAMBERS, P. G. and LYSONS, R. J. *Res. vet. Sci.* **26**, 273 (1979)
44. FISHER, E. W. *Vet. Rec.* **101**, 227 (1977)
45. FISHER, E. W., MARTINEZ, A. A., TRAININ, Z. and MEIROM, R. *Br. vet. J.* **132**, 39 (1976)
46. LAMONT, M. H. *Proc. Br. Cattle Vet. Ass.*, p. 109 (1983–84)
47. GRONSTOL, H., OSBORNE, A. D. and PETHIYAGODA, S. *J. Hyg., Camb.* **72**, 155 (1974)
48. OSBORNE, A. D., NAZER, A. H. K. and SHIMELD, C. *Vet. Rec.* **103**, 233 (1978)
49. COUNTER, D. E. and GIBSON, E. A. *Vet. Rec.* **107**, 191 (1980)

50. HINTON, M. *Br. vet. J.* **130**, 556 (1974)
51. WRAY, C. and SOJKA, W. J. *Res. vet. Sci.* **21**, 184 (1976)
52. OSBORNE, A. D., PEARSON, H., LINTON, A. H. and SHIMELD, C. *Vet. Rec.* **101**, 513 (1977)
53. TAYLOR, S. M. and KILPATRICK, D. *Vet. Rec.* **96**, 342 (1975)
54. JONES, P. W., SMITH, G. S. and BEW, J. *Br. vet. J.* **133**, 1 (1977)
55. McCASKEY, T. A. and ANTHONY, W. B. *J. Dairy Sci.* **58**, 149 (1975)
56. STOTT, J. A., HODGSON, J. E. and CHANEY, J. C. *J. appl. Bact.* **39**, 41 (1975)
57. LINTERMANS, P. and PHOL, P. *Annls Rech. vét.* **14**, 412 (1983)
58. SOJKA, W. J., WRAY, C. and McLAREN, I. *Br. vet. J.* **142**, 371 (1986)
59. WRAY, C. and ROEDER, P. L. *Res. vet. Sci.* **42**, 213 (1987)
60. MINGA, U. M., LICHT, H. H. and SHLUNDT, J. *Br. vet. J.* **141**, 490 (1985)
61. MINISTRY OF AGRICULTURE, FISHERIES AND FOOD, WELSH OFFICE AGRICULTURAL DEPARTMENT, DEPARTMENT OF AGRICULTURE AND FISHERIES FOR SCOTLAND. Animal Salmonellosis Annual Summaries (1985)
62. ANONYMOUS. *Vet. Rec.* **122**, 51 (1988)
63. ANONYMOUS. *Vet. Rec.* **122**, 197 (1988)
64. WRAY, C., McLAREN, I., PARKINSON, N. M. and BEEDELL, Y. *Vet. Rec.* **121**, 514 (1987)
65. CURTIS, P. E., WALTON, J. R. and WARD, W. R. *Vet. Rec.* **121**, 454 (1987)
66. WRAY, C., TODD, J. N. and HINTON, M. *Vet. Rec.* **121**, 293 (1987)
67. PETERS, A. R., WRAY, C. and ALLSUP, T. N. *Vet. Rec.* **121**, 84 (1987)
68. JONES, T. O. *Vet. Rec.* **120**, 399 (1987)
69. GILES, N. and KING, S. C. *Vet. Rec.* **120**, 23 (1987)
70. WRAY, C. and SOJKA, W. J. *J. Dairy Res.* **44**, 383 (1977)
71. JOHNSTON, W. S., HOPKINS, G. F., MacLACHLAN, G. K. and SHARP, J. C. M. *Vet. Rec.* **119**, 201 (1986)
72. CURTIS, P. E., WALTON, J. R. and WARD, W. R. *Vet. Rec.* **119**, 139 (1986)
73. WRAY, C. *Vet. Rec.* **116**, 485 (1985)

Chapter 5

Respiratory infections

With the development of intensive systems of production, there has been an increased incidence of respiratory infection in calves. The cause of the disease is extremely complex, and the relative importance of bacteria, viruses, chlamydia and mycoplasma is still not elucidated.

The disease may vary from a subclinical pneumonia to an acute fatal disease and appears to be the result of the interaction of one or more microorganisms with several predisposing causes, such as the stress of being moved long distances, or through markets, producing the so-called transit fever and shipping fever in the USA, the physical environment in which calves are housed and their nutrition.

The incidence of infection (morbidity) is usually high, but the mortality rate may be very variable. It is usually low, but on occasions may be severe; the severity of the mortality may be associated with the degree of secondary bacterial invasion that has occurred.

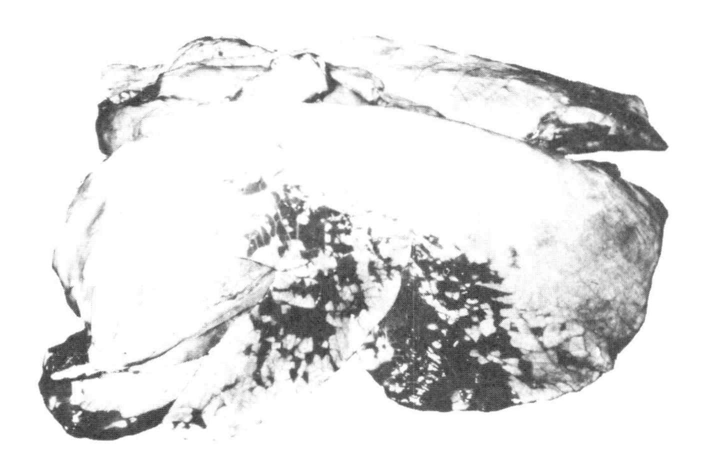

Figure 5.1 Left lateral aspect of the lungs showing lesions in calf infected with parainfluenza type 3 virus[166]. (By permission of the Controller of Her Majesty's Stationery Office. Crown Copyright)

A study of disease incidence in suckled calves, early-weaned calves and veal calves in England revealed that 50 per cent of the treatments for infectious disease were for respiratory conditions[146].

Symptoms

The symptoms are variable, but there is usually a nasal discharge, which is sometimes thin and watery, but at other times may be thick and purulent. A slight discharge may also occur from the eyes. In most outbreaks there is a dry cough, which often persists after the calf has recovered, and is particularly noticeable on exercise. Lesions of variable extent usually occur in the lungs, especially in the apical lobes (see Figure 5.1). In fact, there are a number of discrete pneumonic conditions, with different histopathological changes, which can be ascribed to different infective agents. If obvious signs of respiratory distress are present, the calf will invariably have a high temperature, but not necessarily so in more chronic cases. Diarrhoea is also often associated with the condition.

The organisms

General

In a microbiological study of 50 pneumonic lungs in Denmark *Pasteurella multocida* was isolated from eight, *P. haemolytica* from eight, *Corynebacterium pyogenes* from 13, *Neisseria* sp. from five, *Mycoplasma dispar* from 31, *M. bovirhinis* from 16, *Ureaplasma* from 26 and viruses from 14. Of the viruses, bovine respiratory syncytial virus (RSV) was isolated from four, bovine virus diarrhoea (BVD) virus from three and parainfluenza type 3 (PI3) virus from two, and the remaining viruses were not identified[18].

Of 50 outbreaks of respiratory disease in housed calves in N. Ireland, 47 were characterized by pneumonia. Viruses, particularly PI3 and RSV, were involved in 70 per cent of the outbreaks and in the majority, *Mycoplasma* spp. and *P. multocida* were present in the upper respiratory tract of affected calves. The lungs of surviving calves had a bronchiolitis and alveolitis (inflammation of the bronchioles and alveoli) with necrosis of the bronchiolar mucosa, which progressed to a *bronchiolitis obliterans* and a peribronchiolar fibrosis, i.e. fibrous tissue around the bronchioles. The majority of the calves that died within the first 2 months of these outbreaks had a severe exudative (discharge of fluid) bronchopneumonia with marked vascular damage and necrosis of the pulmonary parenchyma, i.e. the functional part of the lung. The organisms most frequently isolated from pneumonic lungs were *Pasteurella* spp., *Mycoplasma* spp. and PI3[94, 95].

In 1982, a search for new microorganisms that might be responsible for calf pneumonia in the UK revealed the presence of five viruses, four species of mycoplasma and 19 species of bacteria in respiratory material; the only organism not previously associated with calf pneumonia was a coronavirus. Statistical analysis of the results from eight of the outbreaks could not ascribe the disease to a single microorganism or to a particular combination of microorganisms. It was suggested that the failure of a pure culture of an organism to produce disease might be because:

(a) a combination of organisms is required;

(b) the organisms had become attenuated through passage necessary for purification; or
(c) material in the respiratory secretions, other than the organisms identified, is required for disease to be evoked[93].

Viruses

The most important viruses associated with infection of the respiratory tract in calves are bovine herpes virus type 1 (infectious bovine rhinotracheitis (IBR) virus), RSV[28], PI3 virus, bovine adenovirus (nine serotypes)[25–27], bovine rhinovirus (two serotypes), reovirus (three serotypes)[24], BVD virus[5] and enteroviruses. In addition, the psittacosis-lymphogranuloma venereum (PLV) group of organisms, i.e. chlamydia, have also been implicated. Most of these organisms have also been found in groups of healthy calves.

The method adopted to ascertain whether calves had been clinically infected with a particular virus was by the paired serum sample method, in which the first sample was taken at the first signs of the disease and the second 3 weeks later. If there was a four-fold increase in antibody titre, it was considered that the calf was infected with that particular virus at the time of the first sample[4].

Until about 1975, PI3 virus, widespread throughout the world, and adenovirus and reovirus were considered to be the most important viral agents associated with pneumonia in calves in Great Britain. At first sight, it would appear that PI3 virus was of most importance as 84 per cent of serum samples obtained from animals at slaughterhouses had titres greater than 1:32[6] and 23 per cent of serum samples had antibodies against adenovirus[7]. However, by use of the paired sample method, 24 per cent of 150 outbreaks of respiratory infection in calves were associated with PI3, 29 per cent of 133 outbreaks with adenovirus and 26 per cent of 111 outbreaks with reovirus[8]. A similar survey in 1970 showed that 39 per cent of respiratory infections were associated with PI3, 16 per cent with adenovirus, and 15 and 11 per cent with reovirus 1 and 2, respectively. Because of infections with more than one virus, about 50–60 per cent of the outbreaks were probably of viral origin.

In a serological study in Poland of 1067 calves from large breeding farms, 84 per cent had antibodies to PI3 virus, whilst antibodies against IBR virus, adenovirus and RSV were found in 11, 4.7 and 2.3 per cent of the calves respectively. Only 0.5 per cent of the calves had antibodies against reovirus 1 and 2 and BVD virus[92].

In 1974, virus involvement did not seem important in studies of 27 field outbreaks in calves aged 2–8 months[22] or of a large beef rearing farm in the UK[23].

Parainfluenza type 3 (PI3) virus

Pneumonia associated with PI3 has been produced in healthy calves by contact exposure with coughing calves, the incubation period being 7–11 d. It was suggested that prolonged exposure to a low dose of virus was more pathogenic than a single exposure to a larger dose[29]. In a group of 25 calves housed together from birth, PI3 was isolated from the lungs of four calves at 6, 7, 13 and 55 d respectively, but significant amount of antibody was only found in two of the calves[91].

A report in 1979 of four outbreaks of pneumonia in housed calves in N. Ireland where the main clinical symptom was dyspnoea (difficulty in breathing) showed that the pneumonia involved the cranial (apical) lobes and was associated with severe emphysema (dilatation of the alveoli). Histology revealed bronchiolitis, alveolitis with alveolar cell hyperplasia and multinucleate syncytium (a mass of nucleated protoplasm without cell boundaries) formation, intra-alveolar haemorrhage and oedema, and hyaline membrane (a thin transparent membrane beneath the epithelium of the mucosa) formation in terminal bronchioles and alveoli. In each outbreak there was evidence of PI3 and RSV infection with PI3 isolated from all cases[90].

Although PI3 has been isolated from an aborted bovine fetus in Wyoming, it is not considered a common cause of fetal disease, since nearly all cows had antibodies to PI3 and inoculation of such cows with virus failed to show any transplacental transmission. However, inoculation of bovine fetuses *in utero* with PI3 virus did cause disease[89]. The virus has also been isolated from bulls' semen[137].

Respiratory syncytial virus (RSV)

RSV was isolated first in Switzerland and since 1969 has been implicated in respiratory disease with a high morbidity but low mortality rate. The virus is highly contagious and is transmitted by oral or nasal contact. In mild, muggy weather it may have a half-life of about 7 h, but is inactivated by hard frost. Isolation from nasal swabs is only possible within 3–4 d of infection and not after symptoms have appeared[72]. Bovine and human RSV are antigenically similar and both have the ability to multiply in bovine and human cells[96].

Whereas no increase in respiratory disease occurred in Jersey calves infected with PI3 at 58–127 d of age in pens out of doors, infection with RSV and BVD (see below) at an older age, 138–267 d, caused a highly significant difference in mean disease score[30]. Similarly, intranasal infection of gnotobiotic (germ-free) colostrum-deprived calves and conventional calves with RSV produced histological changes in the lungs[31]. Further details of this research showed that RSV caused a biphasic pyrexia with serous nasal discharge. In contrast to the statement above, the virus was recovered from the nasal discharge within 4–10 d and from tracheal and bronchial mucosae and lungs within 7–10 d of inoculation. The virus caused no macroscopic lesions, but a focal degenerative rhinitis (inflammation of the nasal mucous membrane) and a catarrhal bronchiolitis was found with occasional formation of syncytia in bronchioles and alveoli[31].

In a beef suckler herd in Iowa, RSV was isolated from two of seven calves of 45–105 d of age showing signs of acute respiratory tract disease, but no humoral antibody response to the virus was observed[88].

In Scotland, an acute fatal pneumonia which was associated with RSV occurred in multiple-suckled calves that were housed together from 1–2 weeks of age. It was characterized by pyrexia, tachypnoea (rapid breathing), respiratory distress and coughing and histologically showed an alveolitis, with multinucleate syncytia, alveolar epithelial hyperplasia and bronchiolitis. Interstitial emphysema was also present. Although RSV was not isolated from the calves, the majority had neutralizing antibody titres to RSV with some having a rise in titre of four-fold or greater. There was no evidence of PI3 involvement and the adult cows sucked by the calves remained healthy. Six weeks after the outbreak started, six calves had a chronic cuffing pneumonia characterized by lymphocytic bronchiolitis with some also having *bronchiolitis obliterans*. *M. dispar* was isolated from two of the calves[87]. Also in Belgium, naturally-occurring RSV in Friesian calves of between 120 and 160 kg live weight produced an acute obstructive disease since the bronchiolar epithelium was the main target[139]; this condition was progressively replaced by a moderate subclinical restrictive disease[138].

A study of the blood gas values in 12 calves in the acute stage of RSV infection showed that the calves were severely hypoxic. The mean arterial oxygen tension Pao_2 for affected and healthy calves was 7.7 and 14.2 kPa respectively. Hypercapnia (an excess of CO_2 in the blood resulting in overstimulation of the respiratory centre) was present in only three of the diseased calves[141].

Infection of calves with RSV resulted in simultaneous increases of IgA, IgM and IgG_1 in the acute phase of the disease, whereas IgG_2 responses followed at various intervals thereafter. In young animals with maternal antibody, there was no increase in IgG_1 and IgG_2 and only a weak response in IgM and IgA. Since maternal immunity was found to be restricted to IgG_1 antibodies in serum, there was no protection of the mucosal surfaces to RSV infection[140].

A fluorescent antibody test of sera has been developed which provides a rapid and sensitive alternative to conventional tests for PI3 and RSV (i.e. haemagglutination inhibition (HI) for PI3 and virus neutralization for RSV)[143].

Rhinovirus

Rhinovirus has also been isolated from 6–8-month-old calves showing coughing, nasal discharge and pyrexia[32].

Adenovirus

It has been suggested in the USA that adenovirus associated with diseases in calves may be of public health importance, since two strains, antigenically related to human adenoviruses and isolated from the faeces of apparently healthy cows, caused respiratory illness in calves[11].

Infectious bovine rhinotracheitis (IBR) virus

Besides causing enteritis, the IBR virus (bovine herpes virus 1, BHV1) which is related to the herpes virus in man, causes necrosis of the mucosa of the upper respiratory tract rather than of the lower tract, although pneumonia may result from a secondary infection. It also causes necrosis of the lymphoid tissue and of the adrenal cortex[9], and may affect the genital tract, causing infectious pustular vulvovaginitis. Infected bulls may shed BHV1 in semen intermittently at any time and remain latently infected for life[147, 148].

The virus caused severe disease in N. America between 1950 and 1970, spread to Europe in 1972 and caused serious disease in the UK in 1977. It has been suggested that this was associated with the importation of Holstein cattle from Canada[69]. Thus in 1964, IBR virus seemed to be of little importance, antibody titres being found only in 2 per cent of serum samples obtained from animals at slaughterhouses[6], and only one out of 108 outbreaks of respiratory disease could be associated with the virus. However, both this virus and that of BVD have an affinity for lymphoid tissue, and failure to produce antibodies may be the result of destruction of the immunologically competent cells[9, 10].

The first symptom of the disease is the intense reddening of the conjunctiva and mucous membranes of the nose. The discharge from the eyes and nose is clear at first but gradually becomes thick and purulent. The temperature may rise to a high level and the condition may be accompanied by a moist cough. In bought-in calves of 1–2 weeks of age that were sucking dairy cows, IBR was associated with hyperpnoea, pyrexia and conjunctivitis. The morbidity was high but mortality low, and the adults were not infected[69]. Another report showed that neonatal calves were more susceptible than older cattle[149].

Thus, in an outbreak of IBR infection in housed single-suckled calves out of 16 Hereford-cross heifers served by a Charolais bull, 50 per cent of calves developed IBR between 1 and 4 weeks of age. The clinical signs were pyrexia, anorexia, excessive salivation and rhinitis accompanied by unilateral or bilateral conjunctivitis, which was only observed in calves born to the newly-introduced heifers; five of the calves died (31 per cent mortality). After diagnosis of IBR, all surviving calves were given an intranasal IBR vaccine, but they required 2–3 weeks to recover after the vaccination and repeated parenteral antibiotic therapy[147].

Differential diagnosis of the viral agent involved from the lesions in the lungs is difficult in natural outbreaks, because of the much greater effect of the secondary bacterial invaders. However, it appears that a watery discharge and conjunctivitis are associated with IBR virus and that diarrhoea accompanying the pneumonia most frequently occurs with adenovirus, IBR virus and PLV organisms. Serological assays for antibodies to BHV1 have been widely used for the diagnosis of acute infections with IBR[142].

It has been suggested that plasma iron and zinc concentrations should be used as an indicator of infection since in pyrexic cattle or those given *E. coli* lipopolysaccharide endotoxin (O111 B4), plasma iron increased and then declined whilst zinc tended to decline, but in those infected with IBR virus, only plasma iron increased, plasma zinc being unaffected[154].

Reactivation

Latent infection of IBR has been activated in 4-week-old calves by experimental infection with PI3 virus[40]. Other factors that may reactivate IBR are parturition[150], corticosteroid treatment, *Dictyocaulus viviparous* infection[151], transport[152] and oral administration of 3-methyl indole[153] (see Fog fever, p. 149). If reactivation at parturition with nasal re-excretion occurs, a latently infected dam could infect its calf, which would either develop clinical disease if the level of colostral immunity was too low or transmit the virus to other calves[150].

Mycoplasmas

Increased emphasis on mycoplasmas was the result of the isolation of new mycoplasmas, *Ureaplasma* spp. (T-mycoplasmas)[12] and *Mycoplasma dispar*[13,14], and the finding that these two organisms were recovered from 58.5 and 50.8 per cent, respectively, of 65 lungs examined, whilst *M. bovirhinis* was isolated from 23 per cent. Although no viruses were present, chlamydia were isolated from 20 of the lungs[15]. However, these two mycoplasmas have also been found in apparently normal calves[16,17]. *M. bovis* (formerly *M. agalactiae* var. *bovi*) has also been isolated from a pneumonia outbreak of 3–4-month-old Friesian calves, and the disease was reproduced by infection of two Channel Island calves[42]. *M. bovis* was also associated with mastitis, polyarthritis and synovitis, and is pathogenic to the genital tract[42].

In contrast in Poland, *M. bovirhinis* was the commonest species associated with respiratory disease; *M. arginini* and *M. dispar* also occurred but there was no evidence of *M. bovis* or *M. bovigenitalium*[86].

In Louisiana, the nasal flora of healthy calves were positive for mycoplasma on at least one occasion in a year in 80 per cent of the herds and 19 per cent of the calves, but all were negative for chlamydia. Of the isolates, 31 per cent were *M. bovirhinis*, whilst *M. bovis*, *M. arginini* and *Acholeplasma laidlawii* also occurred[76].

In Czechoslovakia only 20 per cent of the nasal swabs from 1000 calves were positive for mycoplasma infection before introduction of the calves in five groups into large-capacity calf houses. After 3 weeks in the calf house, the calves developed respiratory disease and 70–95 per cent of the swabs were positive. Similarly, before introduction 54–100 per cent of the calves had no antibodies to mycoplasma, but after 1–3 months there was a significant increase in antibodies[19].

M. dispar and ureaplasmas have also been associated with so-called cuffing pneumonia, characterized by a peribronchial lymphoid hyperplasia which has the appearance of a peribronchial cuff, in calves housed together for 6 months[20]. Of 20 calves, 12 had cuffing pneumonia, and there was a significantly higher isolation of *M. dispar* and ureaplasmas from these calves than from the unaffected group. *M. bovirhinis* and *A. laidlawii* were isolated from the lungs in both groups. Cuffing pneumonia was found in pneumonic lungs from 75 per cent of clinically healthy calves but only in 5 per cent of the lungs of calves that died[15]. Further indication of the involvement of mycoplasma was the finding that tylosin tartrate but not ampicillin prevented experimental pneumonia in calves given endobronchial inoculation of pneumonic lung homogenates[21].

M. dispar's natural habitat is the respiratory tract, from which it does not spread to the bloodstream, except for irregular occurrence in brain tissue. When given by aerosol to colostrum-deprived calves, it produced pneumonia with an exudative catarrhal purulent bronchiolitis accompanied by moderate interstitial cell proliferation[85].

It is suggested that the extracellular polysaccharide from *M. mycoides* subspecies *mycoides* may cause the symptoms in contagious bovine pleuropneumonia in Australia, the galactans (polysaccharides consisting of galactose units) producing biogenic amines that cause the ill effect[84].

In E. Germany, pneumonia with a concomitant arthritis has been observed in calves originating from herds with *M. bovis* mastitis as well as in large calf rearing units. In most cases of catarrhal and catarrhal purulent bronchopneumonia, *M. bovis* was the only detectable pathogen but in fibrinous necrotizing and abscessed pneumonia, bacteria were also present. *M. bovis* was regularly isolated from joints of calves with polyarthritis although bacteriological findings were negative[101].

A lobular or confluent lobular collateral bronchopneumonia has been induced in 23 calves by intratracheal or endobronchial injection with *M. bovis* although *P. haemolytica* was isolated from five of the calves. One calf, after contracting pneumonia, developed an arthritis caused by *M. bovis*. It was found that intra-articular application of *M. bovis* caused a fibrinous purulent arthritis from all joints[102].

A synergistic effect has been found between *M. bovis* and *P. haemolytica* with calves becoming more severely affected when given both organisms than could be accounted for by the additive effect of each agent separately[103]. Impairment of pulmonary clearance of *P. haemolytica* by *M. bovis* does not appear to be the cause[104].

In further work it was shown that the most severe disease and greatest degree of pneumonic consolidation was induced in conventional calves by intranasal and intratracheal inoculation of 6 h rather than of 18 h cultures of *P. haemolytica* 1 d after the intranasal inoculation of *M. bovis*. It was thought that the 6 h cultures possessed more capsular material and would thus be more resistant to phagocytosis[121]. In gnotobiotic calves it was found that more severe disease and pneumonic consolidation resulted when *M. bovis* was inoculated before *P. haemolytica* rather than *vice versa*[120].

Bacteria

The infection by viral agents is often followed by a secondary bacterial invasion, usually by *Pasteurella haemolytica* or *Corynebacterium pyogenes*. Calves exposed to parainfluenza type 3 (PI3) virus followed by *Pasteurella* organisms 24 h later showed more marked signs of transit fever than those exposed to *Pasteurella* alone[1–3].

In the USSR, a study was made for 4 months of the bacterial flora in the nasal mucus of a group of 54 purchased calves brought into a fattening unit at 15–20 d of age. During the first week, *Alcaligenes* was dominant but staphylococci and streptococci were also present. Between 20 and 40 d, they were replaced by *Pasteurella*. At 105–125 d, the *Pasteurella* declined and a number of other species such as *Neisseria flava*, *Micrococcus luteus*, *Proteus vulgaris* and *E. coli* became established. The appearance of *Pasteurella* was accompanied by mild respiratory disease in most calves, but four calves developed pneumonia. The strain of *Pasteurella* when given by aerosol was pathogenic, but it seemed probable that viral agents were also involved in field outbreaks[64].

In Denmark, a study of calf lungs showed that 44 per cent were bacteriologically sterile, while *P. multocida* was isolated from 28 per cent and *C. pyogenes* from 17 per cent[79].

In E. Germany, *P. haemolytica* infection was considered to be an independent disease with calves showing severe dyspnoea and pyrexia (40.5–41°C)[99]. The peracute form was characterized by a septicaemia, the acute form by a fibrinous pleuropneumonia and the subacute form by a purulent necrotic pneumonia, often with abscess formation. The disease occurred throughout the year and older animals of 1 or 2 years of age might develop the disease. Ninety-eight strains of *P. multocida* and seven strains of *P. haemolytica* were isolated from about 45 per cent of the calves that died. Seven of the strains of *P. multocida* and three strains of *P. haemolytica* were sensitive to chloramphenicol and furazolidone but resistant to chlortetracycline, oxytetracycline and penicillin[100].

Surprisingly, there does not seem to be a great difference in bacterial counts in the lungs of calves with a subclinical or those with a fatal pneumonia. Of lungs from 140 calves with subclinical and also from 60 calves with fatal pneumonia, 48 and 41 per cent respectively contained $>10^4$ colony forming units/g tissue. *P. haemolytica* was associated more with fatal than subclinical cases. Of seven species of bacteria inoculated endotracheally into gnotobiotic calves, only *P. haemolytica* produced any respiratory disease, although some strains of *P. multocida* produced a fatal septicaemia. *C. pyogenes* was either an opportunist invader or the isolates were non-pathogenic variants[105].

Experimental challenge of calves vaccinated with sodium salicylate extracts of the antigens of *P. haemolytica* resulted in clinical disease, possibly because the challenge was too severe, i.e. enhanced by the vaccination[123].

There are two biotypes of *P. haemolytica*, A and T. Biotype A is associated with pneumonia of cattle and sheep and with septicaemia in sucking lambs, whereas biotype T is associated with septicaemia in weaned lambs[122]. Pneumonia, similar to that observed in recently housed, weaned single-suckled calves in Scotland, has been reproduced by infecting calves with a particular strain of *P. haemolytica* biotype A. The major clinical findings were pyrexia, hyperpnoea, tachypnoea, nasal discharge and reduced appetite accompanied by a fibrinous pneumonia at 2–3 d after infection which by 9–10 d after infection had encapsulated into well defined nodules[111].

Anaerobic bacteria may also be associated with pneumonia in cattle. Of 144 lungs of cattle with pneumonia, 45 yielded anaerobes with the number of isolations slightly lower for acute fibrinous or suppurative bronchopneumonia than for chronic pneumonias. The most common isolates were *Peptococcus indolicus, Peptococcus asaccharolyticus, Fusiformis necrophorus, Clostridium perfringens (welchii)* and *Bacteroides fragilis*. There was a significant correlation between the presence of *C. pyogenes* and *E. coli* in the lungs and that of the anaerobes. *B. fragilis* and *Cl. perfringens* showed multiple antibiotic resistance and 33 per cent of *P. indolicus* isolates were resistant to tetracyclines. The remainder were generally susceptible to ampicillin, penicillin G, fefoxitin, cephalothin, clindamycin, chloramphenicol, erythromycin, tetracycline and metronidazole[106].

Site of infection

Initial infections in the respiratory tract occur at specific sites resulting from an interaction between a very complex and effective defence mechanism and the characteristics of the organisms involved. The main site of initial infection for pasteurella, mycoplasma, PI3 and IBR viruses is the upper respiratory tract, whereas for RSV and adenovirus, it is the lower respiratory tract[83].

However, *M. bovis* only induced a subclinical proliferative pneumonia when inoculated intranasally but induced clinical disease when given intratracheally. *P. haemolytica* produced a fatal exudative pneumonia when inoculated intratracheally but *M. dispar, Ureaplasma diversum*, RSV and PI3 viruses produced only a subclinical pneumonia[82]. The response is likely to differ with the strain, which vary greatly in virulence.

Clearance

The presence of large numbers of microorganisms in the lungs clearly results from an imbalance between the numbers inhaled, which is dependent upon the population of organisms in the environment, and the numbers cleared from the lungs. The respiratory tract is kept free from pathogens by the activity of the ciliated and mucus-secreting cells of the respiratory tract, and these latter appear to be susceptible to adverse environmental conditions[53].

A study in Canada of the lungs of calves infected with *P. haemolytica* by aerosol showed that the alveolar macrophage in the phagocytic system provides the major antibacterial action of the lung early in infection. Induction of a flooding alveolar oedema in the lungs by hydrocortisone acetate interfered with phagocytosis, but it was unaffected when a foaming oedema was induced. In calves with haemagglutinating antibodies against *P. haemolytica*, polymorphonuclear leucocytes participated in the phagocytosis. Within 4 h of challenge, an early bronchiolitis and minute areas of atelectasis (shrunken areas devoid of air) occurred at the site of deposit of the organism. Treatment of a calf with an aerosol of histamine impaired clearance of *P. haemolytica* and caused an early fibrinous pneumonia[81].

A major role of RSV may be that of predisposing the lung to a secondary infection. RSV has a catastrophic effect on the ciliated respiratory epithelium by causing extensive damage to the mucous membrane and thus inhibiting clearance and making the lung more susceptible to dust and secondary infection[72].

In a study of clearance, using aerosols of radiolabelled polystyrene spheres of 3.3 μm in diameter[116], 72 per cent of total radioactivity was deposited in lung tissue beyond the major bronchi. After mucocilary clearance, mean alveolar deposition was 34 per cent in a group of bull calves and 52 per cent in a group of heifer calves. When the heifer calves were treated with oestradiol, deposition increased from 47 to 56 per cent. The overall mucocilary clearance rate was 1.28/h with a great variation between calves. Although it is generally considered that temperature, relative humidity (RH) and ammonia concentration affect ciliary action[119], neither mucocilary clearance rate nor alveolar deposition was affected by a change in climate from 14°C and 0.87 RH to 5°C and 0.75 RH[117].

Clenbuterol, the β-adrenergic agonist or more precisely the β_2-receptor sympathomimetic agent, by stimulation of the appropriate receptor produces smooth muscle relaxation of the bronchi, enhances ciliary activity of the ciliated bronchial cells and has serolytic activity[118]. The use of this drug was found to increase the mucociliary clearance of the polystyrene spheres[117].

Another approach to the causes of airway obstruction has been adopted in Belgium where the physiological and mechanical properties of isolated calf lungs have been studied[124–126].

Predisposing factors

Passive immunity from colostrum

In general the outbreaks, which occur mainly in the winter months, do not affect calves younger than 1 month of age, which suggests that the calves are probably protected during this time by the passive immunity received from their dams. Whether the protection afforded is to the secondary invaders or to the virus, or both, is not clear.

The peak incidence of respiratory disease in a group of veal calves occurred at 40–50 d of age, which coincided with the minimum concentrations of systemic Igs[145]. In another survey the peak incidence of respiratory disease occurred when calves were 6–8 weeks of age, several weeks after arrival of purchased calves at the rearing unit[146].

Since high blood Ig levels at a young age are important in protection from arthritis and pneumonia as well as from enteric disorders, the importance of adequate colostrum ingestion cannot be overemphasized. In Denmark, the incidence of pneumonia in the first 6 weeks of life was twice as high in calves with low Ig status and mortality was generally higher[79].

In the USA, the peak onset of pneumonia, which occurred between 2 and 4 weeks of age, was correlated with the lowest IgG_1, IgG_2 and IgA concentration in the serum and the lowest concentration of IgG and IgA in nasal secretions. Most calves developed pneumonia when serum IgG_1 was <15 g/l, IgG_2 <3 g/l, IgA <1 g/l and IgM <2 g/l, and when combined IgG and IgA values in nasal secretions were <0.2 g Ig/g protein. In the first study diarrhoea preceded pneumonia in 63 per cent of 56 calves, and in the second 38 per cent of calves had diarrhoea and/or haemorrhagic faeces before pneumonia occurred[107].

High IgG_1, IgG_2 and IgA, particularly IgG_1, but not IgM levels at 2.5 weeks of age were associated with a reduced susceptibility to pneumonia at 2.5 months of age[39]. Other work has shown a relationship between the incidence of pneumonia up to 5 months of age and low levels of serum Ig[165]. It has been suggested that a good initial immunity may allow an animal to support a gradual and controlled establishment of successive respiratory pathogens so that the immune system can elaborate competent and adequate levels of specific antibody rather than be overwhelmed by a sudden influx of large numbers of pathogens[39].

IgA is the predominant Ig in the respiratory mucosa of healthy calves, but in calves with cuffing pneumonia IgG_1 cells are the major cell type with a marked increase in plasma cells (mature antibody-synthesizing cells that have been differentiated from B lymphocytes in response to an antigenic stimulus) in the bronchial wall. In contrast, in exudative pneumonia, although IgG_1 cells predominate, the total number of plasma cells are similar to those in healthy calves. IgA and to a lesser extent IgM are both transported across the mucous membrane via the bronchial epithelial cells and submucosal glands[163].

Because the interstitial (lymph) and terminal (lavage) compartments of the lungs of sheep contained higher concentrations of IgA and IgM, but not of IgG_1, than did plasma, it appears that there is either local synthesis of these two classes of Ig in the lungs or there exists a class-specific active transport of Ig into the lungs[80]. IgG_2 has been found to neutralize *P. haemolytica* A1 leucotoxin more efficiently than does IgG_1; IgM was ineffective[110].

In neonatal calves of <4 d of age, no Ig-containing cells were found in the tracheobronchial tree, whereas 18-month-old cattle had a greater number of these cells than had 4-month or 8-month-old cattle or cull dairy cows >3 years of age. The relative frequency of each Ig class among the cells of the mucosa was $IgA > IgG_1 > IgG_2 > IgM$ for all cattle of 4 months and older with an IgA to IgG_1 ratio of 3.46. The proportion of cells in each Ig class was similar throughout the airway from trachea to bronchiole, although the density of the cells declined distally[144].

Although after the original introduction of a virus onto a farm mortality may occur, subsequent losses will be limited, because the dams should produce antibodies to protect the fetus during gestation, and antibodies should be concentrated in colostrum to protect the neonatal calf[41].

Housing

It is claimed that the concentration of airborne bacteria colony-forming particles in the air of a calf house can be 100 times that in human dwellings and 1000 times that in fresh air[115].

Sudden changes in climate may affect the colonization of the respiratory tract by microorganisms partly as a result of increased pathogen survival, and thus challenge, and partly as a result of reduced host resistance. The housing of cattle modifies the effect of climate but also increases the density of the pathogens. Thus, any stress that lowers resistance will be magnified when cattle are housed[129].

Although poor ventilation and high humidity of calf houses have been implicated as predisposing causes in outbreaks of the disease, the relative importance of various factors such as air temperature, air movement, relative humidity, ammonia concentration and cubic capacity per calf, has not been elucidated[130].

Even with a very high cubic capacity per calf (20 m^3) and 24 changes of air/h, high humidity at an environmental temperature of 14.5°C, or low humidity at an environmental temperature of 21°C, predisposed Jersey and Friesian calves to respiratory infections. However, no such relationship existed for Ayrshire calves. The ideal temperature and relative humidity for veal calves appeared to be 17°C and 0.65 respectively[33]. Thus cold damp and, to a lesser extent, hot dry conditions seem to be detrimental. It is generally recognized that pneumonia may become a problem under commercial conditions during the cool and damp conditions of late autumn; in contrast, in warm veal units the relative humidity is often artificially increased by water drips, as it is suggested that this reduces problems associated with respiratory infections. A high relative humidity at a high environmental temperature has its parallel in pig 'sweat boxes', in which the sedimentation rate of airborne particles is increased and the bacterial complement of the environment is reduced[56]. In testing stations for growing bulls, more pneumonia occurred during the period October–December[34]. The adverse effect of a particular environment would appear to be associated with the clearance rate of microorganisms from the lung. In mice, cold stress, together with wet fur, has been shown to inhibit the clearance of bacteria from the lungs. In addition, hypoxia and starvation for 24–48 h reduced clearance[54, 55]. Similarly, calves that were chilled when artificially infected with respiratory viruses were more severely affected and showed more extensive lung lesions[43].

In a study of bacteria isolated from nasopharyngeal swabs from veal calves kept either in naturally-ventilated straw yards, or in crates in a fan-ventilated room, or in crates in a controlled environment in the UK, it was found that in one group of calves a significant proportion of the variation in weekly bacterial count was associated with changes in vapour pressure and temperature that occurred between 2 and 4 d previously. With calves kept at a constant temperature of 16°C, the bacterial population was at a minimum between a relative humidity of 0.65 and 0.75 and tended to rise at humidities outside this range. In three groups of calves the number of bacteria at 9 weeks of age was positively correlated with lung lesions at 16–18 weeks of age[127].

In Switzerland[35] the optimum conditions for preventing respiratory infection in veal calves have been proposed as follows:

Environmental temperature:	18°C at beginning and 12°C at end of fattening period
Relative humidity:	0.60–0.80
Maximum tolerable CO_2:	3.5 ml/l
Maximum tolerable NH_3:	25 μl/l
Air movement:	not to exceed 0.2 m/s, except during hot weather when 0.3 m/s is acceptable

In the UK it is recommended[52] that concentrations should not exceed 1 ml/l for CO_2 and 5–10 μg/l for NH_3.

Large diurnal variations in calf house temperature and relative humidity were considered to be more harmful than high values, provided that the temperature and relative humidity did not exceed the upper limit. However, these suggestions are based on experience rather than on experimentation[35]. Thus, in a comparison of calves kept at 5°C and 0.75 RH with those subjected to an abrupt change to 13°C and 0.84 RH directly after exposure to an aerosol of *P. haemolytica* type A1, an increase in respiration rate occurred in the latter calves 8–14 h later and *P. haemolytica* proliferated in the nasopharynx during a 24 h monitoring period.

However, there was no apparent effect of the change in climate on *P. haemolytica* in the trachea[128].

Although the importance of an adequate ventilation rate (0.37–1.87 m^3/kg live weight) and cubic capacity per calf (5.0–7.0 m^3)[52] in reducing the incidence of respiratory infections and improving growth rate has frequently been emphasized[45–51], experiments at Drayton and High Mowthorpe Experimental Husbandry Farms, in which various methods of heating and ventilating calf houses have been compared, have failed to show any marked effect of environmental temperature and relative humidity on the incidence of calf disease[57,58]. In a recent trial in England, there was more respiratory disease in a controlled-environment house than in a naturally-ventilated monopitch building[109].

In Denmark, there was a tendency for an increased incidence of pneumonia in calves given milk substitute, hay and concentrates, when ambient temperature was <10°C and relative humidity was >0.85. In insulated houses, a maximum ventilation rate of 100 m^3/h and a minimum rate of 30 m^3/h per calf were recommended. In a comparison of relative humidities of 0.70 and 0.85 on seven farms, there was no difference in incidence of pneumonia (24 per cent) or mortality rate. Incongruously, there appeared to be fewer cases of pneumonia in an uninsulated house, but mortality was higher (7.7 per cent) than in the insulated house (4.3 per cent)[79].

For IBR, no effect of temperature or humidity on the response of the respiratory tract was found. Moreover, experiments in controlled environments with few calves in climatic chambers have shown little evidence of any effect of the physical environment on susceptibility to IBR[78]. Similarly, calves of 66–158 d of age exposed to aerosols of IBR and then to *P. haemolytica* after an interval of 3–5 d, but not those exposed to *P. haemolytica* alone, developed rhinitis, tonsillitis, laryngitis and tracheitis within 4 d, but the outcome was not affected by the temperature or relative humidity in the environmental chamber either before or after the viral aerosol was given nor by the period that the calves had been exposed to a particular environment[77].

Air filtration

Installation of recirculating air filters that filtered the total volume of air every 2.2 min in sheds, each containing 28 veal calves, within a main building, reduced aerial bacterial contamination by 45 per cent, antibiotic usage by 35 per cent and the mean area of lung consolidation by 38 per cent. The filtration in the plenum-ventilated sheds resulted in increased air movement and was associated with a reduction in both incidence and severity of clinical and subclinical disease. Both the number of treatments for respiratory disease and the area of lung consolidation at slaughter were directly related to the reduction in weight gain[75].

Method of rearing

Time of weaning

Possibly of more importance than the climatic environment is the system of rearing used within a particular climatic environment. In particular, age at weaning appears to affect susceptibility of calves to respiratory infections. Early-weaned calves (weaned at 5 weeks of age) have been shown to have a higher incidence and severity of lung lesions than have calves given both a higher level and longer duration of milk feeding[33,36]. It is not known whether the lower intake of productive energy and the lower fat deposition in the carcass and thinner skins of early-weaned calves may be important, or whether the lower intake and utilization of some micronutrient by calves given dry food from an early age is responsible, or whether inhaled dust and fungal spores from dry food[44] may be the predisposing factor. Moreover, weaning at a young age may be stressful. Repeated and lasting stressor effects, such as by ACTH injections, have been shown to impede humoral antibody production[65].

The effect of early weaning in predisposing calves to infection was suggested in 1964[47], and under range conditions in the USA the first clinical signs of pneumonia appear shortly after weaning, at which time the calves are on a low plane of nutrition[61]. With bulls introduced to a testing station at 6–7 months of age, more pneumonia occurred if they had been weaned within 14 d of entry[34], which suggests an interaction between the stress of weaning and a new environment.

Further evidence[60] on the effect of early weaning and of a high environmental temperature combined with a low relative humidity on the incidence of respiratory infections has come from a study of serum antibody titres in 102 calves. At 3 months of age a four-fold increase in antibody titre over the precolostral serum value was taken as evidence of infection, an assumption that may be unwarranted, since it was not possible to establish how far the titre at 3 months of age represented a residual titre from the passive immunity afforded by feeding 7 kg colostrum and how far it represented the calf's autogenous antibody response to infection. The percentage of calves classified as infected with any one or more of the four viruses PI3, reo 1 and 2 and BVD was greater for early-weaned calves than for those given milk substitutes *ad libitum* and greater for calves reared at a high environmental temperature. Similarly, the incidence of multiple infections and the percentage of calves infected with reo 1 or reo 2 was greater for early-weaned calves and for calves reared at a high environmental temperature, whereas for PI3 no effect of environmental temperature was apparent, although early weaning markedly increased the infection rate. In contrast, infection with BVD virus appeared to be largely limited to calves given a high level of milk substitute feeding.

Level of milk feeding

In contrast to an early-weaning regimen, very high growth rates (1.57 kg/d) and concomitant high body fat deposition resulting from feeding milk substitute diets at high concentration also appear to increase susceptibility to respiratory infections[37]. Moreover, in *ad libitum* milk feeding the large output of urine may result in wet bedding and a high humidity at calf level[62].

More respiratory disease occurred with *ad libitum* machine feeding than with once or twice daily bucket feeding[108] although in one trial a high level of respiratory infection occurred with once daily bucket feeding[109]. In a trial in which calves were allowed to drink milk from a machine as fast as it was dispensed, 50 per cent of calves had respiratory infections[109].

Breed

A breed difference in susceptibility is apparent, with Ayrshire calves being more resistant than Friesian and Jersey calves[38], and Hereford × Friesian being more resistant than purebred Friesian calves. These findings may be associated with the better insulation arising from the thicker skins of the Ayrshire calf and the Hereford × Friesian calf.

Purchase of calves

In a recent survey, the proportion of farmers treating calves for respiratory disease was 5–6 times greater for bought-in than for home-bred calves[146].

Herd size and calf creeps

Pneumonia was found to be more closely related to herd size and building design than to the selenium status of the herd. Herds with >60 calves had a higher incidence. A lower incidence occurred with herds that had separate creeps for the calves of suckler cows than in herds where creeps were part of the cow accommodation[157].

Prevention

Management

In Scotland the problem of pneumonia in young housed suckled calves was overcome by ensuring an adequate intake of colostrum and avoidance of nutritional stress, the mixing of groups of calves, a lack of a calf creep, poor ventilation (the single most important cause), dust and infection with lungworm[74]. Similarly, in Czechoslovakia, minimum morbidity from respiratory infections was obtained by giving calves a controlled amount of colostrum, rearing them in batches on an all-in all-out system and disinfecting the environment before the next batch of calves was introduced[73].

Vaccination

In view of the variety of organisms involved in respiratory disease, vaccination should only be considered where specific agents have been identified. Vaccines against several of the microorganisms involved are available. Some vaccines are more effective when given intranasally, but for some antigens, such as RSV, this is not so. Maternally-derived antibody with a serum neutralization titre >1:8 appears to protect calves against RSV, but interferes with the response to vaccination[72].

Similarly, colostrum-fed calves gave a smaller response to infection by aerosols of PI3 virus than did colostrum-deprived calves, although both groups showed a leucopenia (a decrease in leucocytes) and shed virus from the nasal passages. Titres in serum and nasal secretion of virus-neutralizing antibody at 30 d after viral exposure were greater in the colostrum-deprived calves[98]. Pre-colostral serum of 12 calves from dams vaccinated with PI3 virus contained antibody[59].

In Bulgaria, higher interferon activity in both serum and nasal secretion and higher secretory (local) and serum antibody titres were produced in 6-month-old calves by submucosal inoculation within the nasal opening of a bivalent vaccine, consisting of a mixture of live attenuated PI3 virus and an adenovirus (variant of type 1), than by subcutaneous injection. Adenovirus was a stronger inducer of interferon than was the PI3 virus[71]. Other work showed that two subcutaneous injections of an inactivated PI3 virus would give similar complete protection to that afforded by one subcutaneous injection together with intranasal inoculation[70].

In the USA the IBR virus is considered as one of the inciting agents of shipping fever and a vaccine against it is available. Vaccination by intranasal inoculation of live IBR virus can produce effective immunity[67], but the IBR virus may spread infection to non-vaccinated animals[68]. Thus, vaccination should only be considered where IBR has been identified[69].

The immune response of calves to inoculation with mycoplasma is dependent on the species of mycoplasma and the age of the calf. Whereas IgG_1 response to *M. bovis* varied little with age, that to *M. dispar* increased from 16 to 155 d of age, suggesting that the immune response of young calves to *M. dispar* is impaired or defective[66]. Immunity to colonization by *M. bovis* was induced by both intratracheal and intramuscular injections with a formaldehyde-inactivated organism but not by intramuscular injection alone, indicating that it was necessary to stimulate the local immune system to develop resistance[97].

In Scotland, three different vaccination routines were studied in 44 bought-in calves on a commercial calf rearing unit with a history of virus pneumonia. The routines were intranasal temperature-sensitive IBR vaccine at 3 and 10 weeks, a combined IBR and PI3 vaccine at the same ages and the combined vaccine together with a live attenuated bovine RSV virus given intramuscularly at 7, 10 and 16 weeks. Two outbreaks occurred, one at 3–4 months with RSV and the other at 4–5 months with PI3. The incidence of pneumonia tended to be lower and the number of days of elevated temperature and the number of treatments were significantly lower in groups vaccinated against the virus responsible. However, there was no significant difference in live-weight gain during the outbreaks or during the 10 month period to slaughter[131].

In England, a multicomponent vaccine has been produced at the Institute for Research in Animal Diseases and this is to be developed commercially[132]. The quadrivalent vaccine contains killed antigen of RSV, PI3, *M. dispar* and *M. bovis* emulsified with an oil adjuvant. Calves were vaccinated on three occasions, 3 weeks apart, and starting at either 3, 7 or 12 weeks of age. In a trial period from November 1981 to May 1982, 27 per cent of unvaccinated control calves were treated for respiratory disease compared with 16 per cent of the vaccinated calves. Reduction of non-fatal disease represented a protection rate of almost 40 per cent. Mortality was reduced from 3.4 to 1.9 per cent but this difference was not significant. During a major outbreak of disease associated with RSV the protection rate increased to 69 per cent. In the group of calves vaccinated for the first time at 7 weeks of age, there was a significant reduction in the pneumonic consolidation in the lungs at slaughter at 18 months of age or at death[133]. A comparison of this quadrivalent vaccine with one containing only the RSV antigen was made on a large beef unit during the winters of 1983/4 and 1984/5, involving 736 calves. An outbreak of disease occurred in five of seven batches. In four batches, where disease was associated with RSV or *M. bovis* infection together or singly, the death rates were 9, 2 and 3 per cent respectively for the unvaccinated, quadrivalent vaccine and RSV vaccine groups respectively representing a protection rate of 77 and 68 per cent for the two vaccines respectively. The proportion of calves receiving treatment for respiratory disease was 38, 25 and 27 per cent for the control, quadrivalent and RSV vaccine groups respectively[134].

In the preparation of vaccines against *P. haemolytica* in cattle, it is suggested that serotypes other than A1 should be included, assuming that, as in sheep, protection is serotype specific[136]. Although biotype A serotype 1 is most common, other biotype A serotypes and biotype T serotypes and untypeable strains were also present in cattle[135].

Immediately after the transfer of calves to a rearing unit, the immunological reaction to antigen injections was found to be stronger than when the antigen was applied 3 d after transfer[65].

Antibiotics

The use of antibiotics for prevention is likely to be more effective than their use for therapy. In the USSR antibiotic therapy was considered useless[64]. Treatment is largely aimed at reducing the effect of secondary bacterial invasion.

On the other hand, the mass medication of purchased calves against the shipping fever-bovine respiratory disease complex by intramuscular injection for three successive days of oxytetracycline (11 mg/kg body weight) followed by oral sulphadimethoxine (150 mg/kg body weight) resulted in an 81 per cent reduction in 'treatment days' per calf purchased. This reduction was much greater than that achieved by the use of the drugs individually (20–55 per cent reduction). The use of a long-acting tetracycline and a sustained-release sulphadimethoxine resulted in a 90 per cent reduction and only one handling of the calves was necessary[63].

Similarly, a long-acting oxytetracycline preparation given intramuscularly (20 mg/kg live weight) to beef bulls at performance testing stations of the Meat and Livestock Commission at about 6 months of age tended to reduce the risk of respiratory infection, but the effect was not significant[112].

In the Netherlands, in 23 calves aged 3–8 months that had shown serological evidence of RSV infection and had been treated for 3 d with tetracycline hydrochloride to prevent secondary bacterial infection, the injection intravenously of the anti-inflammatory flunixin meglumine (2 mg/kg body weight) significantly reduced mean body temperature but had no effect on Pao_2 in capillary blood samples or on respiration rate[114].

The efficacy of sulbactam-ampicillin in the treatment of respiratory disease associated with ampicillin-sensitive and ampicillin-resistant strains of *P. haemolytica* and *P. multocida* has been studied. Treatment of 123 Friesian calves, 3–5 weeks old, showing clinical signs of respiratory infections, with either sulbactam-ampicillin or ampicillin alone resulted in the deaths of seven of 59 calves given ampicillin alone compared with only one of 64 calves given sulbactam-ampicillin. Moreover, the combined drug resulted in a greater reduction in the severity of symptoms. With ampicillin, a greater proportion of ampicillin-resistant strains of *P. haemolytica* was isolated subsequently, whereas with sulbactam-ampicillin the proportion of resistant strains that were recovered declined[113].

Fluid

Since dyspnoea and hyperpnoea rapidly lead to dehydration, and the calf may not take water voluntarily, administration of 2 litres fluid intravenously has been recommended[62].

Selenium (see also Chapter 7)

Selenium has been considered to have a preventive effect against diarrhoea and pneumonia, but this has been disputed by other workers who could find no effect of selenium treatment on pneumonia in housed calves[156]. When selenium, as barium selenate, was given to suckler cows, no antibiotic treatment was needed for the calves from the 25 treated cows, whereas of the calves from the 27 untreated cows, 11 had to be treated for diarrhoea or pneumonia and four died[155]. In contrast, in a study of 13 beef suckler units, five dairy rearing units and one beef rearing unit with bought-in calves involving 1341 calves in N.E. Scotland, the occurrence and incidence of pneumonia in housed calves was not related to the selenium status of the herd as measured by glutathione peroxidase activity, nor were they affected by selenium treatment of calves during the neonatal period[157].

Fog fever

This condition causes large economic losses in the USA and other countries and is associated with an atypical interstitial pneumonia together with an acute pulmonary emphysema. It is particularly associated with a change of grazing and the ingestion of aftermath or other lush forage[162]. Ingestion of mouldy sweet potatoes causes a similar condition[164].

Although the disease has been produced experimentally by oral administration of tryptophan[159, 160] or its metabolites indole acetic acid and 3-methyl indole[161], no significant difference has been found in the tryptophan concentration in herbage from two normal pastures and from three pastures on which outbreaks of fog fever occurred. It is suspected that L-tryptophan is converted to 3-methyl indole by the rumen microflora[164].

In a comparison of the pulmonary changes in field cases of fog fever in Canada with those induced by tryptophan or 3-methyl indole, it was found that 48 field cases and the experimentally-produced condition were characterized by a diffuse proliferation of alveolar pneumocytes, whereas 107 other animals had a focal exudative alveolitis comparable to a hypersensitivity reaction. The former condition occurred commonly from September to November, was associated with a change of pasture or diet and occurred in animals aged 2 years and older, particularly suckler cows. The hypersensitivity type occurred in younger cattle throughout the year with a peak incidence in January[158]. Other research has shown that calves are more resistant to fog fever than are adults, since no pulmonary oedema or emphysema was observed when seven young calves were given intraruminally 0.25 g 3-methyl indole (skatole)/kg body weight, although there were mild respiratory signs[164].

References

1. HEDDLESTON, K. L., REISINGER, R. C. and WATKO, L. P. *Am. J. vet. Res.* **23**, 548 (1962)
2. HETRICK, F. M., CHANG, S. C., BYRNE, R. J. and HANSEN, P. A. *Am. J. vet. Res.* **24**, 939 (1963)
3. SORENSEN, D. K., JOHNSON, D. W. and HOYT, H. H. *Nord. VetMed.* **16**, Suppl. 1, 543 (1964)
4. HARBOURNE, J. F. *Vet. Rec.* **78**, 749 (1966)
5. OMAR, A. R. *Vet. Bull.* **36**, 259 (1966)
6. DAWSON, P. S. and DARBYSHIRE, J. H. *Vet. Rec.* **76**, 111 (1964)
7. DARBYSHIRE, H. J. and PEREIRA, H. G. *Nature, Lond.* **201**, 895 (1964)
8. DAWSON, P. S., LAMONT, P. H. and DARBYSHIRE, J. H. quoted by OMAR, A. R. (ref. 5)
9. TYLER, D. E. and RAMSEY, F. K. *Am. J. vet. Res.* **26**, 903 (1965)
10. SHOPE, R. E. Jr. *Diss. Abstr.* **26**, 634 (1965)
11. MOHANTY, S. B. and LILLIE, M. G. *Proc. Soc. exp. Biol. Med.* **120**, 679 (1965)

12. GOURLAY, R. N. *Res. vet. Sci.* **9**, 376 (1968)
13. GOURLAY, R. N. *Vet. Rec.* **84**, 229 (1969)
14. GOURLAY, R. N. and LEACH, R. H. *J. med. Microbiol.* **3**, 111 (1970)
15. GOURLAY, R. N., MACKENZIE, A. and COOPER, J. E. *J. comp. Path.* **80**, 575 (1970)
16. GOURLAY, R. N. and THOMAS, L. H. *J. comp. Path.* **80**, 585 (1970)
17. THOMAS, L. H. and SMITH, G. *J. comp. Path.* **82**, 1 (1972)
18. BITSCH, V., FRIIS, N. F. and KROGH, H. V. *Acta vet. scand.* **17**, 32 (1976)
19. JURMANÓVÁ, K., HÁJKOVÁ, M., KREJČI, J. and ČERNÁ, J. *Vet. Med., Praha* **20**, 583 (1975)
20. PIRIE, H. M. and ALLAN, E. M. *Vet. Rec.* **97**, 345 (1975)
21. GOURLAY, R. N., HOWARD, C. J., THOMAS, L. H. and STOTT, E. J. *Res. vet. Sci.* **20**, 167 (1976)
22. THOMAS, L. H. *Vet. Rec.* **93**, 384 (1973)
23. THOMAS, L. H. and COLLINS, A. P. *Vet. Rec.* **94**, 506 (1974)
24. LAMONT, P. H., DARBYSHIRE, J. H. and DAWSON, P. S. *J. comp. Path.* **78**, 23 (1968)
25. IDE, P. R. and DARBYSHIRE, J. H. *Br. vet. J.* **125**, vii (1969)
26. BETTS, A. O., EDINGTON, N., JENNINGS, A. R. and REED, S. E. *J. comp. Path.* **81**, 41 (1971)
27. REED, S. E., TYRELL, D. A. J., BETTS, A. O. and WATT, R. G. *J. comp. Path.* **81**, 33 (1971)
28. JACOBS, J. W. and EDINGTON, N. *Vet. Rec.* **88**, 694 (1971)
29. LAMONT, P. H. *VetMed. Nauki* **12**, 65 (1975)
30. THOMAS, L. H., STOTT, E. J., COLLINS, A. P. and JEBBETT, N. J. *Res. vet. Sci.* **23**, 157 (1977)
31. JACOBS, J. W. and EDINGTON, N. *Res. vet. Sci.* **18**, 299 (1975)
32. SHIMIZU, Y., NARITA, M. and MURASE, N. *Nat. Inst. Anim. Hlth Q., Tokyo* **14**, 35 (1974)
33. ROY, J. H. B., STOBO, I. J. F., GASTON, H. J., GANDERTON, P., SHOTTON, S. M. and OSTLER, D. C. *Br. J. Nutr.* **26**, 363 (1971)
34. ANDREWS, A. H. *Vet. Rec.* **98**, 146 (1976)
35. MARTIG, J., BOSS, P. H., NICOLET, J. and STECK, F. *Livestk Prod. Sci.* **3**, 285 (1976)
36. ROY, J. H. B. *Span* **16**, 101 (1973)
37. STOBO, I. J. F. and ROY, J. H. B. *J. agric. Sci., Camb.* **93**, 95 (1979)
38. ROY, J. H. B. *Jl R. agric. Soc.* **132**, 81 (1971)
39. WILLIAMS, M. R., SPOONER, R. L. and THOMAS, L. H. *Vet. Rec.* **96**, 81 (1975)
40. MENSÍK, J., POSPÍSIL, Z., SUCHÁNKOVÁ, A., ČEPICÁ, A., ROZKŌSNÝ, V. and MACHATKOVÁ, M. *Zentbl. VetMed.* **B23**, 854 (1976)
41. McCLURKIN, A. W. *J. Dairy Sci.* **60**, 278 (1977)
42. THOMAS, L. H., HOWARD, C. J. and GOURLAY, R. N. *Vet. Rec.* **97**, 55 (1975)
43. JENNINGS, A. R. and GLOVER, R. E. *J. comp. Path.* **62**, 6 (1952)
44. LACEY, J. *J. gen. Microbiol.* **51**, 173 (1968)
45. ESMAY, M. L., WILLIAMS, H. F. and GUYER, B. E. *Res. Bull. Mo. agric. Exp. Stn* No. 527 (1953)
46. MARTIN, B. *Vet. Rec.* **75**, 78 (1963)
47. QUARMBY, W. B. *Vet. Rec.* **76**, 590 (1964)
48. THOMPSON, W. A. *Agriculture, Lond.* **73**, 37 (1966)
49. MARTIN, H. *Vet. Rec.* **81**, 255 (1967)
50. NATIONAL AGRICULTURAL ADVISORY SERVICE. *Rep. Drayton exp. Husb. Fm,* p. 14 (1964)
51. NATIONAL AGRICULTURAL ADVISORY SERVICE. *Rep. Drayton exp. Husb. Fm,* p. 16 (1965)
52. HOSKEN, E. E. Personal communication (1975)
53. PHILLIP, J. I. H. *Vet. Rec.* **20**, 552 (1972)
54. GREEN, G. M. and KASS, E. H. *J. clin. Invest.* **43**, 769 (1964)
55. GREEN, G. M. and KASS, E. H. *Br. J. exp. Path.* **46**, 360 (1965)
56. GORDON, W. A. M. *Br. vet. J.* **119**, 263 (1963)
57. HELPS, M. B. *Agriculture, Lond.* **75**, 557 (1968)
58. BEE, R. *Dairy Fmr, Ipswich* **20**(1), 26 (1973)
59. FRANZ, J., KREJČÍ, J. and MENŠÍK, J. *Zentbl. VetMed.* **B21**, 540 (1974)
60. ROY, J. H. B. and STOBO, I. J. F. *Rep. natn. Inst. Res. Dairy,* p. 68 (1968)
61. BROWN, W. W. *J. Am. vet. Med. Ass.* **152**, 726 (1968)
62. BEE, D. J. *Proc. Br. Cattle Vet. Ass.,* p. 185 (1984–85)
63. LOFGREEN, G. P. *J. Anim. Sci.* **56**, 529 (1983)
64. SIDOROV, M. A., POLYKOVSKII, M. D., SKORODUMOV, D. I. and MUSTAFAEVA, N. I. *Vest. sel'-khoz. Nauki, Mosk.* **9**, 55 (1975)

65. HARTMANN, H., BRUER, W., HERZOG, A. *et al. Arch. exp. VetMed.* **30**, 553 (1976)
66. HOWARD, C. J. and GOURLAY, R. N. *Vet. Microbiol.* **8**, 45 (1983)
67. FRERICHS, G. N., WOODS, S. B., LUCAS, M. H. and SANDS, J. J. *Vet. Rec.* **111**, 116 (1982)
68. NETTLETON, P. F. and SHARP. J. M. *Vet. Rec.* **107**, 379 (1980)
69. PLENDERLEITH, R. W. J. *Proc. Br. Cattle Vet. Ass.*, p. 65 (1983–84)
70. PROBERT, M., STOTT, E. J., THOMAS, L. H., COLLINS, A. P. and JEBBETT, J. *Res. vet. Sci.* **24**, 222 (1978)
71. HARALAMBIEV, H. *Arch. exp. VetMed.* **29**, 397 (1975)
72. McFARLAND, H. J. *Proc. Br. Cattle Vet. Ass.*, p. 69 (1983–84)
73. MENSIK, J., DRESSLER, J. and FRANZ, J. *Vet. Med., Praha* **22**, 463 (1977)
74. LOWMAN, B. G. and WATSON, G. A. L. *Fm Bldg Prog.* **63**, 7 (1981)
75. PRITCHARD, D. G., CARPENTER, C. A., MORZARIA, S. P., HARKNESS, J. W., RICHARDS, M. S. and BREWER, J. I. *Vet. Rec.* **109**, 5 (1981)
76. SPRINGER, W. T., FULTON, R. W., HAGSTAD, H. V., NICHOLSON, S. S. and GARTON, J. D. *Vet. Microbiol.* **7**, 351 (1982)
77. JERICHO, K. W. F. and LANGFORD, E. V. *Can. J. comp. Med.* **42**, 269 (1978)
78. JERICHO, K. W. F. and DARCEL, C. le Q. *Can. J. comp. Med.* **42**, 156 (1978)
79. BLOM, J. Y., THYSEN, I. and OSTERGAARD, V. *Beretn. St. Husdyrbrugsforsog* **570** (1984)
80. GORIN, A. B., STEWART, P. and GOULD, J. *Res. vet. Sci.* **26**, 126 (1979)
81. GILKA, F., THOMSON, R. G. and SAVAN, M. *Zentbl. VetMed.*, **B21**, 774 (1974)
82. ANONYMOUS. *ARC News* p. 17 (Summer 1983)
83. BASKERVILLE, A. *N.Z. vet. J.* **29**, 235 (1981)
84. BUTTERY, S. H., LLOYD, L. C. and TITCHEN, D. A. *J. med. Microbiol.* **9**, 379 (1976)
85. FRIIS, N. F. *Acta vet. scand.* **21**, 34 (1980)
86. PILASZEK, J. and TRUSZCZYNSKI, M. *Bull. vet. Inst. Pulawy* **22**, 6 (1978)
87. PIRIE, H. M., PETRIE, L., PRINGLE, C. R., ALLAN, E. M. and KENNEDY, G. J. *Vet. Rec.* **108**, 411 (1981)
88. LEHMKUHL, H. D., GOUGH, P. M. and REED, D. E. *Am. J. vet. Res.* **40**, 124 (1979)
89. SWIFT, B. L. and TRUEBLOOD, M. S. *Theriogenology* **2**, 101 (1974)
90. BRYSON, D. G., McFERRAN, J. B., BALL, H. J. and NEILL, S. D. *Vet. Rec.* **104**, 45 (1979)
91. ALLAN, E. M., PIRIE, H. M., SELMAN, I. E. and SNODGRASS, D. R. *Res. vet. Sci.* **24**, 339 (1978)
92. BACZYNSKI, Z., MAJEWSKA, H. and ZMUDZINSKI, J. *Bull. vet. Inst. Pulawy* **22**, 14 (1978)
93. THOMAS, L. H., GOURLAY, R. N., STOTT, E. J., HOWARD, C. J. and BRIDGER, J. C. *Res. vet. Sci.* **33**, 170 (1982)
94. BRYSON, D. G., McFERRAN, J. B., BALL, H. J. and NEILL, S. D. *Vet. Rec.* **103**, 485 (1978)
95. BRYSON, D. G., McFERRAN, J. B., BALL, H. J. and NEILL, S. D. *Vet. Rec.* **103**, 503 (1978)
96. PRINGLE, C. R. and CROSS, A. *Nature, Lond.* **276**, 501 (1978)
97. HOWARD, C. J., GOURLAY, R. N. and TAYLOR, G. *Vet. Microbiol.* **2**, 29 (1977)
98. MARSHALL, R. G. and FRANK, G. H. *Am. J. vet. Res.* **36**, 1085 (1975)
99. DAT, DAO TRONG and SCHIMMEL, D. *Arch. exp. VetMed.* **28**, 303 (1973)
100. DAT, DAO TRONG, BATHKE, W. and SCHIMMEL, D. *Arch. exp. VetMed.* **27**, 909 (1973)
101. BOCKLISCH, H., PFUTZNER, H., ZEPEZAUER, W., KUHN, U. and LUDWIG, H. J. *Arch. exp. VetMed.* **37**, 435 (1983)
102. PFUTZNER, H., KIELSTEIN, P., MARTIN, J. and SCHIMMEL, D. *Arch. exp. VetMed.* **37**, 445 (1983)
103. HOUGHTON, S. B. and GOURLAY, R. N. *Vet. Rec.* **113**, 41 (1983)
104. LOPEZ, A., MAXIE, M. G., SAVAN, M., *et al. Can. J. comp. Med.* **46**, 302 (1982)
105. HOUGHTON, S. B. and GOURLAY, R. N. *Res. vet. Sci.* **37**, 194 (1984)
106. CHIRINO TREJO, J. M. and PRESCOTT, J. F. *Can. J. comp. Med.* **47**, 265 (1983)
107. CORBEIL, L.B., WATT, B., CORBEIL, R. R., BETZEN, T. G., BROWNSON, R. K. and MORRILL, J. L. *Am. J. vet. Res.* **45**, 773 (1984)
108. ANDREWS, A. H. and READ, D. J. *Br. vet. J.* **139**, 423 (1983)
109. ANDREWS, A. H. and READ, D. J. *Br. vet. J.* **139**, 431 (1983)
110. O'BRIEN, J. K. and DUFFUS, W. P. H. *Br. vet. J.* **143**, 439 (1987)
111. GIBBS, H. A., ALLAN, E. M., WISEMAN, A. and SELMAN, I. E. *Res. vet. Sci.* **37**, 154 (1984)
112. PETERS, A. R. *Vet. Rec.* **116**, 321 (1985)
113. GRIMSHAW, W. T. R., COLMAN, P. J. and WEATHERLEY, A. J. *Vet. Rec.* **121**, 393 (1987)
114. VERHOEFF, J., WIERDA, A., VAN VULPEN, C. and DORRESTEIJN, J. *Vet. Rec.* **118**, 14 (1986)

115. CURTIS, S. E. and DRUMMOND, J. G. *Handbook of Agricultural Productivity*. Vol. 11 (ed. Miloslav Recheigl, Jr.) Cleveland, CRC, p. 107 (1982)
116. JONES, C. D. R. *Can. J. comp. Med.* **47**, 265 (1983)
117. JONES, C. D. R. and BULL, J. R. *Res. vet. Sci.* **42**, 82 (1987)
118. VERHOEFF, J., HAJER, R., VAN DEN INGH, T. S. G. A. M. and DORRESTEIJN, J. *Vet. Rec.* **119**, 105 (1986)
119. BOURNE, F. J. *Vet. A.* **15**, 23 (1975)
120. GOURLAY, R. N. and HOUGHTON, S. B. *Res. vet. Sci.* **38**, 377 (1985)
121. CORSTVET, R. E., GENTRY, M. J., NEWMAN, P. R., RUMMAGE, J. A. and CONFER, A. W. *J. clin. Microbiol.* **16**, 1123 (1982)
122. CRAFT, D. L., CHENGAPPA, M. M. and CARTER, G. R. *Vet. Rec.* **120**, 393 (1987)
123. GILMOUR, N. J. L., GILMOUR, J. S., DONACHIE, W., JONES, G. E. and GOURLAY, R. N. *Vet. Rec.* **121**, 277 (1987)
124. GUSTIN, P., LEKEUX, P., LOMBA, F. and CLERCX, C. *Res. vet. Sci.* **42**, 272 (1987)
125. GUSTIN, P., LEKEUX, P., LOMBA, F. and CLERCX, C. *Res. vet. Sci.* **42**, 277 (1987)
126. GUSTIN, P., BAKIMA, M., LEKEUX, P., LOMBA, F. and VAN DE WOESTIJNE, K. P. *Res. vet. Sci.* **42**, 313 (1987)
127. JONES, C. D. R. and WEBSTER, A. J. F. *Res. vet. Sci.* **37**, 132 (1984)
128. JONES, C. D. R. *Res. vet. Sci.* **42**, 179 (1987)
129. WEBSTER, A. J. F. *Vet. Rec.* **108**, 183 (1981)
130. HARKNESS, J. W. *Bull. Off. int. Épizoot.* **88**, 3 (1977)
131. THOMPSON, J. R., NETTLETON, P. F., GREIG, A. and BARR, J. *Vet. Rec.* **119**, 450 (1986)
132. ANONYMOUS. *Vet. Rec.* **122**, 49 (1988)
133. STOTT, E. J., THOMAS, L. H., HOWARD, C. J. and GOURLAY, R. N. *Vet. Rec.* **121**, 342 (1987)
134. HOWARD, C. J., STOTT, E. J., THOMAS, L. H., GOURLAY, R. N. and TAYLOR, G. *Vet. Rec.* **121**, 372 (1987)
135. QUIRIE, M., DONACHIE, W. and GILMOUR, N. J. L. *Vet. Rec.* **119**, 93 (1986)
136. GILMOUR, N. J. L., MARTIN, W. B., SHARP, J. M., THOMPSON, D. A. and WELLS, P. W. *Vet. Rec.* **104**, 15 (1979)
137. ABRAHAM, A. and ALEXANDER, R. *Vet. Rec.* **119**, 502 (1986)
138. LEKEUX, P., VERHOEFF, J., HAJER, R. and BREUKINK, H. J. *Res. vet. Sci.* **39**, 324 (1985)
139. INGH, T. S. G. A. M. VAN DEN, VERHOEFF, J. and VAN NIEUWSTADT, A. P. K. M. I. *Res. vet. Sci.* **33**, 152 (1982)
140. KIMMAN, T. G., WESTERNBRINK, F., STRAVER, P. J., VAN ZØAANE, D. and SCHREUDER, B. E. C. *Res. vet. Sci.* **43**, 180 (1987)
141. VERHOEFF, J., WIERDA, A., NIEUWSTADT, A. P. VAN and BUITELAAR, J. W. *Vet. Rec.* **117**, 202 (1985)
142. EDWARDS, S., WOODS, S. B., WESTCOTT, D. G., EMMERSON, M., JONES, P. C. and PHILLIPS, A. J. *Res. vet. Sci.* **41**, 378 (1986)
143. ADAIR, B. M. *Res. vet. Sci.* **41**, 414 (1986)
144. ANDERSON, M. L., MOORE, P. F., HYDE, D. M. and DUNGWORTH, D. L. *Res. vet. Sci.* **41**, 221 (1986)
145. PORTER, P. *Immunology* **23**, 225 (1972)
146. WEBSTER, A. J. F., SAVILLE, C., CHURCH, B. M., GNANASAKTHY, A. and MOSS, R. *Br. vet. J.* **141**, 472 (1985)
147. DREW, T. W., HEWITT-TAYLOR, C., WATSON, L. and EDWARDS, S. *Vet. Rec.* **121**, 547 (1987)
148. BITSCH, V. *Latent Herpesvirus Infections in Veterinary Medicine* (eds. G. Wittman, R. M. Gaskell and H. J. Rhiza), Martinus Nijhoff, The Hague, p. 163 (1984)
149. HIGGINS, R. J. and EDWARDS, S. *Vet. Rec.* **119**, 177 (1986)
150. THIRY, E., SALIKI, J., SCHWERS, A. and PASTORET, P.-P. *Vet. Rec.* **116**, 599 (1985)
151. MSOLLA, P. M., ALLAN, E. M., SELMAN, I.E. and WISEMAN, A. *J. comp. Path.* **93**, 271 (1983)
152. THIRY, E., SALIKI, J., LAMBERT, A.-F., BUBLOT, M., POUPLARD, L. and PASTORET, P.-P. *CEC Seminar on Immunity to Herpesvirus Infections of Domestic Animals, Brussels* (1984)
153. ESPINASSE, J., VISO, M., LAVAL, A., LE LAYEC, C. and MONPETIT, C. *Vet. Rec.* **113**, 15 (1983)
154. DEPELCHIN, B. O., BLODEN, S., HOOREMANS, M., NOIRFALISE, A. and ANSAY, M. *Vet. Rec.* **116**, 519 (1985)
155. HALL, T. J. *Vet. Rec.* **121**, 599 (1987)
156. PHILLIPPO, M., ARTHUR, J. R., PRICE, J. and HALLIDAY, G. J. *Vet. Rec.* **122**, 94 (1988)
157. PHILLIPPO, M., ARTHUR, J. R., PRICE, J. and HALLIDAY, G. J. *Vet. Rec.* **121**, 509 (1987)
158. SCHIEFER, B., JAYASEKARA, M. U. and MILLS, J. H. L. *Vet. Path.* **11**, 327 (1974)

159. JOHNSON, R. J. and DYER, I. A. *Life Sci.* **5**, 1121 (1966)
160. CARLSON, J. R., DYER, I. A. and JOHNSON, R. J. *Am. J. vet. Res.* **29**, 1983 (1968)
161. CARLSON, J. R., YOKOYAMA, M. T. and DICKINSON, E. O. *Science, N.Y.* **176**, 298 (1972)
162. MACKENZIE, A., FORD, J. E. and SCOTT, K. J. *Res. vet. Sci.* **19**, 227 (1975)
163. ALLAN, E. M., PIRIE, H. M., SELMAN, I. E. and WISEMAN, A. *Res. vet. Sci.* **26**, 349 (1979)
164. CORNELLIUS, L. M., COULTER, D., DOSTER, A. and RAWLINGS, C. *Am. J. vet. Res.* **40**, 571 (1979)
165. THOMAS, L. H. and SWANN, R. G. *Vet. Rec.* **92**, 454 (1973)
166. DAWSON, P. S., DERBYSHIRE, J. H. and LAMONT. P. H. *Res. vet. Sci.* **6**, 108 (1965)

Chapter 6

Diseases that may be acquired from the dam and some congenital disorders

Brucellosis (contagious abortion)

Source of infection

Calves born alive from cows suffering from *Brucella abortus* infection may themselves be infected, but such infection is only of a transient nature. As a precaution, such calves should be isolated from susceptible adults. Bull calves intended for breeding should not be fed on milk from infected cows since there is a possibility of their becoming infected in the genital organs.

Vaccination

Before the scheme for eradicating contagious abortion from the UK was begun, it was recommended that, to avoid an outbreak, all heifer calves should be vaccinated with S.19 vaccine at the age of 3–6 months. Under the Ministry of Agriculture, Fisheries and Food Free Calf Vaccination Service, calves were eligible for a *free* vaccination between the 91st and 180th day of life (i.e. from 3–6 months of age). The use of S.19 vaccine was only permitted under the Free Calf Vaccination Service. Vaccination carried out during this period of calfhood would, in most cases, protect against abortion for at least five pregnancies. The Ministry required farmers to undertake not to vaccinate any animals outside the prescribed age without its consent. Vaccinated animals were identified by a tattoo containing a symbol and two figures on the caudal fold under the tail. One figure denoted the quarter and the other the year of the decade.

Eradication schemes

Brucellosis (Accredited Herds) Scheme

With the intention of eradicating contagious abortion, the Brucellosis (Accredited Herds) Scheme was introduced in 1967. This scheme set up a voluntary register of Brucellosis Accredited Herds. To enter this register, a herd was first inspected by the Ministry's staff to ensure that it would be practicable to maintain the herd in a *Brucella*-free state if the required tests were passed.

For dairy herds, three clear milk ring tests at not less than 3-monthly intervals were needed before a herd could become 'supervised'. A negative serum agglutination test of all female cattle and bulls over certain ages would then qualify the herd for full registration. Routine tests would be made at prescribed intervals thereafter.

For beef herds a screening serum agglutination test that proved negative would allow the herd to become 'supervised', and a further clear agglutination test after a prescribed interval would qualify the herd for full registration.

Once a herd had become 'supervised', compensation would be paid for reactors at a rate of 100 per cent subject to a certain maximum with the Ministry retaining the salvage value. 'Supervised' herds would be required to comply with certain conditions; the purchase of cattle from other herds would only be allowed from those of the same status and a permit was required; adequate fencing had to be provided to prevent contact with animals from herds not of the same status; and all premature calvings and abortions had to be notified to the Ministry's Animal Health Division.

Brucellosis Incentive Scheme

To increase the rate of progress in eradicating brucellosis, the Brucellosis (Accredited Herds) Scheme was closed to new applicants in March 1970, and was replaced by the Brucellosis Incentive Scheme. This new scheme was designed for the owner whose herd was free or relatively free from brucellosis and who wished to have his herd voluntarily registered in the British Register of Brucellosis Accredited Herds, before a compulsory eradication programme began in his area. Once the herd had become accredited, the owner received a premium, guaranteed for 5 years based on a gallonage payment (0.176 p/l in 1983) through the Milk Marketing Board (MMB) or a supplement to the Beef Cow Subsidy or the Hill Livestock Compensatory Allowance, provided that the owner had not elected to remain in the original scheme. The majority of dairy herds had completed this period in 1985[2]. There was no compensation for individual reacting animals, so slaughter was at the owner's expense.

After applying to join the scheme, the owner's premises were inspected to see whether the isolation facilities, boundary fences and drainage were suitable. Three free qualifying negative blood tests at a minimum interval of 4 months on all bull calves over 6 months of age and all heifer calves over 1 year old (18 months, if vaccinated) were necessary for registration; steers were not normally tested. If there was a reactor at any stage, it had to be slaughtered and 2 months had to elapse before repetition of the three qualifying tests.

Since 1971 the milk of each herd having an operative wholesale milk contract with the MMB is tested each month with a modified milk ring test, and this method is used for monitoring the status of the accredited herds and was used for encouraging those herds with negative results to join the Brucellosis Incentive Scheme. If a positive test occurs on the milk of an accredited herd, a retest of the herd will be necessary although the incentives will still be paid.

False-positive milk ring tests may occur. Thus, during the year ending March 1987, 340 herds showed a first positive reaction to the milk ring test with 35 (10.5 per cent) of these subsequently being found to contain reactors. The major reasons for false-positive milk ring tests are the inclusion of colostrum, and milk from cows with clinical mastitis, which in small herds of <40 cows can influence the bulk milk test. Further milk samples taken by the Ministry of Agriculture, Fisheries and Food from the herds which had shown the initial positive milk ring tests revealed 42.6 per cent of these herds reacted to a second milk ring test[3].

Since November 1971 the Ministry has allowed certain herds a concession on the blood-testing programme, waiving one of the qualifying blood tests if the owner could produce evidence of negative monthly MMB milk ring tests.

If calves are purchased from non-accredited sources, the owner must obtain a permit and the calves must be isolated for 60 d, during which time heifer calves must have two blood tests and castrated male calves one blood test at their owner's expense. Steers over 6 months of age do not require a permit.

The owner joining the Scheme had to abide by a number of rules. For instance, he had to inform the Ministry immediately that he suspected that an animal had 'aborted' or had calved prematurely (i.e. less than 271 d after service or insemination), and had to keep the animal in isolation. The fetus and placenta had to be retained for veterinary examination. No milk or dairy by-product unless boiled, sterilized, pasteurized or in powder form was allowed to be brought onto the premises without a permit. Owners had to keep records in an approved form, in addition to particulars of movement required by the Movement of Animals (Records) Order 1960. The following details were required:

(a) identification, breed and sex;
(b) date of vaccination with *Brucella* vaccine;
(c) date on which each heifer or cow was served or artificially inseminated;
(d) date on which each heifer or cow calved or aborted;
(e) date of birth, death or disposal of any animal in the herd;
(f) any calving abnormality, such as stillbirth or retention of the placenta 12 h or more after calving.

The only vaccination allowed was that given under the Free Calf Vaccination Service.

Compulsory eradication

Compulsory eradication began in November 1972 with the designation of eradication areas. The scale of compensation for reactors that were slaughtered depended on whether the herd was accredited or not. In non-accredited herds compensation was payable at full market value up to a certain maximum per animal. For accredited herds compensation was limited to three-quarters of market value, up to a lower maximum per animal. The full market value was paid for any healthy animal that was required to be slaughtered as a result of contact with infected animals or reactors.

No vaccination against *Brucella* was allowed in an eradication area except under a licence issued by a veterinary inspector of the Ministry. Cattle could only be moved into an eradication area under licence, except that non-reacting cattle could be moved direct to a slaughterhouse or to a market from which the animals would be slaughtered, and calves under 6 months of age could be taken direct to a market.

In January 1976 a free 45/20 Vaccination Scheme was announced for farmers with infected herds in certain heavily infected areas of the UK and in areas scheduled for compulsory eradication beginning in 1979 and 1980.

The designation of attested areas is the last stage in the programme. N. Ireland was declared a brucellosis-free area in December 1971. Scotland was declared attested on 1 January 1980 followed by England and Wales on 1 November 1981. All herds in the UK are now accredited brucellosis-free. No vaccination is allowed in an attested area and to ensure that herds remain free of infection, all beef and heifer-rearing herds, and heifers and bulls in dairy herds will continue to be blood tested annually and dairy herds will continue to have milk samples tested monthly by the MMB. Although tenders were invited by the Ministry to contract out the blood tests, namely the Rose Bengal plate test, serum agglutination test and complement fixation test, it has been decided to centralize the testing at the Ministry's Central Veterinary Laboratory.

After three years, herds in these attested areas will be Officially Brucellosis Free (OBF) according to EEC Directive 64/432 and stock can be freely traded and can be registered on the OBF register. This is voluntary, but is of particular value to owners of herds with an export potential; in March 1985 there were 819 herds on the register. Within an attested area the controls on farm-to-farm movement are relaxed, but the rules governing movement of stock into the attested area become more rigid.

On 1 October 1985 all herds in the UK were designated Officially Brucellosis Free (OBF).

Present incidence in the UK

In December 1972 just after compulsory eradication began, 17 per cent of herds in England and Wales were free from brucellosis, but by 1975, 70 per cent of herds were accredited and a further 16 per cent of non-accredited herds showed a negative milk ring test.

For the year ending 31 December 1985, there were 373 reactors to *Brucella* in 143 herds out of a total of 152 527 herds in Great Britain, a herd breakdown rate of 0.094 per cent. For the year ending 31 December 1987, 632 herds in England and Wales showed a positive milk ring test of which 32 were subsequently shown to contain serological reactors[132].

Congenital disorders

The number of possible congenital disorders in calves is legion but some information on the more important or recently reported disorders are presented. In the UK they appear to be particularly associated with imported bulls showing extremes of muscle development or milk production, and with the breeding for 'polledness' in cattle.

Achondroplasia (bulldog calves)

The calves are characterized by very short limbs and flattened skulls and frequently have abdominal hernia. Their occurrence is common in the Dexter breed, where 25 per cent of the offspring of Dexters mated together are bulldogs. They have also been reported in the offspring of imported Dutch Friesian bulls, and can occur in other breeds as well.

Arthrogryposis

In 1980, 16 abortions or stillbirths and 26 deformed calves occurred in a beef herd in Tayside, Scotland from 115 Hereford × Friesian cows mated to a Charolais bull. The deformities were principally arthrogryposes (retention of a joint in a flexed position due to muscular contraction or to intracapsular or extracapsular adhesions) with contractures particularly of the forelimbs, shortened legs and various skull abnormalities. These findings were considered to be due to exposure of the fetus to an unidentified teratogen early in pregnancy rather than resulting from heredity, an infectious agent or a nutritional deficiency. The possibility of a mycotoxin in silage being responsible was suggested[129].

Atlanto-occipital fusion

Two cases of fusion due to the failure of the developing arches of the first cervical vertebra to separate from the occiput (back of the head) have been reported in the young calf. It was associated with compressive myelopathy (compression of the spinal cord) and ataxia (loss of control over voluntary movements)[119].

Atresia of ani, recti and/or coli

In Belgium 45 calves were found with intestinal malformation during a 4-year period. Of these 18 were of the East Flemish Red Pied breed, 13 were West Flemish Reds, 12 were Friesians and four were Central and Upper Belgians. Thirty-eight had atresia of ani, recti and/or coli, and successful surgical correction was made on 15 of them[112].

Bovine epidermolysis

This resembles epidermolysis bulbosa simplex in man, where vesicles arise as a result of cleavage between the epidermal cells and heal without scarring. Of 72 calves sired by a Simmental bull, 25 were affected with this skin disease. It appeared to be inherited as an autosomal dominant trait. Affected calves showed hypotrichosis (a deficiency of hair), erythema (redness of the skin) and breaks in the integrity of the skin. Distribution of the lesions and the ease with which excoriations occurred suggested abnormal vulnerability to trauma. Mortality was high, but in affected survivors clinical signs decreased with age[117].

Branched-chain keto acid decarboxylase deficiency

This condition found in Australia in five polled Hereford calves was associated with dullness, recumbency, opisthotonos (a tetanic spasm with head and limbs bent backwards) and severe status spongiosus (multiple vacuoles) in the central nervous system and was considered to be analogous to the same deficiency disease in children, otherwise known as Maple syrup urine disease[125]. It was considered that the original report of neuraxial oedema (see below)[108] included two diseases, i.e. congenital myoclonus (muscular twitch) and branched-chain keto acid (valine, isoleucine and leucine) decarboxylase deficiency. A deficiency of the enzyme complex results in the accumulation of branched-chain keto acids in tissues, plasma and urine and a reduction in their rate of deamination.

Cataracts

Congenital nuclear (central) cataracts at birth have been reported[113,114].

Ectopia cardis

This is defined as an abnormal location of the heart outside the thoracic cavity. Two cases have recently been reported, one in a 3-month-old Friesian heifer calf and the other in a 2-day-old Hereford bull calf. The hearts of both were located in the lower cervical region; the Hereford calf also had a patent foramen ovale[96] (see Chapter 1).

Freemartins

The incidence of twins in cattle is about 2 per cent of births, but the percentage is lower for heifer calvings and greater for older cows[107]. In the Dairy Progeny Testing Scheme of the MMB in 1987–8, twins were produced in 2.6 per cent of all calvings from Friesian-Holsteins[132]. Heterosexual twins account for approximately half of these, so 1 per cent of all calves born might be expected to be freemartins[98]. However, only about 90 per cent of heifers born twin to a bull are sterile (freemartins), owing to fusion of the choriovascular system with a resultant common circulation and the presence of both male and female populations of blood cells in the twin fetuses.

It is suggested that the normal development of the internal female organs is prevented by male hormones entering the female circulation[5]. Whereas the female twins of some other species are protected against the masculinizing effects from the male because their placentae contain an enzyme capable of converting androgens to oestrogens, the enzyme is apparently absent in the bovine placenta[103, 104]. The freemartins and their co-twins usually exhibit sex chromosome chimerism (a mixture of two genetically different populations) and also chimerisms of erythrocytes and blood proteins[99–101].

Diagnosis of a freemartin is generally based on clinical examination, but further evidence may be obtained from a sex-chromatin test and from blood typing and skin grafts. Freemartins are highly tolerant of skin grafts from their male twins[105]. It has been suggested that, in addition to a study of sex chromatin, examination of the polymorphonuclear leucocytes of heterosexual twins might be of use in the early detection of freemartins[97].

The degree of intersexuality in freemartins varies widely. The uterus may be rudimentary and the vagina may be only about one-third of its normal length. The external genitalia look essentially normal, but frequently the clitoris and tuft of hair on the vulva are enlarged. The freemartin may have a fold of skin extending from the navel towards the udder which represents a rudimentary penis. The mammary gland, although underdeveloped, can be palpated, and teats may be shorter than those of normal heifers. Although freemartins will mount other animals on heat, they will not allow themselves to be mounted[106].

In Minnesota, pseudohermaphroditism in a Holstein heifer was observed. Genetically the heifer was a female with female gonads but its external genitalia were partially masculinized[102].

Haemolytic disease of the newborn

This has been produced experimentally in newborn calves by their absorption from colostrum of incompatible blood group antibodies. These were produced in the dam by injecting her with red blood cells from the sire[111].

Inherited parakeratosis

This condition is associated with zinc deficiency, presumably because of a higher than normal requirement. (For further information on zinc, see *The Calf,* 5th edn, vol. 2.) It appears to be inherited as an autosomal recessive trait from heterozygous parents, since in three Friesian calves in Cumbria, it was associated with a single home-bred bull out of dams that were half-sisters or possibly full sisters and in one case was the bull's own dam[120].

In another report on one calf[121], the condition was associated with a very low serum zinc concentration of 0.1 mg/l compared to values of 0.9–2.3 mg/l in four normal calves. Normal serum zinc concentrations are considered to be 0.8–1.2 mg/l[122]. All five calves were marginally hypocupraemic (0.6 mg/l). A year later a further calf that was affected had a plasma zinc concentration of 0.3 mg/l. Both calves were sired by a recently imported Dutch bull, but were from different dams. It was possible to trace this particular bull to bulls that appeared in a list of known carriers of the trait published in the Netherlands in 1974.

Necrotic laryngitis

Incidence of necrotic laryngitis is more important in double-muscled cattle because of the narrower lumen and higher airflow resistance of the larynx[116]. From a study of the mechanics of breathing and of gas exchange in five Belgian double-muscled cattle aged 2–3 months that had a severely reduced appetite, moderate hyperthermia and laborious noisy breathing, especially during inspiration, it was concluded that necrotic laryngitis disturbs pulmonary function to such an extent that it impairs growth and predisposes the infected animals to secondary bronchopneumonia and ventilatory failure due to respiratory muscle fatigue[115].

Neuraxial oedema

This condition (see also branched-chain keto acid decarboxylase deficiency, p. 158) has been reported in the calves of Polled Hereford cattle in the USA[108], Australia[109] and New Zealand[110]. The calf is unable to stand. Any sudden stimulus such as noise or touch produces tetanic spasms with extension of the head, neck and legs, crossing of the hind legs and cessation of breathing. When not in spasm, the calf behaves normally.

Breeding experiments with Polled Hereford cattle indicate that the condition is inherited in an autosomal recessive manner since 25 per cent of calves born to matings of obligatory heterozygotes were affected and both sexes were represented among those affected. The mean gestation length for affected calves was 9 d shorter than that for normal calves. Since clinical signs were observed during parturition, it is presumed that the calves become affected *in utero*[123].

In a further study of the symptoms of the condition in 34 newborn Polled Hereford and Polled Hereford cross calves it was found that all affected calves were unable to stand at birth and when first examined were in lateral recumbency with extension and crossing of the hind legs. All the calves were bright and alert, could lift their heads and apparently could see and hear. When calves were encouraged to stand, spontaneous and stimulus-responsive myoclonic extension spasms with rigidity of the whole body were consistently observed. Of the affected calves, 32 had macroscopic lesions in the coxae (hip bone). No pathological lesions were found in the central nervous system and the water content of the cerebellum was unaffected. It was suggested that, as the central nervous system was unaffected, it should be described as an inherited congenital myoclonus (clonic spasm or twitching of a muscle or group of muscles)[124].

In the UK, 11 Hereford calves with neuraxial oedema with and without hypomyelinogenesis (a reduction in the acquisition of myelin that forms the medullary sheath to the axis cylinder of a nerve fibre) were reported. After birth, the affected calves were recumbent with intermittent extensor spasm and hyperaesthesia (excessive sensitivity to touch, pain or other sensory stimuli). In six calves there was nystagmus (a spasmodic lateral oscillatory movement of the eye), whilst vacuolation of the central nervous system was seen in all calves. In two horned Hereford calves, the vacuolation was restricted to the white matter areas, whilst in the remaining Polled Hereford calves it was distributed in both the white and grey matter and these calves showed hypomyelinogenesis[126].

Polydactyly

Of 69 Friesian cows and heifers served by a pedigree Hereford bull, 22 produced calves with duplication of the inside claw of one or both front feet with considerable twisting and deformity of the foot. The affected animals were reared indoors on soft bedding but owing to difficulties with walking, they failed to grow normally and were slaughtered at 18 months of age[118].

Polymelia

A Hereford × Holstein calf with an extra limb on its left hand side and overlaying its normal left leg has been reported. The calf had gross abnormalities of its pelvis[128].

Primary cardiomyopathy

During a 3.5 year period, 15 newborn Polled Hereford calves with tight curly hair coat died before reaching 6 months of age. Of these calves, seven had exercise interolance, hyperpnoea and dyspnoea from 1–7 d before death. Post-mortem examination revealed focal, diffuse and pale fibrous streaking of the entire myocardium and vascular congestion of the liver, spleen and lungs. On one farm, seven affected calves had a common ancestor of all their sires and dams. The mode of inheritance was considered to be a simple autosomal recessive factor[127].

Siamese twins

Various abnormalities may occur in monozygous (identical) twins. For instance two fully formed ventrally-joined calves with a common umbilical cord were born alive by caesarean[131].

Spina bifida

A case of spina bifida has been reported in a Friesian cross calf. The calf was ambulatory and had no locomotor disturbance, but it had small skin defects in the thoracic region with abnormalities of the ninth and tenth thoracic vertebrae laminae and the associated ribs[130].

Umbilical abnormalities

A survey of the incidence of umbilical hernia and other umbilical abnormalities on 33 dairy farms in S. Wales showed an incidence of 7.5 per cent for hernia and 22 per cent for other defects. Of the 79 hernias, 63 per cent allowed one finger (1F), 35 per cent allowed two fingers (2F) and 1.3 per cent allowed three fingers (3F) to enter the umbilical ring. By 6–9 months, 85 per cent of hernias (all lF and 2F) had disappeared from 20 calves that were reexamined; the remainder had all closed by 22 months.

The incidence of hernias was higher in spring-born than in autumn-born calves, but the reverse was true for other umbilical abnormalities. There was no effect on hernias of umbilical dressing, but other abnormalities were increased in the absence of dressing. The incidence of hernias was considerably higher in group- (9.9 per cent) than individually-housed calves (5.6 per cent). The feeding of whole milk to calves was associated with a reduction of both hernia and other abnormalities[31].

Enzootic bovine leukosis

This virus infection has been introduced into Great Britain in imported cattle. In a survey of cattle imported between 1968 and 1978, the disease was confirmed in 69 herds (67 Holstein, one Brown Swiss and one Charolais)[134]. The infection can exist undetected in a herd for several years[133].

The Enzootic Bovine Leukosis Order 1977 (SI 1838) under the Diseases of Animals Act 1950 made the disease notifiable. Thus, any sign or symptom of the disease in any animal that was suspected of having the disease or any carcass that showed signs had to be reported.

The symptoms on which notification was required were based on the fact that it was difficult to differentiate between enzootic bovine leukosis and leukosis of a sporadic nature. Thus, any animal with swollen, painless lymph nodes or with tumorous changes (other than haemangiomas, papillomas or warts) in any part of the body, or with a lymphocyte count ranging from more than 11 000/ml at 0–1 year of age to more than 5 500/ml for animals over 6 years, or had proved positive to a test for the bovine leucosis virus had to be reported. After notification, the veterinary inspector is obliged to make more tests to confirm whether the disease exists or not.

The Enzootic Bovine Leukosis Order 1978 (SI 975) gave authority for infected animals to be slaughtered and the Enzootic Bovine Leukosis (Compensation) Order 1978 (SI 976) offered full market value as compensation. These two Orders were revoked and replaced by similar Orders in 1980 (SI 79 and 80 respectively). Statutory instrument 79 of 1980 also revoked SI 1838 of 1977 but re-enacted its provisions.

If a farm is declared an infected place, there are requirements for cleaning and disinfection and for infected animals to be marked and kept in isolation. No milk from an affected or suspected to be infected animal is allowed to be fed to a calf without a licence granted by the veterinary inspector unless it has been pasteurized, boiled or sterilized. The fact that boiled milk will predispose the neonatal calf to enteric infection seems to have been overlooked.

Imported animals may be required to be tested within 6 months of importation. If found to be infected, such animals can be ordered to be slaughtered without any compensation.

Infected animals may remain seronegative up to 18–24 months before showing a positive titre, and clinical signs are rarely seen in cattle under 4 years of age. The virus infrequently crosses the placenta to affect the fetus. Serologically-positive cows may become seronegative at parturition and then become positive again a few weeks later. Colostral antibody was found to disappear from calf serum by 6 months of age with a mean of 3.7 months[134].

The virus may be present in colostrum and milk and calves are susceptible to infection from this source. It has also been isolated in semen. The virus can be transmitted between cattle if the closeness of contact is $<2\,m$[134].

In January 1982, the Ministry of Agriculture introduced a voluntary Attested Herds Scheme to encourage the establishment of Enzootic Bovine Leukosis (EBL)-free herds. On 1 April 1987 the scheme was replaced by the Cattle Health Scheme. Members of this Scheme are required to achieve and maintain EBL-attested status; they may also benefit from additional monitoring for Infectious Bovine Rhinotracheitis (see Chapter 5). A prospectus and application form for the Scheme may be obtained from the Divisional Veterinary Officer. By 1 June 1987 there were applications for 647 herds to join the scheme.

Membership of the Cattle Health Scheme involves registration, membership and associated blood sampling and laboratory testing fees. To achieve EBL-attested status, all animals in a herd will have to pass a qualifying test and then a registration test after an interval of 4–12 months. The herd will then be placed on an official register and a first periodic blood test will be carried out within the next 12 months and thereafter at intervals of not more than 24 months. Any reactors found at any test can only be removed from the farm under a licence issued by the Divisional Veterinary Officer. Compensation will not be paid for any reactors and the Divisional Veterinary Officer will advise on restrictions and subsequent testing programmes. Tests for this scheme can only be carried out at a Ministry or MMB Veterinary Laboratory[133].

Of approximately 1000 blood samples tested in each of the years 1986 and 1987 by the MMB, all proved negative[3,132].

Johne's disease

Source of infection

This disease, caused by the organism *Mycobacterium johnei,* is very probably acquired during calfhood although clinical symptoms, which include profuse diarrhoea, sometimes with a bubbly appearance of the dung, and severe emaciation, are not revealed until later in life, generally not until after the first calving. In addition to animals obviously infected, about 17 per cent of apparently normal cattle carry the causal organism.

Infection of the calf can occur *in utero* if the dam has the disease in a very advanced stage[1]. Other calves may thus become infected by contact with the dung of an infected calf. Moreover, the organism has also been isolated from the udder of affected cows. It is not known whether the fetus of a cow with a latent infection may also become infected, but it has been suggested that all calves born of infected cows should be slaughtered.

Resistance to infection

Young animals appear gradually to build up an immunity, but the time that they take to become resistant varies very much between individuals. Poor nutrition of young animals also seems to be important, for the infection can be set up experimentally in undernourished animals more easily than in those on an adequate diet. Dairy cattle, especially those of the Channel Island breeds, seem to be particularly susceptible.

Diagnosis

Although a positive diagnosis of the disease may be made from the faeces of infected animals, since the organisms are found in about 30 per cent of such animals that have diarrhoea, a negative result is not conclusive as the organism is likely to be passed out at irregular intervals. However, in 90 per cent of cases showing a clinical infection antibodies against the organism may be found in the blood. Detection of infected animals in the preclinical stage is much more difficult and cannot be done with certainty.

Prevention

In infected herds direct contact between calves and adult animals must be prevented. The calf should be removed at birth and be fed as described for calves born in herds infected with tuberculosis (see below). Great care must be exercised in avoiding the contamination of colostrum with faeces and in preventing the transfer of faeces on the person or on equipment from the adult stock to the calf pens. Particular attention should also be directed to the cleanliness of the water supply; calves must not be allowed access to a water supply that may have been contaminated by older stock. As the Johne bacilli survive on pastures for up to 1 year, calves should not be turned out to grass used by the adult stock during the previous 12 months.

Mastitis

Unsaleable milk is often offered to calves. In the UK milk may be unsaleable for the following reasons:

(a) compositional quality, e.g. containing colostrum, blood clots, too low concentration of some nutrient;
(b) unsatisfactory hygienic quality;
(c) too high antibiotic concentration;
(d) abnormal amount of sediment; or
(e) maintained at too high a temperature.

The main problem for calves concerns (b) and (c) arising from mastitis in dairy cows and its chemotherapy or prophylaxis.

The official advice[6] is that mastitic milk should preferably be given to pigs and that where it is known that cows are suffering from mastitis and are being given antibiotics, the milk from the cows should not be fed to livestock but be disposed of in farm slurry stores; public authority sewers, if prior consent has been obtained; or as liquid manure on suitable farmland.

In spite of this advice, there can be little doubt that mastitic milk, whether from antibiotic-treated animals or not, is fed to calves on dairy farms. Of 19 dairy farms in S. Dakota, 15 were reported as feeding mastitic milk to calves when it was available[7] and in the UK, it is tacitly assumed that milk unsuitable for sale will be utilized by calves[8,9].

In fact, it is quite common practice to suckle calves on cows affected with clinical mastitis. Calves should not, however, be allowed to suck cows suffering from acute mastitis, as illness and even deaths among the calves may ensue. In other types of mastitis the risk to the calf is probably slight, but where valuable calves are concerned, the practice should be avoided.

A survey of clinical mastitis in 890 herds in England and Wales[10] has shown that on average 71 cases of clinical mastitis occur each year in every 100 cows. In these herds 0.59 per cent of cows were under treatment each day ranging from 0.34 per cent of cows in September up to 0.91 per cent in January. It has been estimated that 40 litres mastitic milk per cow are produced each year[11] and this would amount in England and Wales to approximately 100×10^6 litres annually.

Mastitic organisms found in milk and their relevance to calf health

The proportions of mastitis samples attributed to various organisms and submitted to Veterinary Investigation Centres in 1986 are given in Table 6.1[12]. These proportions would be undoubtedly biased towards the more acute forms of mastitis.

Table 6.1 Proportion of mastitis samples, submitted to the Veterinary Investigation Service in 1986, that were associated with different pathogens[12]

Pathogen	*Percentage*
Corynebacterium bovis	1.1
Corynebacterium pyogenes	9.5
Escherichia coli	21.7
Klebsiella	1.1
Pseudomonas spp.	2.3
Staphylococcus spp. (coagulase positive)	17.7
Staphylococcus (not otherwise specified)	4.1
Streptococcus agalactiae	4.4
Streptococcus dysgalactiae	11.5
Streptococcus uberis	17.3
Streptococcus (not otherwise specified)	1.7
Mycoplasma spp. (not *bovis*)	0.06
Mycoplasma bovis	0.09
Any other cause	7.5

Of the samples from 4400 cows that proved positive to mastitic organisms in the Mastitis Bacteriology Service in 1986, 64 per cent yielded staphylococci, of which just over half were resistant to penicillin, and 18 per cent were so-called environmental organisms, *Streptococcus uberis* accounting for 14 per cent and *E. coli* 4 per cent. *Streptococcus agalactiae* was isolated from 5 per cent of samples[3].

It is probable that cattle with undetected subclinical mastitis will often be excreting a similar number of organisms as excreted by those with clinical disease. Thus calves being reared on whole milk will normally be ingesting a wide range of mastitic organisms.

Although isolated reports of detrimental effects on the calf of ingesting mastitic organisms have been reported, it is by no means a clear-cut situation.

Streptococcus agalactiae (Gram-positive)

It is doubtful if milk containing this organism would have any ill effect providing that the calf did not ingest it before receiving colostrum.

Staphylococcus aureus (Gram-positive)

The position is similar to that for *Str. agalactiae* but the fermentation of milk to degrade antibiotic residues (see below) might increase the amount of enterotoxin produced and this could be detrimental to a calf.

Corynebacterium pyogenes (Gram-positive)

This is the main causative organism of summer mastitis affecting 1.4 per cent of dry cows, 1.6 per cent of pregnant heifers and 0.3 per cent of non-pregnant heifers in the UK in 1986[3]. It certainly may be associated with suppurative lesions in the joints and abscesses in the jaws of young calves.

It has been suggested that calves born of cows that have had summer mastitis fail to thrive. There is evidence of skeletal growth retardation during the time of the fetal crisis and a marked reduction occurs in the weight of the thymus, which may have a deleterious effect on postnatal immune response[13].

Streptococcus uberis (Gram-positive) and Pseudomonas (Gram-negative)

These are both organisms that cause 'opportunist' mastitis and together with *Escherichia coli* are mainly responsible for the so-called 'environmental mastitis'.

Escherichia coli (Gram-negative)

This is also an opportunist invader of the udder but it is potentially the most important pathogenic organism for the neonatal calf. *E. coli* is not regularly present in the faeces of adult cattle, and adult cows are not likely to be a significant source of drug-resistant *E. coli* in man[14]. However, of the 148 'O' group serotypes of *E. coli* that are recognized, 93 were found in healthy calves and 107 in a previous survey in man; 42 of the serotypes were common to both hosts. In this survey 60 per cent of the isolates from calves were resistant to at least one antibiotic. Of these 71 per cent belonged to ten 'O' groups, nine of which were found in man[26].

In general, the number of serotypes of *E. coli* that are associated either with septicaemia in calves, e.g. 'O' groups 15, 78, 86, 115 and 117[27], or with enterotoxaemia, e.g. 'O' groups 8, 9, 26, 101, are quite small and there are few that are common to enteric disorders of calf and infant. To cause a septicaemia in calves, *E. coli* must either be ingested and absorbed by pinocytosis before colostrum ingestion, or the *E. coli* must be a highly virulent serotype producing colicine V and causing tissue invasion e.g. O78:K80[28]. To cause an enteropathogenic condition of diarrhoea and dehydration, the *E. coli* must be of a particular 'O' group which by virtue of its 'K' antigen, particularly K99, allows adherence to the mucosa of the small intestine[29]. Whereas enteric disease in calves is associated with relatively few 'O' groups, a survey of 279 isolates from individual cows with mastitis showed 67 different 'O' groups[30]. Although there is no evidence of adherence of *E. coli* to the epithelium within the udder, the incidence of the various 'O' groups in the udders of the cows in that study[30] were significantly correlated with the incidence of these serotypes in the faeces from healthy calves in an unrelated survey[26, 32].

When O117 isolated as the sole strain from calves dying from colibacillosis under experimental conditions and from field outbreaks was given to calves at $5–8 \times 10^5$/ml in pasteurized milk, only in one of the calves was the organism isolated in the small intestine, whereas another two strains, including O101 present in the calf pens, became dominant in the intestines[33]. Similarly, when calves given raw colostrum containing 10^7 coli/ml was compared with those given the same colostrum but pasteurized, no difference in performance of the calves was observed. In another experiment, either colostrum containing $10–10^6$ coli/ml or pasteurized colostrum, followed either by refrigerated milk that contained 10^2 or less coli/ml, or the same batch of milk in which coli were allowed to multiply for 24 to 36 h so that 79 per cent of samples contained 10^3 or more coli/ml were compared. The raw colostrum again had no effect on mortality or incidence of diarrhoea in surviving calves, whilst the stale milk had very little effect, except to cause a slightly increased incidence of pyrexia[34].

Effect of ingestion of mastitic colostrum or milk

The evidence of detrimental effects is extremely fragmentary. In a small trial, no difference in performance or mortality of calves given milk from a cow suffering from *E. coli* mastitis and that of calves given bulked herd milk was observed[33].

In the USSR, a comparison was made of newborn calves given colostrum from healthy and infected cows; a number of those given mastitic colostrum developed dyspepsia and several died[35]. The pathogenic organisms *Staphylococcus* and *E. coli* isolated from the faeces were identical to those in the mastitic colostrum. Other work from the USSR[36] showed high mortality rates of calves given colostrum from udders infected with streptococcal and staphylococcal mastitis. For staphylococcal mastitis the mortality rate was 71.4 per cent and for *Str. agalactiae* mastitis the morbidity rate was 50 per cent and the mortality rate 20 per cent. Among groups of calves given colostrum from healthy carriers of pathogenic staphylococci and streptococci in the udder the infection rate of calves was between 25 and 27 per cent.

Since inadequate trypsin inhibitor in colostrum from mastitic cows has been observed, it has been suggested that this could impair the assimilation of Ig by newborn calves[37]. That this occurred was supported by the finding that calves that sucked cows with subclinical mastitis had reduced protein and γ-globulin concentration and reduced bactericidal, lysozyme (EC.3.2.1.17) and phagocytic activities in their blood compared with those of calves that sucked healthy cows[38].

In contrast there is evidence from Central America that the sucking of milk from the udder by a calf, even if the calf only sucks the residual milk after milking, reduces the level of clinical and subclinical mastitis[39, 40]. This has been attributed to the mechanical effect of sucking, inhibitory factors in the calf's saliva and better emptying of the udder[41].

In the USA, newborn calves from dams free of staphylococcal udder infection were given, after receiving colostrum for 2 d, pasteurized milk containing *Staph.*

aureus twice weekly for a total of nine feeds. No staphylococci were isolated from the body tissue of the bull calves at post-mortem at about 7 weeks of age[42].

Effect of ingestion of mastitic milk from antibiotic-treated cows

In general, there appears to be little effect on the calf. In a comparison of using 'waste milk' or a milk replacer from 4–42 d of age, there was no difference in performance but a tendency for a higher incidence of diarrhoea in the calves given the 'waste milk'[43]. In Cuba, the feeding of mastitic milk from untreated cows or those treated with oxytetracycline caused no difference in weight gains or incidence of diarrhoea in calves given the milk from 7–90 d of age[44]. Other results from the USA showed that growth of calves to 8 weeks of age given 'waste milk' containing antibiotic residues, where the predominant infection was *Str. agalactiae* and *Staphylococcus*, was equal to or superior to that of the control calves[45]. In Cyprus, no adverse effects of giving mastitic rather than normal milk on weight gain or health of calves have been observed[46]. Similar results have been found in experiments in Louisiana and Florida[47,48].

Long-term effect of ingestion of mastitic milk during calfhood

The evidence of this mainly emanated from the USA in the 1940s. In 1942, it was reported from California[49] that a small proportion of heifers calved with a subclinical infection with *Str. agalactiae*. Field observations indicated that heifer calves became infected through sucking each other's teats after drinking milk containing viable *Str. agalactiae*. In an infected herd where raw milk was fed to calves and they were allowed to suck each other, 57 per cent of heifers were excreting *Str. agalactiae* at parturition. In another infected herd, in which calves were given pasteurized milk and muzzled if they showed a tendency to suck their pen mates, no heifers excreted *Str. agalactiae* at parturition. Of four calves fed infected milk for 7 months and allowed to suck each other at will, one was excreting large numbers of *Str. agalactiae* from two quarters. The only possible source of infection was milk given during calfhood, since the strain of *Streptococcus* was identical to that isolated from milk consumed 2 years previously by the animal.

Other work showed that, when 12 heifer calves that had been given infected milk came into lactation, no *Str. agalactiae* was present but 10 quarters were permanently infected with haemolytic streptococci[50].

More recently a study[51] was made of the tonsils and lymph nodes of 500 calves and of milk samples from 937 heifers until 5 d after calving. There was no support for the hypothesis that *Str. agalactiae* or *Staph. aureus* mastitis originated from ingestion of infected colostrum and subsequent invasion via the tonsils and lymph nodes. The three infections with *Str. agalactiae* that occurred in the heifers at calving were due to manipulations of the teats before infection. Similarly, no difference in the incidence of blind quarters or mastitis during the first lactation was found between heifers that had been given a mixture of colostrum and milk or mastitic milk from antibiotic-treated cows[52]. When pasteurized milk containing a known quantity of *Staph. aureus* was given to heifer calves it had no effect on the subsequent incidence of mastitis[42]. Other limited data also indicated that in the first lactation heifers given mastitic milk in calfhood suffered no more udder trouble than did contempories given normal milk[66].

On the other hand, the incidence of *C. pyogenes* in the tonsils and lymph nodes of 500 calves and heifers was similar to the rate of infection of summer mastitis[53].

Antibiotic residues in milk

The prevention of the marketing of milk containing antibiotics was instituted mainly to prevent the inhibition of starters used in manufactured milk products. However, there was also the possibility of humans becoming sensitized to penicillin from drinking liquid milk or the penicillin having an effect on their oral or intestinal flora.

Since 1 January 1978 and until 1 January 1986, the standard set by the MMB above which farmers were penalized has been 0.02 iu penicillin/ml or the equivalent of other antibiotics. From 1 January 1986, the standard was lowered to >0.01 iu/ml for penicillin G. In terms of tetracycline, the value would be 0.05 μg/ml[54]. On the UK market, there are about 38 intramammary preparations for use in lactating cows and 17 for use in dry cows. Of the 1.2 per cent of milk samples that failed the tests for antibiotic residues in creameries, 62 per cent of the failures were due to 'lactation' products and 31 per cent to 'dry cow' products with a marked seasonal variation in the latter[10].

Most of the products used in the UK are mixtures of penicillin G and streptomycin, or ampicillin or cloxacillin. The insertion of intramammary antibiotics for three consecutive days during lactation means that milk must be withheld for 7 d if penalties are to be avoided.

Effect of low levels of antibiotics in milk

In general, low levels of penicillin have no growth-promoting effect on calves and its use has often significantly decreased growth rate[55]. Since penicillin must be given at very high levels to control Gram-negative organisms such as *E. coli*, the lack of a growth-promoting effect is perhaps not surprising. Similarly, penicillin at the level used was found to have no protective value against *E. coli* septicaemia in calves[56]. On the other hand, ampicillin, which has a range of antibacterial activity closer to that of the tetracyclines has in particular been used as a prophylaxis against salmonellosis[57]. In spite of penicillin having no growth-promoting action in the preruminant calf it does, in contrast to chlortetracycline, result in a reduction in the weight:length ratio, i.e. the thickness of the walls of the small intestine of calves[56].

In Norway, groups of calves were given milk substitutes to which penicillin or streptomycin had been added, or milk from cows recently injected with antibiotics. Rumen juice from these calves had slightly higher bacterial counts than that from untreated controls and a rather higher number of antibiotic-resistant strains. However, it was concluded that feeding milk containing these antibiotics to young calves was a safe procedure[58].

Antibiotic residues in calves at slaughter

In Czechoslovakia, 62 per cent of calves at abattoirs showed chlortetracycline residues in muscles when they had been given a medicated premix without observing withdrawal times. However, the presence of residues did not influence the Gram-negative resistant and R^+ microflora, which was not significantly different from that of calves that had not received antibiotics[59]. No residues were found in liver, kidney and muscle of calves given 40 mg virginiamycin/kg feed until 4 h before slaughter[60]. Similar results have been found in the Netherlands[61].

Degradation of antibiotics by fermentation of mastitic milk

It has been suggested that mastitic milk containing antibiotic should be fermented to reduce the concentration to a level acceptable even for human consumption, namely <0.01 iu/ml.

Penicillin is hydrolysed in 2 d by fermentation, but streptomycin is only slowly degraded with no decline for 10 d. Indeed, streptomycin-resistant *E. coli* might develop during fermentation.

In the USA satisfactory fermentation was achieved if the pH of the milk was reduced to 4.7 within 24 h. This involved an average temperature of 25°C. The infusion of 100 000 iu penicillin G (1 million units ≡ 600 mg[62]) and 150 mg neomycin sulphate into each quarter delayed the achievement of a pH of 4.7 to 53 h. However, at this rate of antibiotic infusion only the first two milkings after the last intramammary infusion had a marked inhibitory effect on fermentation (100 and 73 h to achieve pH 4.7 for the first and second milking respectively)[63].

Inoculation of the milk from cows given 100 000 iu penicillin G and 150 mg neomycin in two quarters, with *Str. cremoris* and *Str. lactis* reduced the time to reach pH 4.7 but was ineffective when milk from the first milking after infusion was used. Similarly, when mastitic milk from cows given 100 000 iu penicillin G and 150 mg sodium novobiocin per quarter was inoculated with commercial buttermilk, fermentation was poor in the first two milkings when the concentration of penicillin was >0.06 iu/ml and that of novobiocin was >0.35 μg/ml. By the third milking, novobiocin was not detectable and the penicillin concentration had decreased to 0.02 iu/ml[64].

In the UK[65], a comparison was made of the fermentation of mastitic milk at 10, 14, 18 and 22°C with the acidification of mastitic milk with propionic acid; the effect of fermentation on the incidence of antibiotic-resistant bacteria was also studied. A temperature of 20–22°C was required to degrade penicillin within 2–4d, the lactic acid and other bacteria reducing the pH to 4.7 or below and producing a stable nutritious food[66]. At lower temperatures lactic acid bacteria could not grow fast enough, and antibiotic degradation was slower. Moreover, the antibiotic inhibited the growth of the lactic acid bacteria so that 'souring' bacteria predominated to produce an unpalatable nutritionally-degraded product. The addition of propionic acid prevented both bacterial activity and antibiotic degradation; other trials showed that lactoperoxidase had a similar effect to that of propionic acid.

However, large numbers of *E. coli* and other Enterobacteriaceae showing resistance to clinically important antibiotics were isolated in the milk before and after fermentation. Moreover, several species of streptococci, including *Str. uberis* and *Str. dysgalactiae* also multiplied in the milk with antibiotic-resistant strains being present in large numbers.

The question arises as to whether it is better to give milk containing small amounts of antibiotics to calves rather than to give milk containing antibiotic-resistant bacteria[65]. However, if the fermentation has been satisfactory, antibiotic-resistant lactic acid bacteria would predominate and these should be harmless whereas resistant Enterobacteriaceae might have serious consequences for the calf. Lactic acid bacteria can produce antibiotic-like substances[67]. On the other hand, the death of a calf has been associated with pure cultures of lactobacillus in the intestine[68]. Even if the resistant organisms were not pathogenic, they might transfer their resistance to pathogenic organisms within the calf.

At extremely high temperatures, fermentation and antibiotic degradation was rapid[65], but the milk looked unpalatable and there was concern that the smell might taint milk in a bulk tank.

The antibiotic preparations used in the experiment in the USA[63,64] contained only 100 000 iu penicillin whereas 'Streptopen' in the UK contains 1 million units of penicillin and 0.5 g dihydrostreptomycin. For satisfactory degradation of antibiotics it appears that 20°C would have to be maintained for 7 d under UK conditions; this would need some form of heating tank. Although 50 iu penicillin G/ml milk could be fermented at 20–22°C with all antibiotic degraded within 2–4 d, some streptomycin residues might still be present since the organism used for assay was more sensitive to penicillin than to streptomycin. Any system of fermentation adopted must be rapid so that the farmer can know when the antibiotic has been completely degraded.

Effect of ingestion of fermented mastitic milk

It has been reported that, although calves given fermented mastitic milk did not differ in mortality rate or weight gain from those given unfermented fresh milk, they had a three-fold higher incidence of diarrhoea and respiratory problems; however, many of the fermented milk samples were above a pH of 4.7[6].

On the other hand, other work has shown that the growth and incidence of health disorders of calves were not affected by feeding milk from cows treated with antibiotics for mastitis, whether offered fresh or fermented. Mastitic milk preserved by propionic acid or formaldehyde was relatively unpalatable[66]. Similarly, no difference in performance from birth to 90 d was found between calves given fermented mastitic milk from antibiotic-treated or untreated cows, fermented colostrum (diluted 1:1 with water) or fresh milk for the first 30 d of life, but a tendency for a lower incidence of diarrhoea to 90 d for calves given the fermented milk was observed[72].

Another report of a comparison of mastitic milk from cows not yet treated with antibiotics, with mastitic milk from the first six milkings after antibiotics were used that had been fermented to pH 4.7 within 15 h, showed that there was less diarrhoea in the calves given the fermented milk. It was recommended that in starting the fermentation, the first two milkings after antibiotics were used should not be included, since fermentation of these two milkings was very slow (100–73 h). In this experiment new batches of milk for fermentation were begun every 2 weeks[73].

A further comparison[69] of calves given fermented or acidified colostrum to 14 d followed by fermented or acidified unsaleable milk to 28 d with those given milk substitute or whole milk showed no differences in weight gain but the acceptability was lower for acidified or fermented milk. More recently, no significant difference in weight gain, food conversion efficiency or incidence of diarrhoea was found between calves given fresh milk, 'waste milk' or fermented milk during the first 42 d of life[70].

Long-term effect of ingestion of fermented mastitic milk in calfhood

Only one case of a blind quarter occurred in 64 first-lactation heifers given, during the first 30 d of life, fermented mastitic milk from cows either treated with antibiotics or not, or fermented colostrum. Moreover, there was no difference in the incidence of mastitis between the three groups of heifers[71].

For the heifers completing their first lactation 7/12, 4/13, 4/9 and 9/10 heifers had mastitis and the number of cases of mastitis was nine, nine, eight and 16 for the fermented mastitic milk containing antibiotic, the fermented milk in the absence of antibiotic, fermented colostrum (diluted 1:1 with water) and fresh milk respectively[72]. Of the infected quarters in the herd, 2 per cent were infected with *E. coli*, 18 per cent with *Str. uberis*, 10 per cent with *Staph. aureus* and 70 per cent with *Staph. epidermidis*.

Summer mastitis

Summer mastitis is an acute suppurative mastitis occurring in non-lactating cows and heifers and in newly calved cows, predominantly in the summer months[18]. Four species of bacteria have been isolated in various combinations from clinical cases. They are *Corynebacterium pyogenes, Peptococcus indolicus, Str. dysgalactiae* and an unnamed micro-aerophilic coccus[19, 20].

On no account should a calf be allowed to suck an udder infected with summer mastitis, even when there are no symptoms of acute mastitis and no gross changes in the appearance of the milk.

The role of flies

Summer mastitis is probably fly-borne as also is keratoconjunctivitis (New Forest Eye)[23]. On six dairy farms in S. England, 2764 muscid flies were netted from in-calf dairy heifers. A total of one or more pathogens were isolated from 27 of 825 *Hydrotaea irritans* (3.3 per cent) examined. *C. pyogenes* and *Str. dysgalactiae* were the most common pathogens detected, but isolations of *Peptococcus indolicus, Staph. aureus* and *Str. uberis* were also isolated from *Hydrotaea* spp. Isolation of more than one pathogen from the same fly occurred more frequently than would be expected by chance and suggested a common source of contamination. *Hydrotaea metorica* was found to be a potential carrier of the pathogen in June[17].

In a survey of summer mastitis in dry cows and heifers in England in 1983 and 1984, *C. pyogenes* was usually involved and the condition occurred predominantly in late gestation. Most cases occurred in a single front quarter, and the incidence was greater in older cows. Only 3 per cent of cases occurred within 3 weeks of cows receiving dry cow therapy, but 50 per cent occurred within 6 weeks of therapy[21].

Fly numbers were reduced by 90 per cent by a single ear tag and by 99 per cent on double-tagged 22–26-month-old heifers; the biting fly *Haemotobia irritans* was totally controlled. The number of face flies was reduced by 69 per cent with one tag and by 80 per cent with two tags. On the udder where *Hydrotaea irritans* predominated, there was no control with one tag but two tags reduced fly numbers by 89 per cent; thus two tags would be necessary for some protection against summer mastitis[22]. Fly infestation was not unexpectedly found to be more severe on river meadows than on chalk downland pasture[24].

Two ear tags produced more even coverage with synthetic pyrethroids than did one tag and reduced fly population around the head much more than in other parts of the body. In fact the tags might even result in a larger proportion of flies around the udder. Thus, impregnated tail strips or legbands might be more efficient in giving protection against the transfer of summer mastitis[25].

Mucosal disease (bovine viral diarrhoea (BVD) virus)

Incidence

Mucosal disease is caused by the bovine viral diarrhoea (BVD) virus-mucosal disease virus complex. In 1962 it appeared to be quite widespread in Great Britain, since 45 per cent of serum samples obtained from animals at slaughterhouses contained antibodies to the condition[77]. By 1987, it was stated that very few cattle herds in the UK were totally free from evidence of infection[89]. Thus, a survey had indicated that 0.4 per cent of cattle in apparently normal herds in the UK were likely to be persistently infected[90].

A review of the literature concerning the disease has recently been published[95].

Symptoms

The disease is characterized by anorexia, enteritis associated with profuse diarrhoea and a leucopenia. Ulcerations of the mucosa occur particularly in the oral cavity, the abomasum and overlying the Peyer's patches in the small intestine. There is also a cellular depletion of the lymphoid tissue associated with the alimentary tract and the thymus.

In the first stage the calf has a high temperature and no apparent symptoms; the temperature usually recedes for about 3 d and then returns to a much higher level with ulcerations in the mouth and reddening of the nostrils and oral mucosa, especially around the teeth. There is also a mucous nasal discharge and sometimes increased lachrymation[78]. In mild cases these are the only symptoms, but in more severe cases diarrhoea is marked and the faeces may become bloodstained. The ulcerations are severe, and besides occurring on the muzzle, lips and mouth, may also occur between the claws of the hoof. The nasal discharge is also profuse, and in the early stages there is much evidence of thirst. Abortions have been associated with some outbreaks of the disease.

It has been postulated that, as the ulcerations in the mouth are similar to those of hyperkeratosis, vitamin A deficiency may possibly predispose calves to the condition[80].

It has been suggested that failure to produce antibodies against the disease may be due to the destruction of lymphoid tissue. There is also a suggestion of genetic susceptibility to the disease, as in one outbreak only calves from one sire were affected[76].

In mild cases of the disease calves recover quickly, but in severe cases the mortality rate may be high.

Immunity

It seems that the infection can occur under two conditions.

Firstly, the bovine fetus becomes infected before 180 d of gestation, i.e. before immunocompetence develops, by a non-cytopathic strain of virus. The virus is tolerated and although there is no specific immune response, there is also no clinical disease. Thus, there is only a low or undetectable level of serum antibodies. If the calves are subsequently infected with cytopathic virus, the virus multiplies rapidly and death invariably follows 2–3 weeks later[81].

Secondly, if the fetus is infected during the last third of pregnancy, it is capable of developing serum neutralizing antibodies against BVD. Thus, of the precolostral sera of 77 newborn calves examined in France for antibodies against adeno, PI3, BVD and IBR viruses, RSV, reovirus 1 and 2, rhinovirus, pustular stomatitis and Aujeszky's disease, two calves had antibodies to PI3 and seven to BVD. IgG was present in five of the samples but there was no trace of IgM[83].

In a study of viral antibodies in fetal fluids, 4 per cent were positive for BVD, 2.4 per cent for BSV (bovine syncytial virus), 0.6 per cent for PI3, 0.3 per cent for both adenovirus and RSV and none for parvovirus and IBR virus. The latter finding was not surprising as IBR virus causes death of the fetus before antibodies can be produced. Nine fetuses (2.7 per cent) were positive for virus (8 BVD; 1 PI3) and three (0.5 per cent) positive for antigen (2 BVD; 1 IBR)[74].

Calves born with a high titre to BVD virus retained a high titre for 1 year, whereas colostrum-derived antibodies to BVD generally decreased during the first 4–6 months of age[86].

If there are antibodies to BVD in colostrum, the condition does not usually affect colostrum-fed calves until they are over 1 month of age. A study of viruses endemic in a group of six calves in New Zealand over a period of 14 months showed that they were infected with a variety of enteroviruses and PI3. Maternal antibodies to IBR (two calves) and BVD (three calves) were present at 1 month but not at 3, 8 and 15 months of age; no acute infection with these two viruses occurred[84].

After ingestion of colostrum most calves developed high antibody titres against reovirus, PI3 and pustular stomatitis. There was no difference in the protein fractions in precolostral serum, whether antibodies were present or not[83].

In a naturally-infected herd in California, the rate of decline of passively acquired antibodies to BVD was exponential varying with initial concentration, and among different calves. The calves that had high initial titres had a faster reduction than those with low initial titres, the titres disappearing at 4–6 months of age. After the titres had disappeared, all calves developed an increase in titre, probably indicating exposure to infection; it is not known what level of titre would protect[82].

It is claimed that there is an essential role for humoral antibody but not for cellular immunity in protection from a systemic BVD infection. Thus, in calves without passively acquired neutralizing antibody, administration of the cortico-steroid dexamethasone to suppress cellular immunity potentiated a systemic BVD infection with the calves developing a fatal uraemia, whereas it had no effect on those with high levels of passively acquired antibodies. It was suggested that there are three types of virus; those controlled by antibody (BVD), those controlled by antibody and cellular immunity (IBR) and those controlled by a complex interaction of cellular and humoral immunity[85].

The fact that calves may be infected with non-cytopathic virus may be the reason why some cattle vaccinated with attenuated BVD virus succumb to mucosal disease[81].

The virus of BVD is sufficiently stable for successful isolation from the blood of persistently infected cattle even with a delay of 5 d between specimen collection and testing; the mean temperature in the dark to which the samples were exposed was 22°C[88].

Interaction with other microorganisms

Simultaneous infection with IBR virus and BVD virus has been shown to produce more severe effects than with either virus singly[75]. Similarly, BVD virus was found to exacerbate the effects of salmonella infection[87] (see Chapter 4).

Introduction of infection

The condition seems to be more severe when it occurs in self-contained herds, probably owing to the low level of herd immunity[79]. The non-cytopathic virus spreads rapidly among cattle in early pregnancy. If calves are exposed to cytopathic virus 6–24 months later, an explosive outbreak occurs[81].

The results of introducing a heifer, persistently infected with BVD virus, into a dairy herd have been reported. The damage was limited to progeny of cows who were in the first 168 d of gestation at that time, and only fetuses up to 81 d of gestation became persistently infected in calfhood. Two of these calves exhibited body tremor but at post-mortem at 3 months of age lacked macroscopic or microscopic lesions in the central nervous system. In contrast, calves at 146 and 153 d of gestation at the time of infection had eliminated the infection but had lesions of cerebellar dysplasia and multifocal retinal atrophy[94].

Embryo transfer

The problem of BVD virus in relation to embryo transfer has been studied in Canada. It was concluded that even if the genetic parents were infected with BVD virus, uninfected calves could be produced provided that the zona pellucida was intact; proper embryo handling procedures between collection and transfer were adopted; and the recipients were free from BVD at the time of transfer and for the appropriate period thereafter[92]. However, a pedigree bull, born as a result of artificial insemination and ovum transplantation, was reported as persistently infected with BVD virus, the virus being cultured from blood, nasal and ocular swabs as well as being present in semen in high titres[93].

Prevention

The crux of the problem of control is to avoid fetal infection in early gestation by immunization of female cattle with a live or inactivated virus vaccine[91].

Treatment

As there is no specific treatment for the disease, the aim is to minimize the effect of any secondary bacterial invasion.

Tuberculosis

Source of infection

Calves may become infected with tuberculosis before birth, although this probably occurs quite rarely. In postnatal calfhood tuberculosis there is little doubt that the infection originates from the feeding of milk containing tubercle bacilli, and from the contact of calves with open cases of bovine tuberculosis.

In the late 1960s, more cases of tuberculosis were occurring in S.W. England than in the whole of the rest of Great Britain. In 1971, *Mycobacterium bovis* was isolated from a badger in Gloucestershire, where bovine tuberculosis had occurred in cattle. Since then infection has been identified in badgers in S.W. England, Surrey, W. Sussex, E. Sussex, Hereford and Worcester, Staffordshire, Dyfed,

Shropshire and Cumbria, and the infection in badgers has been associated with tuberculosis in cattle in E. Sussex, W. Sussex, Staffordshire, Dyfed and Cumbria[15].

In 1985, bovine tuberculosis was found in imported farm deer in Great Britain and two other outbreaks have occurred in 1988. There is thus a risk that the disease may spread to wild deer and to those in deer parks as well as posing a threat to cattle.

Under the Tuberculosis (Deer) Order 1989 (SI 878) the disease in deer became notifiable. The order provided for:

1. The compulsory notification of tuberculosis in deer and in their carcasses;
2. The isolation of affected or suspected of being infected deer;
3. The testing of deer;
4. The prohibition of vaccination and treatment of deer for tuberculosis;
5. Precautions to be taken against the spread of tuberculosis in deer, including the cleansing and disinfection of premises, the marking and identification of deer and the prohibition of exposure of a deer for sale in a market unless correctly marked.

In an amendment (SI 879) to the Movement of Animals (Records) Order 1960, records of all sales of deer by private treaty or in an auction have to be kept by the owner.

From 1 September 1989, under the Tuberculosis (Deer) Notice of Intended Slaughter and Compensation Order (SI 1316) deer affected with tuberculous emaciation or excreting tuberculous material or affected with a chronic cough and showing signs of tuberculosis or a reactor can be ordered to be slaughtered with compensation of an amount equal to 50 per cent of the market value of the deer up to a maximum of £600.

Symptoms

Clinical symptoms of tuberculosis are usually absent and diagnosis depends essentially upon a positive mammalian tuberculin reaction. If clinical symptoms are present, they may only take the form of coughing and unthriftiness, which are common to a variety of calfhood diseases.

Prevention

By careful management disease-free stock may be raised from an infected herd. Each calf should be taken away from its dam at birth; on no account should the dam be allowed to lick or suckle the calf. The calf should then be placed in a thoroughly disinfected calf pen and be given its dam's colostrum, preferably pasteurized, in a clean bucket.

After the colostrum feeding period either the calf should be fed with milk from an attested cow or, if milk from an attested cow is not available, the milk should be pasteurized before feeding.

Pasteurization can be carried out by heating the milk and retaining it at a temperature of not less than 63°C and not more than 66°C for 30 min (Holder pasteurization) or at a temperature of not less than 72°C for at least 15 s (HTST pasteurization). Colostrum can also be effectively pasteurized at the lower temperature (i.e. 63–66°C) without impairing its protective action, although about 15 per cent of the whey proteins will be denatured. However, HTST pasteurization of human colostrum was found to affect the composition of the protein, leading to the appearance of a new large peptide fraction[4].

At no time should the young stock use the same buildings, water troughs or pastures that have been contaminated by the adult stock; contaminated pastures are normally clear of infection after 6 months. Furthermore, the young stock should not use fields adjacent to those used by the adult cattle. If this is impracticable, a double fence, 1.8 m wide, should be erected between the two fields to prevent the adult cattle breathing on or licking the young stock.

Eradication

Although Great Britain is now virtually free from tuberculosis in cattle, the methods that were adopted to achieve this are summarized for the benefit of those in countries where tuberculosis is still endemic.

The young stock were tuberculin-tested at 3 months of age, and thereafter at 6-monthly intervals. If any calves reacted to the test, they were disposed of as soon as possible. When the remaining heifers, reared under clean conditions, were pregnant, the adult cows were tested and any reactors were slaughtered. As soon as two consecutive tests on all cattle on the premises had been carried out and no reactors found, the owner of the herd could apply for registration under the Tuberculosis (Attested Herds) Scheme 1950. These two tests had to be done at an interval of not less than 60 d and not more than 12 months, and the application had to be made within 12 months of the date of the second test. The herd then became a Supervised Herd until an official test had taken place not earlier than 60 d from the date of the second of the two tests mentioned above. If the official test disclosed no reactor, the herd was registered in the Register of Attested Herds.

By 1988, the incidence of reactors in the UK had dropped to 0.015 per cent. However, unaccountable breakdowns still occur. For instance, a 3-yearly test in a previously clear herd disclosed 80 reactors out of 200. Moreover, traces of tuberculosis were found in the lymph glands of slaughtered cattle, which were not reactors[16].

If a breakdown occurs, all reactors are slaughtered and a test must be made on the whole herd after an interval of 60 d. If the disease is confirmed at post-mortem or by the tuberculin test, further tests at 60 d intervals have to be made until the herd is clear of reactors.

References

1. DOYLE, T. M. *Vet. Rec.* **70**, 238 (1958)
2. THE FEDERATION OF THE UNITED KINGDOM MILK MARKETING BOARDS. *Dairy Facts and Figures* (1985)
3. MILK MARKETING BOARD. *Report of the Breeding and Production Organisation* No. 37 (1986–87)
4. NIKOLAEVSKAYA, V. R., GRIBAKIN, S. G., NETREBENKO, O. K. and CHENIKOV, M. P. *Vopr. Okhr. Materin. Det.* **29**, 45 (1984)
5. LILLIE, F. R. *J. exp. Zool.* **23**, 371 (1917)
6. MINISTRY OF AGRICULTURE, FISHERIES AND FOOD. *Methods of Disposing of Unsaleable Whole Milk on Farms.* Adv. Leafl. No. 679 (1981)
7. SCHAFFER, L. V. and McGUFFEY, R. K. *Proc. natn. Mastitis Coun.* **19**, 67 (1980)
8. HOLLINSHEAD, P. *Livestk Fmg* **19**(5), 33 (1981)
9. POOLE, D. E. and EDWARDS, S. A. *Ann. Rev. Boxworth exp. Husb. Farm,* p. 40 (1981)
10. MILK MARKETING BOARD. *Report of the Breeding and Production Organization* No. 32 (1981–82)

11. SWANNACK, K. P. *A Future for Non-Saleable Milk?* Mimeograph, Ministry of Agriculture, Fisheries and Food, p. 1 (1981)
12. MINISTRY OF AGRICULTURE, FISHERIES AND FOOD. *Veterinary Investigation Diagnosis Analysis* **11**, 1986 and 1979–86. The Epidemiology Unit, Central Veterinary Laboratory, p. 5 (1987)
13. RICHARDSON, C., TERLECKI, S. and FRANCIS, P. *Proc. XIIth Wld Congr. Dis. Cattle, Amsterdam, Netherlands,* p. 1063 (1982)
14. HOWE, K., LINTON, A. H. and OSBORNE, A. D. *J. appl. Bact.* **40**, 331 (1976)
15. HEWSON, P. I. and SIMPSON, W.J. *Vet. Rec.* **120**, 252 (1987)
16. ANONYMOUS. *Vet. Rec.* **122**, 26 (1988)
17. BRAMLEY, A. J., HILLERTON, J. E., HIGGS, T. M. and HOGBEN, E. M. *Br. vet. J.* **141**, 618 (1985)
18. MARSHALL, A. B. In *Mastitis Control and Herd Management: NIRD/HRI Tech. Bull.* **4**, 81 (1981)
19. ŚTUART, P., BUNTAIN, D. and LANGRIDGE, R. G. *Vet. Rec.* **63**, 451 (1951)
20. SORENSEN, G. H. *Nord. VetMed.* **26**, 122 (1974)
21. HILLERTON, J. E., BRAMLEY, A. J. and WATSON, C. A. *Br. vet. J.* **143**, 520 (1987)
22. HILLERTON, J. E., BRAMLEY, A. J. and YARROW, N. H. *Br. vet. J.* **141**, 160 (1985)
23. BROWN, J. F. and ADKINS, T. R. *Am. J. vet. Res.* **33**, 2551 (1972)
24. HILLERTON, J. E. and BRAMLEY, A. J. *Br. vet. J.* **142**, 155 (1986)
25. TAYLOR, S. M., MALLON, T., ELLIOTT, C. T. and BLANCHFLOWER, J. *Vet. Rec.* **116**, 566 (1985)
26. HOWE, K. and LINTON, A. H. *J. appl. Bact.* **40**, 317 (1976)
27. FEY, H. *Schweiz. Arch. Tierheilk.* **104**, 1 (1962)
28. SMITH, H. W. and HUGGINS, M. G. *J. gen. Microbiol.* **92**, 335 (1976)
29. ØRSKOV, L. F., ØRSKOV, H., SMITH, H. W. and SOJKA, W. J. *Acta path. microbiol. scand.* **B-83**, 31 (1975)
30. LINTON, A. H., HOWE, K., SOJKA, W. J. and WRAY, C. *J. appl. Bact.* **46**, 585 (1979)
31. MILK MARKETING BOARD. *Report of the Breeding and Production Organisation* No. 35 (1984–85)
32. LINTON, A. H. *R. Soc. Hlth J.* **97**, 115 (1977)
33. ROY, J. H. B. *Vet. Rec.* **76**, 511 (1964)
34. SHILLAM, K. W. G. Studies on the nutrition of the young calf with special reference to the incidence of *Escherichia coli* infections. *PhD Thesis,* University of Reading (1960)
35. KHAKIMOVA, K. M. and ABZOLOVA, A. G. *Uchen. Zap. kazan. vet. Inst.* **122**, 156 (1976)
36. VOLOVENKO, M. A. *Veterinariya* No. 3, 77 (1972)
37. VOLOVENKO, M. A. *Veterinariya* No. 3, 73 (1974)
38. VOLOVENKO, M. A. *Veterinariya* No. 11, 64 (1974)
39. UGARTE, J. and PRESTON, T. R. *Cuban J. agric. Sci.* 9, 15 (1975)
40. ALVAREZ, F. J., SAUCEDO, G., ARRIAGA, A. and PRESTON, T. R. *Trop. Anim. Prod.* **5**, 25 (1980)
41. RIGBY, C., UGARTE, K. and BOUCOURT, R. *Cuban J. agric. Sci.* 10, 35 (1976)
42. BARTO, P. B., BUSH, L. J. and ADAMS, G. D. *J. Dairy Sci.* **65**, 271 (1982)
43. CHICK, A. B., ARCHACOSA, A. S., EVANS, D. L. and RUSOFF, L. L. *J. Dairy Sci.* **58**, 742 (1975)
44. GONZALEZ, F. AND DENIS, I. *Mems Assoc. latinoam. Prod. Animal* **13**, 48 (1978)
45. CHARDAVOYNE, J. A., IBEAWUCHI, J. A., KESLER, E. M. and BORLAND, K. M. *J. Dairy Sci.* **62**, 1285 (1979)
46. ANONYMOUS. *Rep. Cyprus agric. Res. Inst.,* p. 58 (1981)
47. HORN, H. H. VAN, OLAYIWOLE, M. B., WILCOX, C. J., HARRIS, B. Jr. and WING, J. M. *J. Dairy Sci.* **59**, 924 (1976)
48. KEITH, E. A., WINDLE, L. M., KEITH, N. K. and GOUGH, R. H. *J. Dairy Sci.* **66**, 833 (1983)
49. SCHALM, D. W. *Cornell Vet.* **32**, 49 (1942)
50. JOHNSON, S. D. *J. Am. vet. med. Ass.* **110**, 179 (1947)
51. BRAMMER, H. Investigation of the occurrence of *Streptococcus agalactiae* and *Staphylococcus aureus* in the tonsils and lymph nodes of healthy calves and young cattle in relation to periparturient mastitis at first calving. *Thesis,* Tierärtzl. Hochsch., Hannover (1981)
52. HORN, H. H. VAN, MARSHALL, S. P., FLOYD, G. T., OLAKOKU, E. A., WILCOX, C. J. and WING, J. M. *J. Dairy Sci.* **63**, 1465 (1980)
53. BOCK, R. Detection of *Corynebacterium pyogenes* in tonsils and lymph nodes of healthy calves and young cattle and its significance in the occurrence of summer mastitis. *Thesis,* Tierärtzl. Hochsch., Hannover (1980)
54. MINISTRY OF AGRICULTURE, FISHERIES AND FOOD. *Manual of Veterinary Investigation Laboratory Techniques. Part 2. Bacteriology RVG,* 3 p. 53 (1978)

55. ROY, J. H. B., SHILLAM, K. W. G., PALMER, J. and INGRAM, P. L. *Br. J. Nutr.* **9**, 94 (1955)
56. INGRAM, P. L., SHILLAM, K. W. G., HAWKINS, G. M. and ROY, J. H. B. *Br. J. Nutr.* **12**, 203 (1958)
57. KERR, W. J. and BRANDER, G. C. *Vet. Rec.* **76**, 1105 (1964)
58. YNDESTAD, M. and HELMEN, P. *Norsk VetTidsskr.* **92**, 435 (1980)
59. MALIKOVA, M. and KATYAS, Z. *Vet. Med., Praha* **25**, 627 (1980)
60. SCHMIDT, U. *Fleischwirtschaft* **56**, 418 (1976)
61. NOUWS, J. F. M. *Tijdschr. Diergeneesk.* **99**, 1043 (1974)
62. MARTINDALE. The Extra Pharmacopoeia, 28th edn, p. 1102 (1982)
63. KEYS, J. E., PEARSON, R. E. and FULTON, L. A. *J. Dairy Sci.* **59**, 1746 (1976)
64. KEYS, J. E., PEARSON, R. E. and WEINLAND, B. T. *J. Dairy Sci.* **62**, 1408 (1979)
65. MOORE, A. Degradation of antibiotics in mastitic milk for feeding to calves. Mimeograph. *Proc. Microbiologists Tech. Conf. ADAS* (1982)
66. KESLER, E. M. *J. Dairy Sci.* **64**, 719 (1981)
67. MIKOLAJCIK, E. M. and HAMDEN, S. Y. *Cultured Dairy Prod. J.* **10**, 10 (1975)
68. ROY, J. H. B., STOBO, I. J. F., SHOTTON, S. M., GANDERTON, P. and GILLIES, C. M. *Br. J. Nutr.* **38**, 167 (1977)
69. OTTERBY, D. E., JOHNSON, D. G., FOLEY, J. A., TOMSCHE, D. S., LUNDQUIST, R. G. and HANSON, P. J. *J. Dairy Sci.* **63**, 951 (1980)
70. KEITH, E. A., WINDLE, L. M., KEITH, N. K. and GOUGH, R. H. *J. Dairy Sci.* **66**, 833 (1983)
71. KEYS, J. E. Jr. *Proc. natn. Mastitis Coun.* **19**, 51 (1980)
72. KEYS, J. E., PEARSON, R. E. and WEINLAND, B. T. *J. Dairy Sci.* **63**, 1123 (1980)
73. KEYS, J. E. *Hoard's Dairym.* **122**, 1219 (1977)
74. LUCAS, M. H., WESTCOTT, D. G. F., EDWARDS, S., NEWMAN, R. H. and SWALLOW, C. *Vet. Rec.* **118**, 242 (1986)
75. TYLER, D. E. and RAMSEY, F. K. *Am. J. vet. Res.* **26**, 903 (1965)
76. SHOPE, R. E. Jr. *Diss. Abstr.* **26**, 634 (1965)
77. PATERSON, A. B. *Vet. Rec.* **74**, 1384 (1962)
78. BAKER, J. A., YORK, C. J., GILLESPIE, H. J. and MITCHELL, G. B. *Am. J. vet. Res.* **15**, 525 (1954)
79. HUCK, R. A. *Vet. Rec.* **69**, 1207 (1957)
80. LOOSMORE, R. M. *Vet. Rec.* **76**, 1335 (1964)
81. BROWNLIE, J., CLARKE, M. C. and HOWARD, C. J. *Vet. Rec.* **114**, 535 (1984)
82. KENDRICK, J. W. and FRANTI, C. E. *Am. J. vet. Res.* **35**, 589 (1974)
83. MASSIP, A., WELLEMANS, G., LEUNEN, J. and CHARLIER, G. *Annls Rech. vét.* **5**, 397 (1974)
84. BURGESS, G. W. *N.Z. vet. J.* **25**, 178 (1977)
85. SHOPE, R. E., MUSCOPLAT, C. C., CHEN, A. W. and JOHNSON, D. W. *Can. J. comp. Med.* **40**, 355 (1976)
86. CORIA, M. F. and McCLURKIN, A. W. *Can. J. comp. Med.* **42**, 239 (1978)
87. WRAY, C. and ROEDER, P. L. *Res. vet. Sci.* **42**, 213 (1987)
88. RAE, A. G., SINCLAIR, J. A. and NETTLETON, P. F. *Vet. Rec.* **120**, 504 (1987)
89. EDWARDS, S., DREW, T. W. and BUSHNELL, S. E. *Vet. Rec.* **120**, 71 (1987)
90. HOWARD, C. J., BROWNLIE, J. and THOMAS, L. H. *Vet. Rec.* **119**, 629 (1986)
91. ROEDER, P. L. and HARKNESS, J. W. *Vet. Rec.* **118**, 143 (1986)
92. HARE, W. C. D. *Vet. Rec.* **118**, 544 (1986)
93. BARLOW, R. M., NETTLETON, P. F., GARDINER, A. C., GREIG, A., CAMPBELL, J. R. and BONN, J.M. *Vet. Rec.* **118**, 321 (1986)
94. ROEDER, P. L., JEFFREY, M. and CRANWELL, M. P. *Vet. Rec.* **118**, 44 (1986)
95. DUFFELL, S. J. and HARKNESS, J. W. *Vet. Rec.* **117**, 240 (1985)
96. WEST, H. J. and PAYNE-JOHNSON, C. E. *Vet. Rec.* **121**, 108 (1987)
97. BHATIA, S. and SHANKER, V. *Br. vet. J.* **141**, 42 (1985)
98. HARVEY, M. J. A. *Vet. Rec.* **98**, 479 (1976)
99. DATA, S. P. and STONE, W. H. *Proc. Soc. exp. Biol. Med.* **113**, 756 (1963)
100. KOSAKA, S., KANAGAWA, H. and NISHIDA, S. *Tohoku J. agric. Res.* **18**, 207 (1967)
101. GERNEKE, W. H. *S. Afr. vet. Med. Ass. J.* **40**, 279 (1969)
102. BURTON, M. J. and MOMONT, H. W. *Vet. Rec.* **119**, 155 (1986)
103. BENIRSCHKE, K. and BROWNHILL, L. E. *Cytogenetics* **1**, 245 (1962)
104. RYAN, K. J., BENIRSCHKE, K. and SMITH, O. W. *Endocrinology* **69**, 613 (1962)

105. BILLINGHAM, R. E. and LAMPKIN, G. H. *J. Embryol. exp. Morph.* **5**, 351 (1957)
106. HAFEZ, E. S. E. and JAINUDEEN, M. R. *Anim. Breed. Abstr.* **34**, 1 (1966)
107. ORTAVANT, R. and THIBAULT, C. *Annls Biol. anim. Biochim. Biophys.* **10**, 9 (1970)
108. CORDY, D. R., RICHARDS, W. P. C. and STORMONT, C. *Pathologia vet.* **6**, 487 (1969)
109. BLOOD, D. C. and GAY, C. C. *Aust. vet. J.* **47**, 520 (1971)
110. DAVIS, G. B., THOMPSON, E. J. and KYLE, R. J. *N.Z. vet. J.* **23**, 181 (1975)
111. DIMMOCK, C. K., CLARK, I. A. and HILL, M. W. M. *Res. vet. Sci.* **20**, 244 (1976)
112. STEENHAUT, M., DE MOOR, A., VERSCHOOTEN, F. and DESMET, P. *Vet. Rec.* **98**, 131 (1976)
113. FRANCE, M. P., BARLOW, R. M. and BARNETT, K. C. *Vet. Rec.* **121**, 528 (1987)
114. ASHTON, N., BARNETT, K. C., CLAY, C. E. and CLEGG, F. G. *Vet. Rec.* **100**, 505 (1977)
115. LEKEUX, P. and ART, T. *Vet. Rec.* **121**, 353 (1987)
116. GUSTIN, P., BAKIMA, J., LEKEUX, P. and LOMBA, F. *Proc. XIVth Wld Congr. Dis. Cattle, Dublin, Ireland,* p. 697 (1986)
117. BASSETT, H. *Vet. Rec.* **121**, 8 (1987)
118. MATHER, D. B. *Vet. Rec.* **120**, 487 (1987)
119. BOYD, J. S. and McNEIL, P. E. *Vet. Rec.* **120**, 34 (1987)
120. DYSON, D. A. *Vet. Rec.* **119**, 635 (1986)
121. O'BRIEN, J. K. *Vet. Rec.* **119**, 206 (1986)
122. MILLS, C. F., DALGARNO, A. C., WILLIAMS, R. B. and QUARTERMAN, J. *Br. J. Nutr.* **21**, 751 (1967)
123. HEALY, P. J., HARPER, P. A. W. and BOWLER, J. K. *Res. vet. Sci.* **38**, 96 (1985)
124. HARPER, P. A. W., HEALY, P. J. and DENNIS, J. A. *Vet. Rec.* **119**, 59 (1986)
125. HARPER, P. A. W., HEALY, P. J. and DENNIS, J. A. *Vet. Rec.* **119**, 62 (1986)
126. DUFFELL, S. J. *Vet. Rec.* **118**, 95 (1986)
127. MORROW, C. J. and McORIST, S. *Vet. Rec.* **117**, 312 (1985)
128. JOHNSTON, A. *Vet. Rec.* **116**, 585 (1985)
129. NICOLSON, T. B., NETTLETON, P. F., SPENCE, J. A. and CALDER, K. H. *Vet. Rec.* **116**, 281 (1985)
130. BOYD, J. S. *Vet. Rec.* **116**, 203 (1985)
131. JOHNSTON, R. W. *Vet. Rec.* **116**, 84 (1985)
132. MILK MARKETING BOARD. *Report of the Breeding and Production Organisation* No. 38 (1987–88)
133. THE FEDERATION OF THE UNITED KINGDOM MILK MARKETING BOARDS. *Dairy Facts and Figures,* 25th edn (1987)
134. WATSON, W. A. *Proc. Br. Cattle Vet. Ass.,* p. 21 (1979–80)

Chapter 7

Miscellaneous infections, metabolic disorders, nutritional deficiencies and poisoning

In this chapter only the more important deficiencies of major minerals, trace minerals and vitamins are considered. For the less important deficiencies and toxicities, and for nutritional requirements of major minerals, trace minerals and vitamins, the reader should refer to *The Calf*, 4th edn or, when published, *The Calf*, 5th edn, vol. 2.

Abomasal ulceration

Whereas duodenal ulcers are rare in calves[89], abomasal ulceration was found in 264 of 304 veal calves at slaughter at 3–5 months of age. The incidence and severity were greater in loose-housed calves with access to straw and given milk substitute *ad libitum* (97 per cent) than in those kept in crates and given milk by bucket (66 per cent). However, the calves in crates were mainly Friesian bulls whereas those in loose housing were mostly Hereford-cross females. The majority of lesions were located in the distal pylorus. There was no evidence that the abomasal erosions and ulcers in the majority of veal calves affected their growth rate or were deleterious to their welfare[90].

It was thought that the straw caused a partial blockage of the pylorus, delaying abomasal emptying and rendering the mucosa susceptible to ulceration, as a result of muscular constriction and mucosal compression causing localized hypoxia[91]. The quality and concentration of the powder in a milk substitute diet might also affect susceptibility[92]. It has been shown that, before there is overt mucosal damage, an increase in the depth of the mucosa with loss of mucin in the regions of the erosions and ulcers occurs[93].

Black quarter (quarter-ill or blackleg)

Incidence

Blackleg, caused by the organism *Clostridium chauvei*, has in exceptional cases been responsible for the death of calves under 2 months of age, which have been kept in pens since birth. However, it is more usual for calves to be affected when sucking their dams out at grass, in those areas of the country where the incidence of blackleg is high. Low-lying damp permanent pastures seem to be associated with the disease, which is contracted by animals eating food infected with clostridial spores or by infection through a wound.

Symptoms

The affected animal has a fever and shows inappetence, followed by the appearance of swellings in the muscles of the body, which are at first hot and painful but later become cold and painless. The skin over the affected muscles becomes hard and stiff, and a crackling noise occurs when pressure is applied, owing to the stiffness of the skin and to the gas produced by the bacteria beneath the skin. The organism produces a toxin, which will rapidly lead to death. If an animal dies from the disease, the carcass should be burnt, deeply buried or removed from the field intact, as otherwise it may be a source of infection due to release of bacteria and spores.

Treatment

Treatment with antibiotics and serum may be effective in early cases. As a preventive measure, calves on farms where blackleg occurs are usually immunized when 4–12 weeks of age or a few weeks prior to turning out to grass.

Bloat (tympany)

Bloat occurs as a result of excessive gas production in the abomasum or reticulo-rumen.

Abomasal bloat

It seems probable that on occasions this condition is one of distension with fluid due to malfunction of gastric emptying. If a large amount of fluid from a liquid feed passes into the rumen, then the subsequent distension of the rumen may restrict the emptying from the abomasum. However, abomasal displacement to the left-hand side, as a result of the abomasum being full of gas and curdled milk substitute, has been reported in a 9-week-old calf. When the gas was released a normal fibre-filled rumen was behind it and the rumen returned to the normal position. A similar case had been reported some 20 years earlier of a 6-week-old calf being found dead after being blown for 2 d. A distended abomasum was displaced to the left-hand side with occlusion of the pylorus by the ventral sac of the rumen[113]. If the abomasum is displaced to the right, the condition is more serious and the calf is often found dead[112].

Abomasal bloat may also be associated with the proliferation of an undesirable or excessive flora in the abomasum, possibly of *Escherichia coli* that has spread to the abomasum after having multiplied in the duodenum as a result of excessive undigested protein passing from the abomasum, but more probably of lactobacilli as a result of too high a level of soluble carbohydrate in the diet. In Canada, the addition of 0.5 or 1 g formalin (370 ml HCHO/l) per litre to milk substitutes reduced gas production *in vitro* by 79 and 96 per cent respectively. In *ad libitum* feeding of lambs formalin addition effectively controlled abomasal bloat which was associated with Gram-positive rods[111]. Also in lambs, there was no relationship between protein source, i.e. skim milk, rape seed and soyabean, in a milk substitute and abomasal bloat, which occurred between the 15th and 18th day of *ad libitum* feeding of warm milk twice daily. It was considered that there was no support for a forward or backward movement of gas from the rumen or from the small intestine to the abomasum[110].

Certainly, the feeding of milk substitutes of low fat content seems to be one of the predisposing causes. Treatment lies mainly in reducing milk intake, and thereafter using a diet that does not predispose to the condition. Release of gas can be affected by means of a stomach tube, but the process will usually be necessary after each meal. Bloat of milk-fed calves is also associated with nutritional myopathy (see below)[60].

Ruminal bloat

Milk substitute diets

Ruminal bloat may be caused by the passage of milk substitute diets into the rumen, either as a result of failure of oesophageal groove function or possibly from backflow into the rumen from the abomasum. This is particularly associated with diets containing certain non-milk proteins, and may be accompanied by the multiplication of *Clostridium perfringens* and other gas-forming organisms[119].

Thus, a high incidence of ruminal bloat occurred as early as the first week of life when single-cell protein (Toprina) replaced 40 or 70 per cent of milk protein, and the incidence of mortality was fairly high. It was associated with a low pH of ruminal contents (4.6–5.1) compared with 7.2–7.4 for healthy calves. Of the permitted antibiotics for addition to milk substitutes, i.e. zinc bacitracin, flavomycin and virginiamycin, only virginiamycin reduced gas production *in vitro*. The diet tended to be rather unpalatable and calves tended to drink the milk slowly, which rather suggests a failure in oesophageal groove function.

A second type of ruminal bloat appeared to be the result of backflow from the abomasum. It occurred in 3-month-old preruminant calves, where small hairballs about 25 mm in diameter were blocking the pylorus.

A third type has been associated with keeping cattle as preruminants up to 350 kg live weight. This bloat may also result from hairballs. Cattle showing this condition typically only consume one rather than two liquid feeds daily. Supplementation with roughage seems to aggravate the situation.

Dry diets

Ruminal bloat does not usually occur in calves fed conventional rations, but is restricted to regimens where roughage is not included in the diet, particularly where all-concentrate diets are given in restricted amounts rather than *ad libitum*[61] and where the fibrous characteristics of the concentrate mixture are lacking. Thus, diets containing large amounts of ground maize are more likely to cause bloat than are those containing crushed barley[62].

Cattle that are susceptible to bloating on pastures in New Zealand have high concentrations of several proteins in their saliva and of 'yellow bubbles' in their rumen contents. The induction of this protein synthesis arises from a psychic stimulus, i.e. the sight of food, on the ingestion of different feeds and from acid stimulation. The least susceptible cattle had the highest salivation rate. The heritability of bloat is not greater than about 0.6[107–109].

Proliferation of *E. coli* has been associated with bloat. In 24–84-week-old steers, few *E. coli* were present in the duodenum compared to those in the more distal intestine. However, exceptionally large numbers of *E. coli* occurred at all intestinal sites of one steer which had bloat 3 weeks before slaughter. Only 32 per cent of 331 *E. coli* isolates were serotypable and represented 18 'O' groups with the most frequently isolated being O2A . Serotypes, with which the steers were inoculated at birth, were not recovered at the time of slaughter[106].

The effect of sodium chloride

In Canada, the inclusion of 40 g rather than 5 g NaCl/kg in a concentrate diet of feed-lot cattle with the intention of reducing bloat resulted in increased water intake and number of protozoa, whilst cell-free rumen fluid viscosity, acidity and carbohydrate in cell-free rumen fluid decreased and there was less accumulation of cytoplasmic polysaccharide granules[105].

In a comparison of lucerne hay with all-concentrate diets supplemented with either 40 g or 5 g NaCl/kg, it was found that 40 g NaCl/kg increased the flow of material from the rumen and appeared to alter rumen fermentation in such a way as to oppose the development of conditions that led to bloat. The low-salt diet was associated with bloat, lower rumen pH, lower protozoal counts, higher viscosity, higher soluble carbohydrate and dry matter concentrations than occurred with the hay or high-salt concentrate diets. With the low-salt diet, there was massive lysis of bacterial cells and the intact cells were full of reserve carbohydrate granules and surrounded by thickened fibrillar capsules. The flow of water through the rumen was highest on the lucerne hay diet and least with the low-salt concentrate diet[104].

Bovine spongiform encephalopathy

The first clinical signs of this disease were observed in April 1985. It occurs nearly always in Friesian-Holstein cattle aged between 3 and 6 years in widely separate geographical locations of England[191]. The condition has an insidious onset in healthy cattle in good bodily condition; the afflicted cattle show mild incoordination of gait, muscular weakness and become apprehensive and hyperaesthetic, i.e. show increased sensitivity. Thus the affected cattle exhibit fear and aggressive behaviour whilst auditory stimuli produce an exaggerated response, even causing falling-over. Eventually frenzied behaviour and unpredictability in handling or recumbency necessitate slaughter in 1–6 months after the first clinical signs. On rare occasions, death occurs unexpectedly. Histological examination shows a bilaterally symmetrical degenerative change involving vacuolation in certain localities in the brain stem grey matter. The condition is thought to be caused by an unconventional transmissible agent – a prion protein[195] – similar to that causing scrapie[194] in sheep and to have a very long incubation period of several years; it is claimed that it can be transmitted from cows to the calf at birth[192].

A prion protein is a small '*pro*teinaceous *in*fectious' particle which is resistant to inactivation by most procedures that modify nucleic acids[195]. Replication of the prion protein may occur as a result of unmasking of a binding site. Prion proteins have also been implicated[198] in Creutzfeldt Jakob disease, a progressive pre-senile dementia of man in which 15 per cent of cases were found to be familial[196] and also in Gerstmann-Strausler syndrome[197], which is a hereditary transmissible dementia in man.

In June 1988, under the Animal Health Act 1981, the Bovine Spongiform Encephalopathy Order (SI 1039) was issued. This made the disease notifiable to the Divisional Veterinary Officer by any person in charge of an affected or suspected to be affected animal or the carcass of such an animal and required the animal or carcass to be retained on the premises until examined by a veterinary inspector. The carcass was defined as meat, blood, bones, hair, hide, horn, hooves, offal and the intestinal contents of a bovine animal.

The Order gives powers to restrict movement from the premises except under licence of other bovine animals that might risk the spread of the disease. The sale, supply and use of feed containing animal protein for ruminating animals is prohibited as is the feeding of such feedstuff. 'Animal protein' is defined as derived from a carcass of an animal, but did not include milk or milk products or dicalcium bone phosphate. Fishmeal appears to be excluded from the prohibition and it would seem that poultry carcasses are also excluded. 'Ruminating animals' was all embracing and included cattle, sheep, goats, deer and all other ruminating animals. Cleaning and disinfection of the premises and equipment where a suspected animal or carcass had been present within the previous 56 d could be ordered and had to be carried out at the owner's expense.

In December 1988, to reduce the risk to human health of the disease, bovine spongiform encephalopathy was designated a zoonosis under the Zoonoses Order 1988 (SI 2264) of the Animal Health Act 1981 and was subject to its provisions (see also Zoonoses Order 1975 in Chapter 4).

On 30 December 1988, SI 1039 Order was revoked and re-enacted as Bovine Spongiform Encephalopathy (No. 2) Order 1988 (SI 2299). This new Order included the provisions of the old Order but also enabled animals to be slaughtered on account of the disease.

A new Article prohibited the sale or supply of milk for human consumption or feeding to an animal, or use in manufacture of any product for sale or supply for human consumption from milk of an affected or suspected to be affected animal. A recipient animal in this context was any kind of mammal, except man, and any kind of four-footed beast, which is not a mammal, and any bird.

However, suckling its own calf by an affected or suspected to be affected animal was not prohibited nor was the use of such milk for feeding to animals in research establishments under the authority of a licence issued by a veterinary inspector.

In June 1989, it was reported that the use of bovine brain, thymus, spleen, spinal cord, tonsils and intestines in foods for humans would be banned. By October 1989, the statutory instrument enacting the ban under the Food Act 1984 had not been promulgated because responses from the food industry were being considered.

Calf diphtheria

Symptoms

This disease, caused by the organism *Fusiformis necrophorus*, attacks calves usually before they are 6 weeks old, and may occur as early as 4 d from birth. It is characterized by the formation of ulcers on the mucous lining of the mouth and pharynx. The base or sides of the tongue, the inside of the cheeks adjacent to the teeth and the lips are infected first, and the lesions may then spread throughout the mouth and nasal passages, and down the trachea to the lungs. The destroyed tissue is greyish-yellow in colour, and around the edge of an ulcer the mucous lining is slightly raised, reddened and granulated.

The infected calf will tend to slobber, and swellings may be seen at the side of the cheek or in the throat region. Where the lesions involve the nasal and respiratory passages, laboured breathing, coughing and a sticky, greenish-yellow discharge from the nostrils will be apparent. The swollen tongue may also protrude from the mouth and an offensive odour is often emitted. In addition, the calf loses its appetite, goes back in condition and may show signs of diarrhoea. The mortality rate is usually high, acute cases dying within a week, whereas a calf suffering from a milder attack may survive for a number of weeks.

Treatment

When the disease has been diagnosed on a farm, the sick animals should be segregated, and the calf pens and utensils thoroughly disinfected. All healthy animals should be examined each day for the early detection of new cases. Veterinary advice should be obtained for the treatment of the sick animals.

Cerebrocortical necrosis

The condition of cerebrocortical necrosis occurs on a variety of diets, including pelleted grass meal, pasture with additional concentrates and also, apparently, on conventional hay and concentrate diets. Some of the symptoms are similar to those of hypomagnesaemia. The first signs are anorexia, followed by apparent blindness, staggering and tetanic spasms. Within 12–72 h of the onset of the symptoms the calf collapses, the ears droop and in the final stages the limbs and head are extended. General twitching of the musculature of the ears and eyelids, weaving of the head and neck and grinding of the teeth with groaning may occur; there may also be a brown diarrhoea. Death usually occurs within 3–6 d. The main lesions in such calves are necrotic areas in both cerebral hemispheres[69,70,74].

Injections of thiamin appear to cure the condition if given at an early stage[71,75,76]. It appears that there is no failure in the synthesis of thiamin in the rumen in affected animals, and thus the condition must be associated with poor absorption of thiamin or interference in its metabolism within the tissues[72]. It has been suggested that there is a similarity between the symptoms and those of salt-poisoning, affected calves having lower serum potassium values (3.7 mM) than normal (4.8 mM)[73].

The condition has been produced in preruminant calves by oral or parenteral administration of the thiamin antimetabolite amprolium, whilst with simultaneous administration of thiamin the calves remained healthy[77]. The rumen contents of affected calves have been shown to contain thiaminase[78], which is thought to be of bacterial origin. The insertion of a thiamin antagonist in the rumen has also produced the condition. It has been suggested that the condition may possibly be associated with the consumption of mouldy feeding stuffs[193]. One outbreak was associated with calves having access to mouldy bales of straw which were being used as a calf pen wall. The main fungus isolated was *Acrospeira macrosporoides* (Berk.), which has been shown to produce thiaminase[162].

Because of the high dosage of thiamin antagonist necessary, the condition is considered by some workers not to be a simple thiamin deficiency[79,80] but a thiamin analogue-induced deficiency[79]. It has also been suggested that cattle may have a greater need for thiamin when fattened on diets of high carbohydrate content[74]. Feeding large amounts of molasses was claimed to cause cerebrocortical necrosis, especially when there was no forage in the diet, but the condition did not respond to thiamin supplementation either orally or parenterally[82, 83]. It was suggested that owing to a change in rumen fermentation, resulting in low rates of propionic acid production, damage to the brain resulted from an inadequate supply of glucose[83].

Cobalt deficiency

Symptoms

Cobalt is necessary for the synthesis of vitamin B_{12} by rumen bacteria. A deficiency of cobalt thus causes a deficiency of vitamin B_{12}, a condition which prevents the animal from metabolizing propionate into succinate and causes the accumulation of methyl malonate[94]. Vitamin B_{12} is also associated with iron and copper in haematopoiesis (the formation of blood cells). A serum vitamin B_{12} concentration between 38 and 76 μmol/l would indicate only a marginal deficiency, whereas a concentration of methyl malonic acid in plasma >5 μmol/l would be a better guide to diagnosis of a B_{12} deficiency.

Cobalt deficiency occurs in many parts of the world, including Australia, Kenya, New Zealand, Scotland and the USA. The deficiency results in a decline in appetite, in severe emaciation and wasting of the musculature (marasmus or 'pining'), and in anaemia, which may be of the same type as in iron deficiency[95]. Impaired immunity to *Ostertagia* infection in cobalt-deficient cattle has been reported (see Chapter 8).

A level of 40–60 μg Co/kg dry matter in the livers of cattle indicates a cobalt deficiency compared with normal values of 80–120 μg/kg or more[96]. However, it is possible to maintain cattle on cobalt-deficient pastures by injection with vitamin B_{12}, without raising the liver cobalt content to normal. A better indicator of cobalt deficiency is therefore the vitamin B_{12} level of the liver, which in normal animals should contain more than 0.3 μg/g fresh liver to ensure optimum growth[97].

In deficiency areas the cobalt status of the pasture and soil is low (e.g. for herbage under 80 μg/kg dry matter). Calves born to cobalt-deficient cows are weak and do not generally survive for more than a few days[98]. Supplementation of the dam increases the cobalt content of the tissues of the calf, especially the liver and the kidneys. The cobalt content of newborn calves is from 0.1 to 0.2 mg/kg dry matter, although skin and hair have a much higher level of 1 mg/kg dry matter[99]. However, a cobalt level in black hair of cattle of 5 μg/kg has been considered to indicate a borderline supply[100]. Supplementation of the dam increased the cobalt content of colostrum about 100-fold[99].

Prevention and treatment

When a cobalt deficiency is diagnosed it can be overcome by feeding calves 3.5 ml (one teaspoonful) of a cobalt sulphate solution (6 g/l) or, alternatively, by top-dressing the pastures with 2.2 kg/ha of cobalt chloride or sulphate mixed in with normal fertilizers. Another method of counteracting cobalt deficiency is the supplementation of cattle by the use of cobaltic oxide pellets of high specific gravity ('bullets') which usually remain in the reticulum[97]. If given in early pregnancy, the bullets have been shown to increase the liver storage of cobalt and vitamin B_{12} in the newborn calf, to improve growth rate during the first 7 weeks of life, but not to increase birth weight[101]. These bullets were found to be unsuitable for calves under 6 weeks of age, as the bullets were not satisfactorily retained. With a diet of sorghum grain and silage containing less than 20 μg Co/kg dry matter, an intraruminal cobalt oxide pellet improved weight gain from 1.22–1.38 kg/d[102].

More recently boluses of controlled-release glass containing about 60 g cobalt have been tested on 22 steers, which were slaughtered between 17 and 145 d after dosing. Boluses released about 0.85 mg cobalt daily. In both the treated and untreated animals, serum and liver B_{12} concentration was at the upper end of the normal range[16].

For further information on cobalt, see *The Calf,* 5th edn, vol. 2.

Copper deficiency

Copper deficiency tends to occur on fenland and sandy and peaty soils throughout Great Britain. It has been identified in such areas as Caithness[19], Cheshire, Lancashire[20], Lincolnshire[159], Shropshire, S. Humberside[159], Staffordshire[57], the Fens and the Thames Estuary.

For more information on copper the reader should refer to *The Calf,* 5th edn, vol. 2.

Symptoms

The symptoms of copper deficiency when calves are at pasture vary somewhat in different areas, but the dams are usually copper-deficient during pregnancy, and in beef herds the symptoms appear in the calves at 2–3 months of age, or about 1 month after they are at pasture. Dairy calves that have received concentrates do not usually show symptoms until they are at pasture in their first or second year.

Figure 7.1 Copper deficiency in an Aberdeen Angus calf showing pining and depigmentation of hair around the eyes[19]. (By courtesy of Cambridge University Press and by permission of the Controller of Her Majesty's Stationery Office. Crown Copyright)

Copper deficiency has also been reported in veal calves, and has been associated with the chewing of wooden veal crates, and a depression in growth rate after 70 d of age[23].

The first sign of copper deficiency is a stilted gait, particularly of the hind legs, followed by a progressive loss of condition during the next 4 or 5 months, which may result in extreme emaciation and death. In black- or brown-coloured breeds the hair around the eyes becomes grey or dun-coloured, and the pigmentation frequently extends down the forehead and lower jaw (see Figure 7.1). This depigmentation occurs as a result of the failure of melanin formation, for which the copper-containing enzyme tyrosinase is required. In some cases, but not in all, diarrhoea occurs.

Concentrations of copper in liver and blood

Normal calves have a liver copper concentration of about 100 mg/kg dry matter, whereas deficient calves may have liver copper concentrations ranging from 2–16 mg/kg dry matter. The copper concentration in the liver of newborn calves, the main storage organ, is unevenly distributed throughout the lobes[34] and is about 4–8 times greater than that in the liver of the dam[25]. During the first 6 weeks of life there is a greater selective accumulation of absorbed copper in the liver than occurs in the older animal.

Similarly, deficient calves have blood copper levels below 0.6 mg/l[26], the normal range being 0.7–1.3 mg/l. Plasma copper concentration in the newborn calf (0.27 mg/l) is lower than in the dam (1.00 mg/l), but increases to near-adult values by 1 week. A plasma copper concentration of <9 μmol/l (0.6 mg/l) is an indicator of a marginal deficiency but the concentration may have to fall to <3 μmol/l (0.2 mg/l) before there is a risk of dysfunction and loss of productivity in cattle[14].

Single-suckled calves with normal blood copper concentrations but showing reddish coat colour, anaemia and louse infestation have been reported as responding markedly to oral supplementation with copper sulphate solution[36]. Thus, the test of whether the diagnosis of copper deficiency is correct rests on whether there is an improvement in growth and health when a copper supplement is given.

In Lincolnshire and S. Humberside, where hypocupraemia is common, higher concentrations of copper in the serum occurred in dairy than in beef herds. Moreover, there was a relationship with rainfall; the higher the monthly rainfall, the lower the serum copper concentration. Mean serum copper concentrations were frequently below 0.7 mg/l[159].

Plasma copper concentration, which reflects changes in rate of ceruloplasmin (a blue copper-containing α-globulin in plasma) synthesis by the liver, is considered a more sensitive criterion of copper status than is whole blood copper[35]. Recently, the diagnosis of copper deficiency has been improved by the assay of erythrocyte superoxide dismutase (EC.1.15.1.1)[199] as well as that of copper status[158].

Although there is a wide genetic variation in copper absorption between different breeds of sheep, this has not so far been reported for different breeds of cattle[158].

Concentration of copper in hair

The copper concentration in white and red hair was found to increase with increase in copper intake but that of black hair gave inconsistent results[33]; white hair was considered the most useful for determining mineral status. However, less than 7 mg Cu/kg black hair is considered to indicate a borderline copper intake[100]. Although a positive correlation exists between the concentration of copper in blood and in hair[24], it has been suggested that copper and manganese levels in hair are too variable to be effective for diagnosis[37].

Veal calves

With intensively reared preruminant calves in Germany, daily intakes of 2.2 mg copper resulted in signs of copper deficiency after 70 d of age, the liver reserves at 140 d of age being 17.6 mg/kg dry matter. A daily intake of 7.7 mg copper was considered sufficient to meet the requirement of the calf up to this age. A reduced copper concentration in kidney, blood and hair was found at the lower intake of copper, but no difference in the copper concentration in the muscle[23]. In the USA, with calves given whole milk diets, a supplement of 6 mg copper daily maintained normal serum values, but had no effect on the haemoglobin levels of anaemic and normal calves[32], in spite of the known interaction between copper and iron in haemoglobin synthesis in other species.

When veal calves were given no supplementary copper, and the milk powder contained 0.8 mg Cu/kg powder, the calves became copper-deficient. With a supplement of 30 mg Cu/d, about 50 per cent was retained in the liver[38]. In contrast, in a further experiment a supplement of 5 mg Cu/kg to a milk powder, which contained 0.5 mg Cu/kg, had little effect on growth or haematological status, although there was a significant interaction between copper and iron intake on mean corpuscular volume and on mean corpuscular haemoglobin concentration. There was no evidence of copper deficiency as measured by blood and liver copper concentration or by ceruloplasmin and cytochrome oxidase (EC 1.9.3.1), but the calves did not start on the experiment until 16 d of age[39].

Calves at pasture

Copper deficiency is most likely to occur in grazing animals because of the low availability of copper in lush grazed pasture compared with that in conserved forage. Small increases in herbage molybdenum and sulphur affect absorption of copper and there is also antagonism from iron ingested in the soil[158].

Signs of copper deficiency after the first month of life are associated in Australia and New Zealand with diets containing less than 5 mg Cu/kg dry matter and with those which contain an excess of molybdenum or insufficient phosphorus. Similarly, in Florida yearlings and heifers suckling calves suffer from a nutritional anaemia associated with a combined copper and iron deficiency, when grazing the natural vegetation[21]. In Great Britain levels of less than 5 mg Cu/kg dry matter are rarely found in pasture. However, when grass is growing fast in the spring, its copper content tends to be at its lowest, whereas its molybdenum content is at its highest. Moreover, with diets containing high levels of inorganic sulphate, molybdenum, even at a low concentration, will interfere with the assimilation of copper[22].

In Lincolnshire and S. Humberside, an area of marginal copper status, the herbage samples contained 4.4–9.8 mg Cu/kg dry matter, which may be compared with the official recommendation of the Agricultural Research Council[160] that dietary herbage should contain at least 8 mg Cu/kg dry matter to avoid deficiency.

Calculated from the formula that takes into account the molybdenum and sulphur in the herbage[161], the available copper was 0.19–0.37 mg/kg dry matter.

Nevertheless, it is suggested that adoption of preventative measures should be prompted by biochemical evidence from the cattle rather than that from the soils or pasture[14].

Moreover, high intakes of molybdenum (e.g. 50 mg/kg dry matter) by heifers will give no symptoms of molybdenum toxicity, but addition of 0.3 per cent sulphate sulphur will cause molybdenum toxicity, which can be corrected by the addition of copper[40]. It is thought that molybdenum and sulphate depress copper storage[41] by the formation of a Cu–Mo complex unavailable for metabolism[42, 43]. Severe clinical copper deficiency has been found in N. Staffordshire associated with an excess of molybdenum in the pasture and soil[57].

The diarrhoea which occurs in cattle on 'teart' pastures of high molybdenum content may result from the formation of thiomolybdates by the rumen microflora; thiomolybdates are known to cause diarrhoea in rats. The curative effect of copper sulphate could result from the production of insoluble copper thiomolybdates[56].

A very high iron intake (1 g/kg dry matter) was found to depress absorption of copper and decrease liver reserves of copper[44]. When cattle are markedly copper-deficient, liver iron increases greatly[45] and it has been suggested that copper is required for release of iron from storage sites in the liver[46]. High levels of copper in the ration have been shown to increase the manganese concentration in hair, but they did not affect sodium, zinc, calcium, phosphorus, magnesium, iron or potassium concentration. With veal calves a depressing effect on copper status by supplementing the diet with zinc sulphate has been reported[47]. This could be prevented by addition of soyabean oil meal to the milk substitute.

Copper and immunity

Increased susceptibility to infection and growth retardation in lambs and infertility in cattle have been associated with copper deficiency[158]. In particular, neutrophils from cattle with copper deficiency induced by a high concentration of molybdenum (0.052 mmol/kg diet) or of iron (8.95 mmol/kg diet) had an impaired ability to kill ingested *Candida albicans*. A restricted intake of diet (0.8 of *ad libitum*) also decreased neutrophil candidacidal activity. Ingestion of *C. albicans* by neutrophils was impaired by molybdenum and iron but not by the restricted intake. The changes in neutrophil function and the severity of the copper deficiency induced by molybdenum and iron were greater than those induced by diets of low copper content[156].

Prevention

Normal development of calves in copper-deficient areas has been obtained by administering copper. If it is suspected that calves are showing symptoms of copper deficiency, veterinary advice should be obtained. In the past it was usually recommended that a drench containing 2 g copper sulphate should be given weekly, although some authorities advocated 1 g copper sulphate daily. The amount and frequency of dosage depends on the age of the animal and how severely the animal is depleted.

The most suitable supplement is cupric sulphate, but in the ruminant animal uptake of copper has been found to be higher for cupric chloride than for the sulphate or nitrate. Cupric carbonate compared favourably in its absorption, but cupric oxide was poorly absorbed[54]. Although cupric carbonate had a high absorption, there was also a high excretion in faeces and urine. Cuprous oxide was fairly well absorbed[55].

Nevertheless, a comparison of giving 24 g copper oxide needles, presumably cupric oxide, or a subcutaneous injection of 100 mg copper as calcium copper edetate at the start of the grazing season showed that plasma copper concentration was significantly higher at the end of the season when the copper oxide needles were used. The needles also gave an adequate, but not an excessive, liver copper concentration[163].

Different doses of cupric oxide particles have been compared in crossbred steers of 220 kg live weight given a diet of barley and hay *ad libitum*. They received a single oral dose of either 0, 5, 10, 20 or 40 g cupric oxide particles. A dose of 5 g increased liver copper stores for 240 d and higher doses increased liver stores for longer but 40 g was no more effective than was 20 g (85 mg cupric oxide pellets/kg live weight). The variation among individuals was large, but the highest liver copper concentration (7.59 mmol/kg dry matter, i.e. 482 mg/kg dry matter) produced no evidence of copper toxicity.

The shape of the particles – clumps, short rods and long rods – when given at a rate of 5 mg/kg live weight to steers of 173 kg live weight did not affect the retention in the alimentary tract or the accumulation of copper in the liver[157].

Alternatively, prevention of copper deficiency in calves has been achieved by drenching the dam with 10 g copper sulphate every 2 weeks during pregnancy. In these circumstances there should be no need to supply copper to the calves during the subsequent grazing season[27,28]. Spring-born calves from cows given copper supplementation during the preceding winter gained weight at 1 kg/d, compared with 0.7 kg/d for calves of untreated cows[29].

Copper may be given in the form of copper glycine by subcutaneous injection into the brisket. It has been recommended that an injection of 120 mg copper should be given to the dam in the last 2 months of pregnancy, followed thereafter by injection of the calf with the same amount at 4 months of age, and if necessary a further injection 4–6 months later[26]. Copper glycine (120 mg Cu), given in a single injection to suckling calves at pasture, has been shown to increase weight gain during the summer on a farm where there was a previous history of copper deficiency[30]. Occasionally, large swellings occur both in calves and in adults after injection with copper glycine, reaching a maximum after 4–5 d and subsiding within 3–5 weeks after injection, but they apparently leave no after-effects[26]. After a single injection of 100 mg copper as copper calcium edetate at the beginning of the grazing season in dairy young stock, mean weight response was 10–70 per cent, representing an increase in weight of 14–32 kg per animal over a 6-month grazing period[58]. Alternative methods that have been used for administering copper to calves include access to copper-rich mineral mixtures and the spraying of pastures with 6–11 kg copper sulphate/ha. This latter method is effective when the pasture is deficient in copper, but not always satisfactory if the copper deficiency is a 'conditioned' one.

Soluble glass boluses containing copper given to cattle of 160–350 kg live weight 29 d before turn-out to a copper-deficient pasture resulted in a significantly higher mean plasma copper concentration of 16.1 μmol/l (1.0 mg/l) at 81 d after turn-out compared to 12.8 μmol/l (0.8 mg/l) for the control cattle. The treated group obtained sufficient copper for the 240 d trial, but their increase in weight gain was not significant[103].

Copper toxicity

Before copper is administered to calves, it is essential that copper deficiency should be confirmed, owing to the risk of poisoning. If spraying of pastures is adopted, care must be taken if sheep also graze the land because of their greater susceptibility to copper poisoning. Some 'denaturation' processes used on skim milk powder to enable an EEC subsidy to be obtained include addition of copper sulphate and these skim milk powders should not be used in calf diets.

Chronic copper poisoning occurs in two phases: a period of accumulation of copper in the liver with no toxic symptoms, which may vary from a few weeks onwards; and a toxic phase with acute illness and death in 2–6 d with jaundice, methaemoglobinaemia and haemoglobinuria[48]. Tissue copper values in excess of 150 mg/kg fresh liver or 15 mg/kg fresh kidney are considered to be indicative of poisoning[49].

In preruminant calves, copper toxicity occurred when a milk substitute containing 115 mg Cu/kg dry matter was fed[31]. However, with newborn Holstein calves receiving a restricted quantity of milk substitute, up to 200 mg Cu/kg dry matter was included without ill-effect for 35 d. No significant increase in weight gain occurred and blood haemoglobin concentration appeared to be depressed[53]. In ruminant calves very much higher amounts (up to 900 mg Cu/kg dry matter) have been fed as a growth promoter. Significant increases in weight gain have been reported with amounts varying from 250 to 900 mg Cu/kg dry matter in some but not in all experiments[50–52]; liver copper levels were markedly increased yet no toxicity was reported[50–52].

Ergot poisoning

This arises from the ingestion of the sclerotia of the fungus *Claviceps purpurea* which is a parasite of cereals, particularly rye, and also of many species of grass. The alkaloids present in the fungus cause either convulsant symptoms or vasoconstrictive lesions leading to gangrene, usually in the hind limbs causing necrosis as evidenced by an indented ring below the knee or hock.

In serious cases the hooves may be shed and occasionally lesions may be observed on the tips of the ears and tail.

Mycotic abortion

In Scotland, 11 out of 36 suckler cows in late pregnancy aborted their calves 7–10 d after introduction to a ryegrass heavily infested with ergot[124].

Footrot

Footrot is normally associated with *Fusiformis necrophorus,* which causes necrosis of the skin and the underlying tissue in the cleft between the two claws of the hoof. However, more than one organism may be involved. In Missouri, the predominant organisms were *Fusiformis necrophorus* (*Sphaerophorus necrophorus*) and *Bacteroides melaninogenicus.* A mixed inoculum applied to scarified interdigital skin or inoculated intradermally into interdigital skin induced the typical lesions[59].

Hypomagnesaemia

Incidence

It has been known for a number of years that whole milk supplemented with iron, copper, manganese and vitamin D is inadequate as the sole diet of the calf. This inadequacy was found to be associated with a shortage of magnesium in the diet, resulting in hypomagnesaemia and tetany.

The condition usually occurs with calves growing at a rapid rate on very large amounts of milk or milk substitute and having no access to other supplementary feeds. The higher the rate of gain, the greater the depression in serum magnesium[64]. On a high level of liquid food, access to fibrous matter may exacerbate the risk of hypomagnesaemia because of the increased saliva production resulting in a greater loss of endogenous magnesium[84–86]. Moreover, hypomagnesaemia is considered to be an increasing problem in beef calves that are grazing during the 'spring flush'. The condition usually occurs during the first 3 months of life, during which time the magnesium concentration in the blood may be steadily falling. The ruminant calf, except when grazing, does not usually show signs of hypomagnesaemia because, although the availability of magnesium from dry food is lower than that from milk, the magnesium concentration in dry diets is usually about five times that of milk.

Symptoms

Affected calves show symptoms similar to those of cows suffering from lactation tetany[63]. The calf shows much excitement, the head is retracted, the whites of the eyes are very apparent, owing to an upward rotation of the eyeballs, and there is often continual movement of the ears. Walking is uncoordinated and the calf tends to exaggerate the lifting of its feet. When touched, the animal seems to be hypersensitive. In the final stages convulsions occur; the animal falls on its side with neck stretched out, the jaws clenched and all respiration movements stopped. Control of urination and defaecation is lost and the eyes alternately bulge and sink in their sockets (see Figure 7.2).

The normal plasma magnesium concentration in the calf is about 22–27 mg/l, and convulsions develop when the concentration has fallen to 3–7 mg/l[87]. Even at these low plasma levels of magnesium, only slight clinical signs may occur at first, because depletion of magnesium from the bone is occurring. Normal calf bone contains about 7–8 g magnesium/kg and convulsions are likely to occur when the concentration of magnesium in the bone has fallen to below 3.7 g/kg. The best criterion for confirming that a calf has died of hypomagnesaemia is given by chemical examination of the bone. Ratios of calcium to magnesium in the bones of affected animals are 90:1 or greater, compared with 60:1 or less in normal calves[88].

Prevention

Prevention of the condition can be achieved by feeding hay and concentrates or by the addition of 5 g magnesium oxide or 8 g magnesium carbonate/d. In one experiment[64] 0.5 g magnesium as the carbonate (i.e. 1.75 g magnesium carbonate) given daily largely prevented tetany in calves given large quantities of milk, but 2 g magnesium carbonate/d did not appear to be enough for animals over 2 months of age.

(a)

(b)

Figure 7.2 Calf showing first signs of magnesium deficiency. (*a*) Change in the carriage of the head and ears; and (*b*) opisthotonos (arching of the spinal column so that the head is thrown backwards)[63]

Treatment

In treating this condition, it is necessary to restore the depleted stores of magnesium in the bone as well as to increase the blood magnesium. The blood magnesium content can be restored by intravenous injection of about 100 ml of a solution containing 14 g calcium borogluconate and 3.6 ml magnesium lactate followed by 14 g magnesium sulphate in 50 ml of water given by subcutaneous injection. To restore the bone magnesium it is necessary to feed 14 g magnesium oxide/d in the milk or concentrate mixture for several weeks.

Lead poisoning

Symptoms

Cattle, and especially calves, seem to be particularly susceptible to lead poisoning. The commonest sources of lead are from painted woodwork and metalwork, discarded paint tins, old lead plates from batteries, painted tarpaulins and vegetation in orchards recently sprayed with lead arsenate solution.

The lethal dose for young calves is about 156–250 mg/kg body weight, but older calves may survive doses of this order. Calves tolerated 6 mg lead/kg body weight daily for a period of 3 years, but a single dose of 200–400 mg/kg may produce symptoms within 24 h[67, 68]. However, a time lag of several days may occur before the onset of symptoms, the animal then dying within a few hours. A healthy calf may be suddenly found dead, or it may have a fit from which it does not recover. In less acute cases of lead poisoning, the calf is dull and dejected and has no appetite. Following this the eyes become sunken and the coat staring, and there is evidence of abdominal pain and teeth grinding with salivation. There is often severe constipation and cessation of urination, and in some cases the constipation is followed by diarrhoea. In the next phase, which may be the first in acute cases, the calf becomes greatly excited, attempting to push against the walls of its pen, appears to be blind and staggers around with rolling eyes and frothing mouth. After collapse, muscular spasms and tetany occur before death.

Treatment

In acute cases there is no treatment that is likely to be entirely effective, but in subacute cases a purgative should be used. Calcium versenate* given daily for two periods of 6 d as a 25 per cent solution injected subcutaneously or intravenously, together with saline solution to reduce dehydration, has been partially successful in the cure of calves exhibiting the nervous symptoms resulting from lead poisoning[81]. The lead becomes complexed with the EDTA* and is eliminated in the bile and urine without apparently causing damage to the kidneys. Magnesium sulphate (Epsom salts) may also be of particular benefit by forming insoluble lead sulphate in the alimentary tract of the calf.

* Calcium versenate is the commercial name of the calcium disodium salt of ethylenediaminetetraacetic acid (Ca-EDTA).

Leptospirosis

The organism

In the USA, this condition in cattle is associated with the spirochaetes, *Leptospira pomona, L. grippotyphosa* and *L. hardjo* and is known as redwater of calves. *L. icterohaemorrhagiae,* a common parasite of rats, is usually associated with leptospirosis in dogs but has been responsible for outbreaks of the disease in calves in the UK[121–123].

Symptoms

The symptoms are pyrexia, prostration, inappetence, dyspnoea, jaundice, haemoglobinuria and anaemia. Haemolytic jaundice and haemoglobinuria occur in 50 per cent or more of the affected calves. *L. icterohaemorrhagiae* infection appears to be much less severe in calves than in dogs. In one outbreak, although the calves between 3 weeks and 3 months of age were jaundiced and emaciated, there was no evidence of diarrhoea or haemoglobinuria and the mortality rate was low[121]. Since *L. icterohaemorrhagiae* is pathogenic for man (Weil's disease), hygienic precautions must be taken by the person tending the sick calves.

Abortions

Leptospira interrogans has been associated with abortions and stillbirths in cattle. In N. Ireland infection by the organism (almost entirely caused by serovar *hardjo*) was diagnosed in 57 per cent of 505 calves (472 aborted fetuses, 20 stillborn calves and 13 perinatal deaths) during a 6-year period. There was a seasonal increase in September, October and December and the incidence of infection was significantly higher in fetuses aborted by dairy than by beef cows. The majority of abortions occurred from 6 months of gestation onwards. Cows that aborted infected fetuses had not previously shown overt signs of agalactia, i.e. absence of milk. There was no association between leptospiral infection and retention of fetal membranes[117].

Listeriosis

Listeriosis caused by *Listeria monocytogenes,* a Gram-positive small rod, is an uncommon cause of septicaemia and meningitis in older calves. The organism is pathogenic to man.

It has been reported that outbreaks are increasing both in frequency and size. Silage was found to contain the organism in excess of 12 000 organisms/g silage at the edge of a heavy-duty plastic sheet covering the silage. The top few centimetres of the silage were heavily contaminated[120].

Navel-ill or joint-ill

Symptoms

In this disease the infective agent appears to enter via the navel. The mortality rate is high, and in acute cases death is rapid with no specific symptoms. In less acute cases there is usually some swelling of the navel, with abscess formation, or the infection may spread to the liver with more serious effects. In addition, the calf loses its appetite, has a slightly elevated temperature and tends to be prostrated. The joints, especially the hocks and knees, usually become swollen and painful. Over a period of 8 years, 10 dairy calves varying in age from 3–12 months (mean approximately 4 months), either Ayrshire or Friesian × Ayrshire calves of both sexes, were reported from Scotland to have suffered from internal umbilical abscesses. It was considered that infection entered the umbilicus at birth and contaminated the urachus (the portion of the fetal membrane that forms the middle umbilical ligament postpartum), producing an abscess between the bladder and umbilicus. Most umbilical abscesses have external manifestations[116].

A number of different bacteria are associated with this condition; an organism of the psittacosis-lymphogranuloma venereum (PLV) group, i.e. a chlamydia, and a species of mycoplasma have been associated with polyarthritis in calves[65,66]. Gangrene in the hind limbs of calves with navel infection bacteraemia and also in systemic infection with *S. dublin* has been reported[114]. Similarly, a diffuse gangrene in the hind limbs with navel infection has been found in Tanzania[115].

Prevention

Where navel disease is prevalent on a farm, the buildings used for calving should be thoroughly disinfected and all navel cords should be dressed at birth.

Nutritional myopathy

Interactions between selenium, vitamin E (α-tocopherol) and other compounds

Calves may be affected by nutritional myopathy in one of three ways: first as a result of being suckled by cows that are deficient in selenium or vitamin E, a condition designated 'enzootic nutritional myopathy'; second, when older, receiving a diet deficient in selenium or vitamin E; and third, being given milk substitute diets in which the fat contains a large proportion of polyunsaturated fatty acids.

Thus, nutritional myopathy may develop spontaneously or be induced by substrates for peroxidation, such as unsaturated fatty acids or substances such as antibiotics, carbon tetrachloride, copper and paraquat, which establish peroxidation as part of the pathological process[13]. Excessive consumption of monensin (an ionophore antibiotic widely used to improve the efficiency of food use in cattle by altering the fermentation pattern in the rumen) through its interaction with selenium and vitamin E and its action on intracellular ion concentrations may have the potential to induce nutritional myopathy[12].

The interrelationship between vitamin E and other nutrients is complex, since there is a delicate balance between the synergistic effects of vitamin E, selenium, sulphur-containing amino acids and certain antioxidants, on the one hand, and the antagonistic effects of polyunsaturated fats and vitamin A in the diet, on the other.

For further details concerning selenium and vitamin E, see *The Calf,* 5th edn, vol. 2.

Symptoms

Nutritional myopathy or muscular dystrophy, is locally known by names such as 'white muscle disease', 'white flesh' and 'waxy muscle degeneration'. The factors affecting the incidence of nutritional myopathy have been reviewed by several workers[176,177]. The symptoms are those of a much weakened musculature, but vary according to which muscles are affected. The myopathy is bilaterally symmetrical[145]. There is often a general instability of gait, a tendency for the hind-legs to cross when the calf walks, 'winging' of the shoulder blades, straightening of the pastern joints and spreading of the claws of the feet, general weakness and inability to stand. Breathing, swallowing and the heart may also be affected. Death may be sudden, or the calf may linger on for some time. Wasting of the muscles does not, however, occur. The muscle tissues may be dark red and interspersed with greyish-white areas, and slight darkening of the body fat may also occur[176,178]. Very high concentrations of iron are found in the bloodstream of affected animals due to the massive transfer of iron from broken cells of the affected muscle to the bloodstream[179]. Red coloration of the urine may occur in older animals[180].

The level of serum aspartate amino transferase (AST) (EC 2.6.1.1)[199] activity has been used as a criterion for diagnosis of nutritional myopathy. In one experiment 31 per cent of untreated calves had AST activity greater than 2500 units/l compared with 2–5 per cent of the selenium-treated calves[141]. Similarly, in Canada all sick calves had values greater than 1100 units/ml but some calves from treated cows, although apparently healthy, also showed very high levels[137].

Clinically normal growing cattle may have selenium concentrations in the blood below that of calves showing clinical symptoms (i.e. <30 μg Se/l blood). These low blood selenium concentrations are associated with low blood glutathione peroxidase (GSH-Px) activities; this enzyme contains selenium[15].

The selenium status of a calf is considered adequate when blood concentrations are >0.1 mg/l, equivalent to 14 000 GSH-Px units/l, whereas calves are selenium-deficient when blood concentrations are <50 μg/l (6500 GSH-Px units/l)[1].

GSH-Px (EC 1.11.1.9)[199] is the only known active form of selenium in mammalian tissues. Its concentration in erythrocytes is positively correlated with the selenium concentration in whole blood[8]. A kit is available so that non-specialist laboratories can perform the assay for GSH-Px[9]. However, blood selenium and GSH-Px concentrations may be unreliable for diagnosis of a selenium-responsive condition, because other nutrients may determine what is adequate[14].

Although the selenium content of the hair of cows may indicate the likelihood of their progeny developing nutritional myopathy (>0.25 mg Se/kg for normal hair; <0.12 mg Se/kg for deficient animals), there is not such a good relationship for calves, which have higher values than those of cows[137].

Enzootic nutritional myopathy

It appears that enzootic nutritional myopathy is associated with areas of the world in which the soil content of selenium is low. The deficiency can be prevented by the administration of either small amounts of selenium or amounts of vitamin E that are large in relation to the normal requirement of the animal[136], the daily requirement for either one being increased if the other is deficient in the diet. In this condition the benefits of vitamin E do not seem to reside in its antioxidant properties[91], nor does the feeding of cystine to calves born from cows fed hay of low selenium content have any effect in preventing nutritional myopathy[181]. Similarly the feeding of linoleic acid (C18:2) to calves given a selenium-deficient diet does not affect the incidence of myopathy or inhibit the protective power of selenium[182]. In fact, it has been reported that the milk consumed by affected calves may contain less unsaturated fat than that consumed by healthy calves[151–153], although this is not always the case. An increased concentration of linoleic (C18:2) and arachidonic acids (C20:4) has been found in the muscle phospholipids of affected calves[164] and the vitamin E content of the muscles was reduced[165]. A high level of dietary sulphur may increase the requirement for selenium although this has not been confirmed[137]. Arsenic (1 mg/kg dry matter) in the form of sodium arsenate given to ewes on a selenium-deficient diet was found to reduce the incidence of nutritional myopathy in their lambs[166].

Effect of selenium deficiency in the dam

The vitamin E requirement of the calf can thus be increased by a selenium deficiency in the soil or crops. For instance, in N. Ontario, the incidence of the disease is severe where the forage contains less than 0.1 mg Se/kg dry matter[137]. Similarly, in an area of N.E. Scotland a high incidence of the condition occurred in single-suckled calves turned out to pasture in the spring and was associated with dams given a diet consisting mainly of straw and turnips during late pregnancy[138]. Serum tocopherol concentrations were 0.40 mg/l for affected calves, compared with 1.50 mg/l for normal calves[143]. Crops cut for silage in an immature state in selenium-deficient areas and fed to cows prepartum appear to give better protection to calves in spite of the calves' low plasma selenium levels than do crops cut at maturity, especially if they have a high legume content; this may be associated with the higher vitamin E content of the immature crop[167, 168]. Thus, it appears to be difficult to produce nutritional myopathy in calves from dams given a selenium-deficient diet unless the calves receive a low intake of vitamin E[167]. Supplementation of the dam appears to be equally as effective as supplementing the calf. In Canada supplements of both selenium and vitamin E were superior to a supplement of selenium alone[137]. Calves born from outwintered cows or from those given silage were rarely affected. Under conditions which resulted in inadequate amounts of vitamin E in colostrum, 150 mg α-tocopheryl acetate given daily to calves prevented the condition, whereas 10 mg was insufficient[139]. Further experiments showed that not less than 20 mg α-tocopheryl acetate daily would protect the majority of calves, whereas 0.25 mg Se/d was slightly more effective[140]. In testing out other methods of selenium administration, it was found that a single subcutaneous injection of 15 mg selenium at birth or 5 mg at 3-weekly intervals from birth afforded considerable but not as complete protection as 0.25 mg Se/d[141].

In certain areas of the world, particularly in Nebraska, grazing animals suffer from selenium poisoning[142]. Owing to the high toxicity of selenium, prophylactic treatment for nutritional myopathy should be given preferably in the form of vitamin E. By feeding 1 g α-tocopheryl acetate (about 2 mg tocopherol/kg live weight) prepartum to cows in N.E. Scotland, enzootic nutritional myopathy in the calves was largely prevented[139, 143]. Administration of selenium by implantation in the ear of a slow-releasing compound has been suggested[155]. A soluble glass bolus containing selenium raised the selenium status of cattle of 160–350 kg live weight, based on GSH-Px activity in the blood, between 81 and 240 d after insertion[103].

A study of the effect of an injection of barium selenate, which has a very low solubility, at 1 mg/kg live weight on GSH-Px activity of 13 Angus × Friesian heifers, aged 18–24 months and given a complete diet containing 0.07 mg Se/kg dry matter, has been made. The treated heifers and seven control heifers were slaughtered 119 d after injection. GSH-Px activity in blood increased within 4 weeks of administration and remained high thereafter, but there was no increase in activity in liver, kidney or muscle. The concentration of selenium in blood, liver and muscle increased significantly from 30 to 119 d. Between 76 and 99 per cent of the selenium injected remained at the site of injection[11].

Treatment of cereals with propionic acid or sodium hydroxide

Acute myopathy in yearling cattle between 13 and 20 months of age and weighing 225–320 kg after being turned out to pasture in the spring, has been associated in the UK with a deficiency of vitamin E and selenium in the food during the winter months[154]. The feeding of cereal grains preserved with propionic acid, which is known to reduce their vitamin E content, together with the stress of unaccustomed exercise and inclement weather, were considered to be predisposing causes of the condition. A marked increase in plasma creatine kinase (CK) (EC 2.7.3.2)[199] levels and a decrease in CK level in affected muscles occurred with animals showing clinical symptoms. The winter diet contained about 8 mg tocopherol/kg dry matter and about 40 μg Se/kg dry matter, resulting in intakes of about 0.17 mg vitamin E and 0.9 μg Se/kg body weight. However, exercise does not appear to contribute to the development of the disease[169], and in lambs controlled exercise delayed and reduced the severity of the condition[170].

In N. Ireland, a diet low in vitamin E and selenium and based on sodium hydroxide-treated barley, given to 14 14-week-old calves, resulted in a rapid decline in plasma α-tocopherol concentration, a gradual decline in GSH-Px activity, elevated plasma CK activity, teeth grinding and in three cases electrocardiographic changes. Necropsy at 127–137 d revealed skeletal myopathy in six calves and pale muscle in two calves. The control calves given the same diet together with α-tocopherol and selenium had normal plasma α-tocopherol concentrations and erythrocyte GSH-Px activities and no clinical abnormalities.

When protected linseed oil as a source of polyunsaturated fatty acids was given to the depleted calves, there was a rapid increase in plasma CK activity and linolenic acid associated with cardiopulmonary and locomotory signs; electrocardiographic changes and myoglobinuria occurred within 6–11 d. Necropsy revealed widespread severe skeletal myodegeneration and myocardial lesions with preferential involvement of the left ventricular myocardium[4]. In further work in N. Ireland, housed yearling cattle given a complete diet based on sodium hydroxide- treated selenium-deficient barley developed nutritional degenerative myopathy[10].

Selenium and ‘weak calf syndrome’ (see also Chapter 3)

A double blind control trial on four commercial dairy herds was made in N. Ireland to test the effect of a 50 mg injection of selenium 10 d prepartum on the incidence of weak calves. Although the treatment marginally increased the selenium status of treated calves, it did not decrease the incidence of ‘weak calf syndrome’[7].

It has been suggested that the reason why injections of sodium selenite 2 weeks prepartum had no effect on ‘weak calf syndrome’ was because the sodium selenite is soluble and gives only a short boost to the cow’s selenium status. It has been claimed that on farms with a history of high calf mortality, mortality was significantly reduced when cows were treated in mid-pregnancy with 500 mg selenium as barium selenate[6].

Nutritional myopathy induced by diets containing unsaturated fats

The inclusion of cod liver oil, lard and other fats, such as maize and linseed oil, containing a large proportion of polyunsaturated fatty acids in diets based on skim milk may cause nutritional myopathy[144, 145, 171]. Unlike enzootic nutritional myopathy, it is not preventable by supplementation with selenium[146, 176] although the disease may be postponed and the severity diminished[180]. On the other hand, certain antioxidants are effective, namely methylene blue, DPPD (N-diphenyl-*p*-phenylenediamine) and ethoxyquin (6-ethoxy-2,2,4-trimethyl-l,2-dihydroquinoline), while ascorbic acid also has some effect[176]. Butylated hydroxyanisole (5 mg/kg dry matter) did not prevent nutritional myopathy in calves given a milk substitute containing 20 per cent lard, or 10 per cent lard and 10 per cent beef dripping[149]. However, the addition of 20 mg D,L-α-tocopheryl acetate/kg dry matter prevented the condition and maintained similar serum tocopherol values in calves given diets containing lard to those in calves given diets containing beef dripping. Even so, the serum tocopherol levels resulting from these various diets may not have been really adequate for maximum performance, since a positive relationship was found between serum tocopherol level and weight gain, irrespective of the source of fat or of differences in initial live weight and food consumption.

In contrast also to the enzootic forms of the disease, the lipid fractions of the tissues of calves reared on a milk substitute containing maize oil and supplemented with vitamin E had decreased amounts of linoleic acid (C18:2) but increased amounts of arachidonic acid (C20:4) compared with supplemented calves, and it is suggested that vitamin E has an inhibiting effect on desaturation but not on chain elongation of fatty acids[172]. Heat production of affected calves is increased and is probably related to excessive nitrogen catabolism[145].

Where a shortage of vitamins A or D in early life is suspected, these vitamins should be given in the form of a proprietary concentrate rather than in the form of fish liver oil. Although cod liver oil contains a considerable amount of vitamin E, it is insufficient to protect the calf against the effects of the unsaturated fatty acids present in the oil. Cod liver oil given at a rate of 15–18 ml/d was sufficient to overcome the protective action of as much as 50 mg D,L-α-tocopheryl acetate. Similarly, to prevent nutritional myopathy in calves given a tocopherol-deficient diet supplemented with 30 ml cod liver oil, 200 mg water dispersible D-α-tocopheryl acetate was necessary[146]. It has been suggested that the antagonism between vitamin E and the polyunsaturated fatty acids of cod liver oil may arise because of the replacement of the vitamin E in the cells by such acids, or, alternatively, that vitamin E may be excreted through the bile or may remain in the tract longer and be destroyed in animals supplemented with cod liver oil[173].

Cod liver oil toxicity probably does not occur to the same extent in the ruminant calf, owing to the hydrogenation of unsaturated fats in the rumen[147]. However, a considerable amount of vitamin E may be destroyed in the rumen, especially when diets containing large amounts of maize are given. With a diet containing 80 per cent maize, 42 per cent of the vitamin E was destroyed, compared with an 8 per cent loss when the diet contained 20 per cent maize[174].

There is possibly some relationship between the magnesium status of the calf and nutritional myopathy. Cod liver oil when given to Guernsey calves as a supplement to whole milk resulted in a depletion of the body reserves of magnesium. Although this may have been due to the formation and excretion of magnesium soaps in the faeces, it was found that addition of either α-tocopherol or magnesium restored reserves to normal[148]. Similarly, it has been reported from Norway that the maintenance of normal serum magnesium levels by supplementation with 30 mg Mg/kg live weight had some effect in preventing nutritional myopathy in calves[150], although other results showed no relationship between serum magnesium levels and the incidence of the disease[64, 175].

Selenium or vitamin E and immunity

Vitamin E is involved in the immune response of the body. Moreover, selenium has been considered to have a preventive effect against diarrhoea and pneumonia[2], but this has been disputed[3, 5] (see Chapter 5, p. 148).

In the hope of reducing the impact of calfhood disorders on performance, supplementation of milk substitutes with vitamin E has tended to rise from 20 mg to 50 mg/kg powder, even when the diets contain a low proportion of polyunsaturated fats.

An experiment has been made in which veal calves were given 7 kg colostrum containing either a low or high Ig concentration followed by milk substitute based on whey and an antigen-free soya concentrate, supplying 50 per cent of the protein, supplemented with either 20 or 60 mg vitamin E/kg powder. Milk intake, live-weight gain (1.15 kg/d to 12 weeks) and food conversion rate were unaffected by Ig intake or level of vitamin E inclusion. Calves given 145 g IgG in their first feed of colostrum had a mean value of 26 ZST units at 2 d of age compared with 12 ZST units for those given 44 g IgG, the incidence of pyrexia to 3 weeks of age being greater for the calves given the low IgG intake. Plasma vitamin E concentrations were higher after 7 d of age for calves given the diet containing 60 mg vitamin E/kg and by 77 d the concentrations were 0.85 and 2.1 mg/l for calves given the diets containing 20 and 60 mg vitamin E/kg respectively. Since the additional vitamin E showed no benefit, it is considered that, under the good environmental conditions under which the experiment was made, 20 mg vitamin E/kg powder is sufficient in a milk substitute containing only a small proportion (20 g/kg) of polyunsaturated fatty acids[118].

Organophosphorus poisoning

Poisoning from an organophosphate pour-on lice medication has been reported; it arose because the operator did not follow the maker's instructions. The mean acetyl cholinesterase activity was 42.6 iu/l compared to 83 iu/l blood for normal cattle[18].

Photosensitization

Photosensitization, including skin lesions, has been observed in 13 suckling calves aged 15–40 d in Brazil. The dams were grazing guinea grass and *Brachiaria decumbens*. The calves were housed separately at night with access to perennial star grass. Lesions were observed in a dam of one of the calves. Photosensitivity was of a hepatogenic type (derived from the liver) and was thought to be caused by ingestion of *Pithomyces chartarum* spores by the cows which produced a toxin that passed into the milk. The symptoms disappeared when calves were protected from sunlight and did not recur in calves >60–70 d of age[17]. See also Chapter 8, p. 224.

Rickets

This disease of the young calf is characterized by a failure of the bones to calcify normally. The cause of the condition is a deficiency of vitamin D, or faulty proportions or insufficient amounts of calcium and phosphorus.

Vitamin D deficiency in the young animal causes rickets, even in the presence of adequate calcium and phosphorus[125–128]. The symptoms include a diminished appetite, the occurrence of digestive disturbance and a certain amount of tetany. These are followed by swelling and stiffness of the joints, a tendency for the long bones to be curved resulting in a typical stilted walk, and a humped back (see Figures 7.3 and 7.4)[129, 133].

On hill land there is a possibility that calves receiving no supplementary food may suffer from calcium and phosphorus deficiency resulting in rickets. Furthermore, a deficiency in phosphorus alone may also cause rickets, especially if the calf obtains excessive quantities of calcium at the same time. Moreover, a shortage of phosphorus in the presence of adequate calcium and vitamin D will also give rise to rickets[130]. Even by increasing the calcium intake of calves on diets deficient in vitamin D but adequate in phosphorus, the effect in reducing rickets is slight[128].

A deficiency of phosphorus, besides being associated with rickets, results in anorexia and even a depraved appetite, and the animal thus becomes emaciated.

It has been suggested that magnesium has a sparing effect on vitamin D[131] and that, if large amounts of vitamin D are given to preruminant calves before hypomagnesaemia develops, the concomitant appearance of hypocalcaemia with the fall in blood magnesium will be prevented[132].

A deficiency of vitamin D has also been shown to reduce the digestibility and

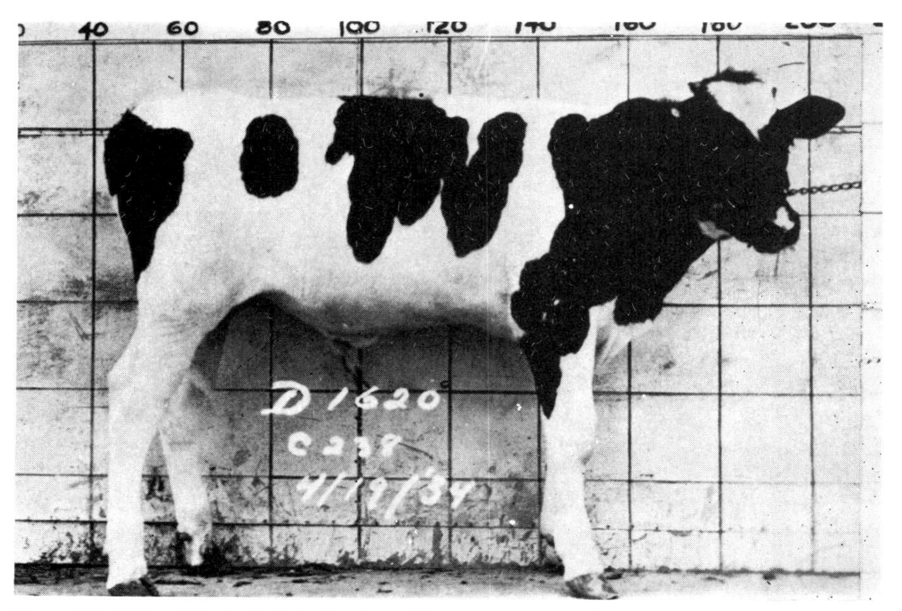

Figure 7.3 This calf had subclinical rickets. There were severe changes in the ribs and blood analysis but the leg bones show no clinical evidence as yet (after reference 129)

Figure 7.4 Calf suffering from severe rickets. Note emaciation, humping of back, swelling of joints, knuckling of pasterns and bowing of legs (after reference 129)

retention of protein as well as of ash and to increase basal metabolic rate[133]. High rates of weight gain may also increase the requirement of vitamin D per unit of live weight.

A deficiency of vitamin D in pregnant cows may result in bone abnormalities of calves at birth[134, 135].

For further details, including requirements of calcium, phosphorus and vitamin D, see *The Calf*, 5th edn, vol. 2.

Urinary calculi

In calves under native range conditions in S. Alberta, Canada, the commonest mineral constituents of urinary calculi are phosphates, oxalates and silica. Obstruction of the urethra by calculi composed largely of silica is an important cause of death among range calves[183, 184]. Even when in confinement, a high proportion of steer calves when given native range grass, as hay, developed siliceous calculi[185, 186].

Whereas dietary ammonium chloride has been shown to reduce the formation of phosphatic calculi in cattle[188], it does not affect the deposition of silica and appears to increase the deposition of oxalate[189]. However, provision of large amounts of sodium chloride caused an increase in water intake of up to 10 kg/d and prevented the formation of siliceous calculi[186], the increased water intake resulting in an increase in urine volume and a decrease in the concentration of silicic acid in the urine.

Provision of a barley-based supplement containing 20 per cent sodium chloride when given up to the time of weaning, or one containing 12 per cent sodium chloride given up to 90 d after weaning, increased the water turnover and reduced the weight of calculi in range calves compared to those of control calves that had access to block salt. However, although calculus formation was invariably low in calves with a high water turnover, it was not always high in calves with a lower water turnover[187].

Of the silica present, 7 per cent in prairie hay and 16 per cent in alfalfa hay were taken into solution in the reticulo-rumen. Of this, 12 and 45 per cent for prairie and alfalfa hay respectively were absorbed, and of that absorbed, 60–75 per cent of silicic acid from prairie hay and 40–60 per cent from alfalfa hay were excreted in the urine[190].

References

1. ANDERSON, P. H. *In Practice* (Suppl. to *Vet. Rec.*) **1**, 25 (1979)
2. HALL, T. J. *Vet. Rec.* **121**, 599 (1987)
3. PHILLIPPO, M., ARTHUR, J. R., PRICE, J. and HALLIDAY, G. J. *Vet. Rec.* **122**, 94 (1988)
4. KENNEDY, S., RICE, D. A. and DAVIDSON, W. B. *Res. vet. Sci.* **43**, 384 (1987)
5. PHILLIPPO, M., ARTHUR, J. R., PRICE, J. and HALLIDAY, G. J. *Vet. Rec.* **121**, 509 (1987)
6. CAWLEY, G. D. *Vet. Rec.* **120**, 47 (1987)
7. RICE, D. A., McMURRAY, C. H., KENNEDY, S. and ELLIS, W. A. *Vet. Rec.* **119**, 571 (1986)
8. RICE, D. A. and BLANCHFLOWER, W. J. *Vet. Rec.* **118**, 479 (1986)
9. ANDERSON, P. H., BERRETT, S. and PATTERSON, D. S. P. *J. comp. Path.* **88**, 181 (1978)
10. RICE, D. A. and McMURRAY, C. H. *Vet. Rec.* **118**, 173 (1986)
11. MALLINSON, C. B., ALLEN, W. M. and SANSOM, B. F. *Vet. Rec.* **117**, 405 (1985)
12. HOSIE, B. D. and ROLLO, D. G. *Vet. Rec.* **116**, 132 (1985)
13. McMURRAY, C. H. and RICE, D. A. *Ir. vet. J.* **36**, 57 (1982)
14. SUTTLE, N. F. *Vet. Rec.* **119**, 148 (1986)
15. ALLEN, W. M., PARR, W. H., ANDERSON, P. H., BERRETT, S., BRADLEY, R. and PATTERSON, D. S. P. *Vet. Rec.* **96**, 360 (1975)
16. ALLEN, W. M., SANSOM, B. F., MALLINSON, C. B., STEBBINGS, R. J. and DRAKE, C. F. *Vet. Rec.* **116**, 175 (1985)
17. FAGLIARI, J. J., PASSIPIERI, M. and OLIVEIRA, J. A. DE. *Archos bras. Med. Vet. Zootec.* **35**, 479 (1983)
18. WHELAN, H. M., MUSKETT, B. D., THOMPSON, T. R. and KENNEDY, B. W. *Vet. Rec.* **116**, 354 (1985)
19. JAMIESON, S. and ALLCROFT, R. *Br. J. Nutr.* **4**, 16 (1950)
20. MORGAN, D. E. and CLEGG, A. *N.A.A.S. q. Rev. 41*, 38 (1958)
21. NEAL, W. M., BECKER, R. B. and SHEALY, A. L. *Science, N.Y.* **74**, 418 (1931)
22. DICK, A. T. *Aust. vet. J.* **29**, 233 (1953)
23. ANKE, M. *Arch. Tierernähr.* **16**, 199 (1966)
24. RUSSOFF, L. L. *Bull. Fla agric. Exp. Stn* No. 356 (1941)
25. COMAR, C. L., DAVIS, G. K. and SINGER, L. *J. biol. Chem.* **174**, 905 (1948)
26. ALLCROFT, R. and UVAROV, O. *Vet. Rec.* **71**, 797 (1959)
27. FIELD, H. I. *Vet. Rec.* **69**, 788 (1957)
28. FIELD, H. I. *Vet. Rec.* **69**, 832 (1957)
29. FIELD, H. I. *St. vet. J.* **8**, 1 (1953)
30. MORGAN, D. E., CLEGG, A., BROOKSBANK, N. H. and McCREA, C. T. *Anim. Prod.* **4**, 303 (1962)
31. SHAND, A. and LEWIS, G. *Vet. Rec.* **69**, 618 (1957)
32. MATRONE, G., CONLEY, C., WISE, G. H. and WAUGH, R. K. *J. Dairy Sci.* **40**, 1437 (1957)
33. O'MARY, C. C., BELL, M. C., SNEED, N. N. and BUTTS, W. T. Jr. *J. Anim. Sci.* **31**, 626 (1970)
34. BINGLEY, J. B. and DUFTY, J. H. *Res. vet. Sci.* **13**, 8 (1972)

35. BINGLEY, J. B. and DUFTY, J. H. *Clin. chim. Acta* **24**, 316 (1969)
36. PEEK, I. S. *Vet. Rec.* **88**, 318 (1971)
37. CHAUVAUX, G., LOMBA, F., FUMIÈRE, I. and BIENFET, V. *Annls Méd. vét.* **109**, 174 (1965)
38. BREMNER, I. and DALGARNO, A. C. *Br. J. Nutr.* **29**, 229 (1973)
39. BREMNER, I. and DALGARNO, A. C. *Br. J. Nutr.* **30**, 61 (1973)
40. VANDERVEEN, J. E. and KLEENER, H. A. *J. Dairy Sci.* **47**, 1224 (1964)
41. HOGAN, K. G., MONEY, D. F. L. and BLAYNEY, A. *N.Z. Jl agric. Res.* **11**, 435 (1969)
42. DOWDY, R. P. and MATRONE, G. *J. Nutr.* **95**, 191 (1968)
43. DOWDY, R. P. and MATRONE, G. *J. Nutr.* **95**, 197 (1968)
44. STANDISH, J. F., AMMERMAN, C. B., PALMER, A. Z. and SIMPSON, C. F. *J. Anim. Sci.* **33**, 171 (1971)
45. CHAPMAN, H. L. Jr. and KIDDER, R. W. *Bull. Fla agric. Exp. Stn* No. 674 (1964)
46. MARSTON, H. R. and ALLEN, S. H. *Nature, Lond.* **215**, 645 (1967)
47. VAN LEEUWEN, J. M. and VAN DER GRIFT, J. *Versl. landbouwk. Onderz.* 731 (1969)
48. TODD, J. R. *Proc. Nutr. Soc.* **28**, 189 (1969)
49. BUCK, W. B. *J. Am. vet. med. Ass.* **156**, 1434 (1970)
50. FELSMAN, R. J., WISE, M. B., BARRICK, E. R. and HARVEY, R. W. *J. Anim. Sci.* **34**, 358 (1972)
51. FELSMAN, R. J., WISE, M. B., HARVEY, R. W. and BARRICK, E. R. *J. Anim. Sci.* **36**, 157 (1973)
52. GILL, R., LEE, D. D. Jr. and MARION, G. B. *J. Anim. Sci.* **37**, 343 (1973)
53. ST LAURENT, G. J., AMER, M. A., BARETTE, D. C. and BRISSON, G. J. *J. Anim. Sci.* **35**, 1135 (1972)
54. LASSITER, J. W. and BELL, M. C. *J. Anim. Sci.* **19**, 754 (1960)
55. CHAPMAN, H. L. Jr. and BELL, M. C. *J. Anim. Sci.* **22**, 82 (1963)
56. DICK, A. T., DEWEY, D. W. and GAWTHORNE, J. M. *J. agric. Sci., Camb.* **85**, 567 (1975)
57. THORNTON, I., KERSHAW, G. F. and DAVIES, M. K. *J. agric. Sci., Camb.* **78**, 157 (1972)
58. THORNTON, I., KERSHAW, G. F. and DAVIES, M. K. *J. agric. Sci., Camb.* **78**, 165 (1972)
59. BERG, J. N. and LOAN, R. W. *Am. J. vet. Res.* **36**, 1115 (1975)
60. BLAXTER, K. L., WOOD, W. A. and MACDONALD, A. M. *Br. J. Nutr.* **7**, 34 (1953)
61. WHITELAW, F. G., PRESTON, T. R. and MACLEOD, N. A. *Anim. Prod.* **5**, 227 (1963)
62. PRESTON, T. R., AITKEN, J. N., WHITELAW, F. G., MACDEARMID, A., PHILIP, E. B. and MACLEOD, N. A. *Anim. Prod.* **5**, 245 (1963)
63. BLAXTER, K. L., ROOK. J. A. F. and MACDONALD, A. M. *J. comp. Path. Ther.* **64**, 157 (1954)
64. BLAXTER, K. L. and SHARMAN, G. A. M. *Vet. Rec.* **67**, 108 (1955)
65. HUGHES, K. L., EDWARDS, M. J., HARTLEY, W. J. and MURPHY, S. *Vet. Rec.* **78**, 276 (1966)
66. STORZ, J., SHUPE, J. L., SMART, R. A. and THORNLEY, R. W. *Am. J. vet. Res.* **27**, 987 (1966)
67. ALLCROFT, R. *J. comp. Path. Ther.* **60**, 190 (1950)
68. ALLCROFT, R. and BLAXTER, K. L. *J. comp. Path. Ther.* **60**, 209 (1950)
69. TERLECKI, S. and MARKSON, L. M. *Vet. Rec.* **73**, 23 (1961)
70. SPENCE, J. B., STEVENS, A. J., SAUNDERS, C. N. and HARRIS, A. H. *Vet. Rec.* **73**, 28 (1961)
71. DAVIS, E. T., PILL, A. H., COLLINGS, D. F. VENN, J. A. J. and BRIDGES, G. D. *Vet. Rec.* **77**, 290 (1965)
72. LEWIS, G., TERLECKI, S., MARKSON, L. M., ALLCROFT, R. and FORD, J. E. *Proc. Nutr. Soc.* **26**, xiii (1967)
73. CLEGG, F. G. *Vet. Rec.* **78**, 505 (1966)
74. EDWIN, E. E. and LEWIS, G. *J. Dairy Res.* **38**, 79 (1971)
75. HENTSCHL, A. F., WALTON, J. F. and MILLER, E. W. *Mod. vet. Pract.* **47**, 72 (1966)
76. PILL, A. H. *Vet. Rec.* **81**, 178 (1967)
77. MARKSON, L. M., LEWIS, G., TERLECKI, S., EDWIN, E. E. and FORD, J. E. *Br. vet. J.* **128**, 488 (1972)
78. EDWIN, E. E., SPENCE, J. B. and WOODS, A. J. *Vet. Rec.* **83**, 417 (1968)
79. LITTLE, P. B. *Diss. Abstr. Int.* **B30**, 4449 (1969)
80. PASS, M. A. *Aust. vet. J.* **44**, 562 (1968)
81. HOLM, L. W., RHODE, E. A., WHEAT, J. D. and FIRCH, G. *J. Am. vet. med. Ass.* **123**, 528 (1953)
82. GEERKEN, C. M. and FIGUEROA, V. *Rev. cubana Cienc. agric.* **5**, 205 (1971)
83. LOSADA, H., DIXON, F. and PRESTON, T. R. *Rev. cubana Cienc. agric.* **5**, 369 (1971)
84. IVINS, L. N. and ALLCROFT, R. *Br. vet. J.* **126**, 505 (1970)
85. SMITH, R. H. *J. agric. Sci., Camb.* **56**, 105 (1961)
86. SMITH, R. H. *J. agric. Sci., Camb.* **56**, 343 (1961)
87. SMITH, R. H. *Nord. VetMed.* **16**, Suppl. 1, 143 (1964)
88. BLAXTER, K. L. and ROOK, J. A. F. *J. comp. Path. Ther.* **64**, 176 (1954)

89. JENSEN, R., PIERSON, R. E., BRADDY, P. M. *et al. J. Am. vet. med. Ass.* **169**, 524 (1976)
90. WELCHMAN, D. de B. and BAUST, G. N. *Vet. Rec.* **121**, 586 (1987)
91. DAMMRICH, K. In *Indicators Relevant to Farm Animal Welfare* (ed. D. Smidt), Martinus Nijhoff, The Hague, p. 143 (1983)
92. UNSHLEM, J., ANDREAE, U. and SMIDT, T. In *Welfare and Husbandry of Calves* (ed. J. P. Signoret), Martinus Nijhoff, The Hague, p. 70 (1982)
93. PEARSON, G. R., WELCHMAN, D. DE B. and WELLS, M. *Vet. Rec.* **121**, 557 (1987)
94. MARSTON, H. R., ALLEN, S. H. and SMITH, R. M. *Nature, Lond.* **190**, 1085 (1961)
95. FILMER, J. F. *Aust. vet. J.* 9, 163 (1933)
96. McNAUGHT, K. J. *N.Z. Jl Sci. Technol.* **30A**, 26 (1948)
97. SKERMAN, K. D., SUTHERLAND, A. K., O'HALLORAN, M. W., BOURKE, J. M. and MUNDAY, B. L. *Am. J. vet. Res.* **20**, 977 (1959)
98. MARSTON, H. R. *Physiol. Rev.* **32**, 66 (1952)
99. WARD, G. M., BENNE, E. J., WEBSTER, H. D., DUNCAN, C. W. and HUFFMAN, C. F. *J. Anim. Sci.* **8**, 632 (1949)
100. ANKE, M. *Arch. Tierernähr.* **17**, 1 (1967)
101. SKERMAN, K. D. and O'HALLORAN, M. W. *Aust. vet. J.* **38**, 98 (1962)
102. MORRIS, J. G. and GARTNER, R. J. W. *J. agric. Sci., Camb.* **68**, 1 (1967)
103. DRIVER, P. M., CARLOS, G. M., EAMES, C. and TELFER, S. B. *Anim. Prod.* **42**, 472 (1986)
104. CHENG, K. J., BAILEY, C. B., HIRONAKA, R. and COSTERTON, J. W. *Can. J. Anim. Sci.* **59**, 737 (1979)
105. HIRONAKA, R., CHENG, K. J. and BAILEY, C. B. *Can. J. Anim. Sci.* **54**, 733 (1974)
106. WARD, A. C. S., WALDHALM, D. G., FRANK, F. W., MEINERSHAGEN, W. A. and DUBOSE, D. A. *Am. J. vet. Res.* **35**, 953 (1974)
107. REID, C. S. W., GURNSEY, M. P., WAGHORN, G. C. and JONES, W. T. *Proc. N.Z. Soc. Anim. Prod.* **35**, 13 (1975)
108. COCKREM, F. R. M. *Proc. N.Z. Soc. Anim. Prod.* **35**, 21 (1975)
109. McINTOSH, J. T. *Proc. N.Z. Soc. Anim. Prod.* **35**, 29 (1975)
110. ARSENAULT, G., BRISSON, G. J., SEONE, J. R. and JONES, J. D. *Can. J. Anim. Sci.* **60**, 303 (1980)
111. GORRILL, A. D. L., McQUEEN, R. E., MacINTYRE, T. M. and NICHOLSON, J. W. G. *Can. J. Anim. Sci.* **54**, 727 (1974)
112. MACLEOD, N. S. M. *Vet. Rec.* **76**, 223 (1964)
113. MACLEOD, N. S. M. *Vet. Rec.* **83**, 101 (1968)
114. HOLLANDS, R. D. *Vet. Rec.* **118**, 646 (1986)
115. MBASSA, G. *Vet. Rec.* **116**, 662 (1985)
116. SHEARER, A. G. *Vet. Rec.* **118**, 480 (1986)
117. ELLIS, W. A., O'BRIEN, J. J., BRYSON, D. G. and MACKIE, D. P. *Vet. Rec.* **117**, 101 (1985)
118. STOBO, I. J. F., ROY, J. H. B., GANDERTON, P., PERFITT, M. W., ANDERSON, P. H. and DUNCAN, A. L. *Anim. Prod.* **40**, 533 (1985)
119. JAYNE-WILLIAMS, D. J. *J. appl. Bact.* **47**, 271 (1979)
120. FENLON, D. R. *Vet. Rec.* **118**, 240 (1986)
121. INGRAM, P. L., JACK, E. J. and SMITH, J. E. *Vet. Rec.* **64**, 865 (1952)
122. FIELD, H. I. *Proc. R. Soc. Med.* **42**, 719 (1949)
123. FIELD, H. I. and SELLERS, K. C. *Vet. Rec.* **62**, 311 (1950)
124. APPLEYARD, W. T. *Vet. Rec.* **118**, 48 (1986)
125. THOMAS, J. W. and MOORE, L. A. *J. Dairy Sci.* **34**, 916 (1951)
126. RUPEL, I. W., BOHSTEDT, G. and HART, E. B. *Res. Bull. agric. Exp. Stn Univ. Wis.* No. 115 (1933)
127. GULLICKSON, T. W., PALMER, L. S. and BOYD, W. L. *Tech. Bull. Minn. agric. Exp. Stn* No. 105 (1935)
128. THOMAS, J. W. *J. Dairy Sci.* **35**, 1107 (1952)
129. BECHTEL, H. E., HALLMAN, E. T., HUFFMAN, C. F. and DUNCAN, C. W. *Tech. Bull. Mich. agric. Exp. Stn* No. 150 (1936)
130. THEILER, A. *Vet. J.* **90**, 143 (1934)
131. HUFFMAN, C. F. and DUNCAN, C. W. *J. Dairy Sci.* **18**, 605 (1935)
132. SMITH, R. H. *Biochem. J.* **67**, 472 (1957)
133. COLOVOS, N. F., KEENER, H. A., TERRI, A. E. and DAVIS, H. A. *J. Dairy Sci.* **34**, 735 (1951)
134. WALLIS, G. C. *J. Dairy Sci.* **21**, 315 (1938)

135. WALLIS, G. C. *Bull. S. Dak. agric. Exp. Stn* No. 372 (1944)
136. BLAXTER, K. L. *Proc. Nutr. Soc.* **21**, 211 (1962)
137. HIDIROGLOU, M., CARSON, R. B. and BROSSARD, G. A. *Can. J. Anim. Sci.* **45**, 197 (1965)
138. SHARMAN, G. A. M. *Vet. Rec.* **66**, 275 (1954)
139. BLAXTER, K. L. and McGILL, R. F. *Vet. Revs Annot.* **1**, 91 (1955)
140. SHARMAN, G. A. M., BLAXTER, K. L. and WILSON, R. S. *Vet. Rec.* **71**, 536 (1959)
141. BLAXTER, K. L., McCALLUM, E. S. R., WILSON, R. S., SHARMAN, G. A. M. and DONALD, L. G. *Proc. Nutr. Soc.* **20**, vi (1961)
142. FRANKE, K. W. and POTTER, VAN R. *J. Nutr.* **10**, 213 (1935)
143. BLAXTER, K. L. and SHARMAN, G. A. M. *Nature, Lond.* **172**, 1006 (1953)
144. BLAXTER, K. L. and BROWN, F. *Nutr. Abstr. Rev.* **22**, 1 (1952)
145. BLAXTER, K. L., WATTS, P. S. and WOOD, W. A. *Br. J. Nutr.* **6**, 125 (1952)
146. MAPLESDEN, D. C. AND LOOSLI, J. K. *J. Dairy Sci.* **43**, 645 (1960)
147. HOFLUND, S., HOLMBERG, J. and SELLMANN, G. *Cornell Vet.* **46**, 51 (1956)
148. DEHORITY, B. A., ROUSSEAU, J. E. Jr., EATON, H. D. *et al. J. Dairy Sci.* **44**, 58 (1961)
149. ROY, J. H. B. *Vet. Rec.* **76**, 511 (1964)
150. TOLLERSRUD, S. *Nord. VetMed.* **15**, 543 (1963)
151. DOTTA, U. and CANALE, A. *Atti Soc. ital. Sci. vet.* **17**, 671 (1963)
152. GARTON, G. A., DUNCAN, W. R. H., BLAXTER, K. L., McGILL, R. F., SHARMAN, G. A. M. and HUTCHESON, M. K. *Nature, Lond.* **177**, 792 (1956)
153. HIDIROGLOU, M. and JENKINS, K. J. *Can. J. Anim. Sci.* **52**, 385 (1972)
154. ALLEN, W. M., BRADLEY, R., BERRETT, S., PARR, W. H., SWANNACK, K., BARTON, C. R. Q. and MACPHEE, A. *Br. vet. J.* **131**, 292 (1975)
155. ALLEN, W. M. and SANSOM, B. F. *Rep. Inst. Res. Anim. Dis.*, p. 32 (1977)
156. BOYNE, R. and ARTHUR, J. R. *Res. vet. Sci.* **41**, 417 (1986)
157. SUTTLE, N. F. *Res. vet. Sci.* **42**, 224 (1987)
158. SUTTLE, N. F. *Vet. Rec.* **119**, 519 (1986)
159. BAIN, M. S., SPENCE, J. B. and JONES, P. C. *Vet. Rec.* **119**, 593 (1986)
160. AGRICULTURAL RESEARCH COUNCIL. *The Nutrient Requirements of Ruminant Livestock.* Commonwealth Agricultural Bureaux, Slough (1980)
161. SUTTLE, N. F. and McLAUCHLAN, M. *Proc. Nutr. Soc.* **35**, 22A (1976)
162. DAVIS, E. T., PILL, A. H. and AUSTWICK, P. K. C. *Vet. Rec.* **83**, 681 (1968)
163. RICHARDS, D. H., HEWETT, G. R., PARRY, J. M. and YEOMAN, G. H. *Vet. Rec.* **116**, 618 (1985)
164. POUKKA, R. *Br. J. Nutr.* **20**, 245 (1966)
165. POUKKA, R. *Br. J. Nutr.* **22**, 423 (1968)
166. MUTH, O. H., WHANGER, P. D., WESWIG, P. H. and OLDFIELD, J. E. *Am. J. vet. Res.* **32**, 1621 (1971)
167. HIDIROGLOU, M., CARSON, R. B. and BROSSARD, G. A. *Can. J. Physiol. Pharmac.* **46**, 853 (1968)
168. HIDIROGLOU, M., JENKINS, K. J., WAUTHY, J. M. and PROULX, J. E. *Anim. Prod.* **14**, 115 (1972)
169. HIDIROGLOU, M. and JENKINS, K. J. *Can. J. Anim. Sci.* **48**, 7 (1968)
170. GODWIN, K. O. *Aust. J. exp. Agric. Anim. Husb.* **12**, 473 (1972)
171. SOLIMAN, K. N., WAHBY, M. M., AYOUB, M. H. and ISKANDER, M. *Br. vet. J.* **120**, 535 (1964)
172. POUKKA, R. and OKSANEN, A. *Br. J. Nutr.* **27**, 327 (1972)
173. HIDIROGLOU, M., JENKINS, K. J., LESSARD, J. R. and BOROWSKY, E. *Can. J. Physiol. Pharmac.* **48**, 751 (1970)
174. ALDERSON, N. E., MITCHELL, L. G. E. Jr., LITTLE, C. O., WARNER, R. E. and TUCKER, R. E. *J. Nutr.* **101**, 655 (1971)
175. HIDIROGLOU, M. and JENKINS, K. J. *Can. J. Anim. Sci.* **51**, 803 (1971)
176. BLAXTER, K. L. *Vitams Horm.* **20**, 633 (1962)
177. LAMAND, M. *Annls Nutr. Aliment.* **20**(3), 13 (1966)
178. LOHR, J. E. and VAN DER WOUDEN, M. *N.Z. vet. J.* **19**, 222 (1971)
179. LAMAND, M. *Annls Biol. anim. Biochim. Biophys.* **5**, 309 (1965)
180. OKSANEN, H. E. *Acta vet. scand.* **6**, Suppl. 2 (1965)
181. JENKINS, K. J., PROULX, J. G. and HIDIROGLOU, M. *Can. J. Anim. Sci.* **51**, 237 (1971)
182. JENKINS, K. J., HIDIROGLOU, M., MACKAY, R. R. and PROULX, J. G. *Can. J. Anim. Sci.* **50**, 137 (1970)
183. CONNELL, R., WHITING, F. and FORMAN, S. A. *Can. J. comp. Med.* **23**, 41 (1959)

184. PARKER, K. G. *J. Range Mgmt* **10**, 105 (1957)
185. BAILEY, C. B. *Science, N.Y.* **155**, 696 (1967)
186. BAILEY, C. B. *Can. J. Anim. Sci.* **53**, 55 (1973)
187. BAILEY, C. B. *Can. J. Anim. Sci.* **56**, 745 (1976)
188. CROOKSHANK, H. R., KEATING, F. E., BURNETT, E., JONES, J. H. and DAVIS, R. E. *J. Anim. Sci.* **19**, 595 (1960)
189. BAILEY, C. B. *Can. J. Anim. Sci.* **56**, 359 (1976)
190. BAILEY, C. B. *Can. J. Anim. Sci.* **56**, 213 (1976)
191. WELLS, G. A. H., SCOTT, A. C., JOHNSON, C. T. *et al. Vet. Rec.* **121**, 419 (1987)
192. STUTTAFORD, T. *The Times*, June 9, p. 18 (1988)
193. EDWIN, E. E., LEWIS, G. and ALLCROFT, R. *Vet. Rec.* **83**, 176 (1968)
194. PATTISON, I. H. *Vet. Rec.* **123**, 661 (1988)
195. PRUSINER, S. B. *Science, N.Y.* **216**, 136 (1982)
196. MASTERS, C. L., GAJDUSEK, D. C. and GIBBS, C. J. Jr. *Brain* **104**, 535 (1981)
197. MASTERS, C. L., GAJDUSEK, D. C. and GIBBS, C. J. Jr. *Brain* **104**, 559 (1981)
198. TAYLOR, D. M. and WATERS, A. M. *Lancet* **ii**, 460 (1986)
199. INTERNATIONAL UNION OF BIOCHEMISTRY. In *Enzyme Nomenclature*, Academic Press, New York and London (1984)

Chapter 8

Parasitic infections

Babesiasis

The organisms

Babesiasis is a haemoprotozoan disease that particularly affects both exotic and indigenous cattle in subtropical and tropical regions of the world and causes serious economic loss. In those regions the organism mainly involved is *Babesia bigemina* causing 'Texas fever'. In more temperate areas of Europe, Africa, Asia and the E. Indies *B. bovis* is the responsible organism whilst in N. Europe, including the UK, a small babesia, *B. divergens* causes the disease of 'redwater', although a large babesia *B. major* has been detected in British cattle.

Redwater is associated with rough pastures and moorland districts where ticks, which harbour and transmit the parasite, can find shelter and suitable breeding places.

Symptoms

In some cases there is only a slight fever and haemoglobinuria. In more acute cases rumen function is affected and there is emaciation, jaundice, diarrhoea followed by constipation and anaemia with a marked reduction in the red cell count. Death may result in 3–4 d.

Prevention

Immunization against babesiasis consists of transmitting the disease to susceptible animals by the inoculation of infective blood with, if necessary, subsequent treatment with a babesicidal drug to prevent severe illness or death.

A comparison of untreated 4-month-old control calves has been made with those that had been immunized against *B. bigemina* by treating them after experimental infection and during the prepatent, i.e. incubation, period with one or two doses of diminazene aceturate; after 45 d, the calves were challenged with the homologous strain of *B. bigemina*-infected blood. The four untreated control calves had severe symptoms and two died, whilst the calves of both groups that recovered resisted the challenge infection with a mild relapse in the immunized group, thus indicating that immunity had been conferred by the vaccination. One dose of the drug was sufficient to suppress clinical disease[55].

The effect of continuous oxytetracycline administration on the development of parasitaemia by *B. divergens* in natural and artificial infections, has been studied. During natural exposure to grazing that was heavily infected with the tick, *Ixodes ricinus*, seven out of 42 cattle (18 months–2 years of age) with no previous exposure to tick-borne diseases were injected every 4 d with a long-acting oxytetracycline at 20 mg/kg live weight. During a 6-week grazing period, 21 untreated cattle developed a patent parasitaemia of *B. divergens* and all became seropositive to a fluorescent antibody test. In contrast, no parasites were observed in the treated cattle and antibody titres remained low.

In a comparison of different amounts of long-acting oxytetracycline, groups of cows given 10^8 infected erythrocytes were injected every 4 d with a long-acting oxytetracycline at 20, 10 and 5 mg/kg. The highest level inhibited parasite replication and antibody formation. The same was true for one animal given 10 mg/kg, but the remainder and those given 5 mg/kg developed a low parasitaemia and a high antibody titre, whereas untreated cows developed a severe babesiasis. A further control group was then added and 3 weeks after the end of oxytetracycline treatment, all cattle were injected with 10^9 erythrocytes infected with the homologous isolation of *B. divergens*. The control animals and those in which previous infection had been completely inhibited, i.e. the group given 20 mg/kg, developed severe clinical babesiasis but the remainder were refractory to parasite development[53].

In another experiment, 40 11–13-month-old calves were inoculated with live *B. bovis* vaccine. Of these, 20 were treated with the same antibiotic 7 and 15 d after receiving the vaccine. Parasites were found in nine of the treated calves and in all 20 of the untreated group. Treated calves were less febrile and had higher PCV, but all calves in both groups developed considerable antibody titres to *B. bovis*[52].

After vaccination of 1-year-old calves simultaneously with live *B. bovis* and *B. bigemina* followed by treatment with a long-acting oxytetracycline from 6 d after vaccination at the rate of 20 mg/kg in 2, 3 or 4 doses, the calves showed no evidence of disease when challenged with virulent parasites of both species 5 months later[54].

Whilst application of synthetic pyrethroids to cattle by means of impregnated ear tags is a useful method for reducing the number of flies on cattle, it appears to be less effective against the tick *Ixodes ricinus*[56]. The concentration of pyrethroids may not be high enough or the ambient temperature too high. Some insects are susceptible at a temperature <15°C, but withstand pyrethroids at a higher ambient temperature[57].

Coccidiosis

The organism and its incidence

Coccidiosis is caused by a protozoan parasite belonging to the genus *Eimeria*, which is specific to cattle. The coccidia, of which there are a number of different species, form spores which are capable of surviving on the ground for many months. Cattle harbouring this parasite produce large numbers of the spores, or 'oocysts', which are passed out in the faeces and, after a minimum ripening period of 2 d under conditions of warmth and moisture are capable of infecting other animals.

Calves up to 1 year of age are most susceptible, and the disease may become widespread among a group. Calves become infected by swallowing infective oocysts with their food or drinking water. On entering the digestive tract, the spores develop into active coccidia which penetrate the walls of the intestine and multiply, damaging the mucosal lining. In due course, oocysts are produced and released in the faeces.

Coccidiosis is most common in marshy districts during the summer and early autumn, although acute outbreaks may occur as early as 4–8 weeks of age in housed stock[1]. Most healthy calves are infected with a few coccidia but show no clinical symptoms.

In a study of coccidiosis infection in a large calf house in Czechoslovakia, the calves were infected with 10 species of *Eimeria*. The proportion of calves infected increased from 23 per cent at 2 months to 100 per cent at 5 months of age, with a concomitant increase in the number of oocytes/g faeces. Weaning onto dry food was accompanied by a marked increase in the proportion of calves infected. The main species were *E. zuernii, E. bovis, E. ellipsoidalis, E. subspherica, E. auburnensis*, with sporadic isolations of *E. alabamensis, E. cylindrica, E. wyomingensis, E. bukidnanensis* together with oocysts of *Isospora* spp.[27].

Symptoms

Affected animals often have foul-smelling diarrhoea containing blood either as liquid or in clots, and stand straining with backs arched in a characteristic posture for several minutes. The animals gradually become anaemic, weak and emaciated. The disease frequently occurs where a number of young animals are crowded together. If an epidemic breaks out, the infection becomes less severe as the epidemic progresses, possibly owing to the development of an immunity in the affected calves. Adult cattle are resistant, but may act as carriers and infect susceptible young calves.

Prevention

Young calves should never be introduced into crowded quarters occupied by a group of calves of various ages. As most animals carry coccidia, the pens occupied by the original group of calves will carry a substantial number of infective oocysts available for reinfection and, ultimately, large numbers of oocysts will be discharged by most of the animals. Calves subsequently introduced are exposed to a heavy infestation of oocysts, and the process is repeated until the parasite population is raised to a dangerous level.

Prevention of reinfection is essential. This can be achieved by cleaning out the building every 2 d, as this period is insufficient for the oocysts to become infective. Alternatively, the calves should be tied up in stalls so that they cannot take food or water contaminated by faeces. Attempts to inoculate calves against the condition, including the use of irradiated oocysts, have not been successful[2].

When young Holstein-Friesian calves in the USA were inoculated with *E. bovis* on the third day after the start of a 31 d period of feeding a ration containing either 16.5 g or 33 g monensin/t, the calves remained free of clinical signs of coccidiosis whereas uninoculated control calves given no monensin developed diarrhoea and had an excessive oocyte excretion from 19 to 28 d after they were given *E. bovis*. Although not significant, weight gains and feed efficiency in monensin-treated calves were equal to or superior to those of non-inoculated control calves. Moreover, inoculated control calves had significantly lower weight gains to 28 d than did those on the other two treatments[28].

Affected calves should be isolated, treated by a veterinary surgeon and, after treatment, removed to clean quarters. Bedding used by affected cattle should be carried right away from the building and should be buried deeply in a manure heap. Ammonia and other products of decomposition will effectively destroy the oocysts. The walls of the pens should be washed with hot soda solution, followed by rinsing and spraying with a 10 per cent solution of ammonia.

Fascioliasis (liver fluke disease)

Life cycle

This is caused by the trematode *Fasciola hepatica*. The intermediate host is a species of snail, *Limnaea truncatula*. The snail becomes infected by small ciliated larvae that have developed from eggs excreted by sheep or cattle, and in turn these animals can be infected by cercariae excreted by the snail. The cercariae may be swallowed with drinking water or encysted on grass. After ingestion the infective cysts hatch in the small intestine and the juveniles burrow through the gut wall, aided by their backward pointing spines, into the peritoneal cavity where they browse for a few days before penetrating the liver. After a few weeks migrating in the liver they enter the bile ducts and mature to the adult form.

Resistance to infection

In experiments with rats[39], it was found that 3 weeks after infection they became resistant to reinfection with *F. hepatica*. In resistant rats, the newly hatched flukes were coated with antibody whilst still in the lumen of the small intestine and 20–30 per cent of flukes failed to penetrate the intestinal wall. Those that penetrated into the peritoneal cavity acquired a coating of antibody from the peritoneal fluid, and eosinophils and neutrophils adhered to and degranulated on the surface of the flukes. This resulted in total erosion of the surface of the flukes. The dead flukes then attracted macrophages which disposed of them. It was thought that a similar cycle of events might occur in cattle.

However unfortunately, unlike rats, cattle sensitized by intraperitoneal injection of fluke antigens showed low resistance to infection; the cattle produced antibody, but no lymphocyte proliferation occurred[48].

Symptoms

The infected animal may at first show an improvement in condition due to stimulation of the liver by the flukes, but thereafter the calf becomes progressively more emaciated and anaemic with diarrhoea and depressed appetite and rumination.

In calves infected with 1000 metacercariae of *F. hepatica* followed by a further 500 3 d later, no effect was found on digestibility of dietary protein, but there was a marked reduction in nitrogen retention, due to the inflammatory reactions in the liver reducing protein formation and increasing nitrogen excretion in the urine[29].

After oral infection with *F. hepatica*, immature flukes were found to have penetrated the liver in large numbers by day 7, but there did not appear to be a corresponding reduction of the numbers in the body cavity. The liver showed progressive tissue disruption with a neutrophil and lymphocyte infiltration by day 7 and a marked eosinophil response by day 14[46]. It has been suggested that the increase in serum γ-glutamyl transpeptidase (γGT) (EC 2.3.2.2)[125], which occurred 56 d after the first inoculation of 6-month-old calves with 200 metacercariae during a 72 h period, coincided with the penetration of the bile duct by migrating flukes and could be used as a measure of biliary damage; γGT remained at a high concentration for at least 83 d after infection[47].

Treatment

Drug therapy is not entirely satisfactory since liver damage will have occurred before treatment begins and treated animals become infected again when they graze contaminated pasture and therefore regular dosing is necessary. Moreover, most drugs have a withdrawal period during which meat and milk cannot be sold for human consumption.

A comparison of triclabendazole, nitroxynil and rafoxamide treatment in cattle naturally infected with predominantly immature stages of *F. hepatica* indicated, from faecal egg counts 9 weeks after treatment, an efficiency in the first trial of 100 per cent for triclabendazole and 95 per cent for nitoxynil, and an efficiency, in a second trial 15 weeks after treatment, of 98 per cent for triclabendazole and 53 per cent for rafoxamide. Compared with a control group, the reduction in fluke burdens were 97 per cent for triclabendazole and 76 per cent for the nitroxynil group[49]. Other work has shown that triclabendazole is highly effective against early immature, immature and mature infections of *F. hepatica*[50].

Routine dosing with triclabendazole is normally done in the autumn followed by further treatment in January, but it has been shown that infection can be picked up after January. In 1986 there was clinical evidence of significant pasture contamination lasting well into April in certain parts of the UK. Under these circumstances, a further dose should be given in early summer to reduce snail infection and metacercarial production during the following autumn and winter[51].

Ox warble flies

Species

Two species of fly, *Hypoderma bovis* and *H. lineatum*, attack cattle in the UK. The adult insects are fairly large, dark, hairy flies with bands of yellow or orange which give them a superficial resemblance to humble bees. *H. lineatum* is on the wing earlier (May) than is *H. bovis,* and is more prevalent in W. England. Like many flies, their activity is restricted during periods of rain and high winds and, moreover, their cruising range appears to be limited to about 400 m. As warble flies fly at a low altitude and appear to dislike crossing water, high hedges and rivers help to restrict the area of their activity.

Life cycle

The flies lay their eggs on the hair of the legs and bellies of cattle, those of *H. bovis* being laid singly and those of *H. lineatum* in rows of up to 20. After 4 or 5 d the eggs hatch into maggots, which penetrate the skin of the leg by means of their sharp mouth hooks and in the case of *H. lineatum* move up to the wall of the oesophagus, where they spend the winter. In February they leave the oesophagus and pass to the loins, where they encyst until the pure maggot stage is reached. Some of the *H.*

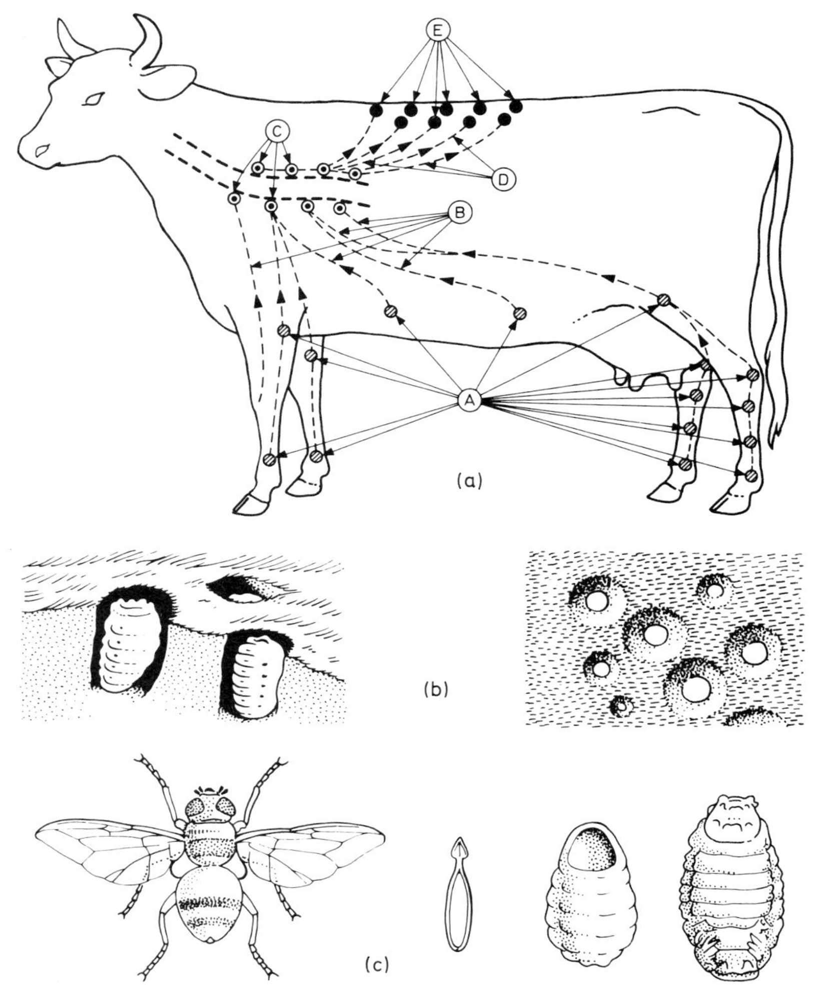

Figure 8.1 Life cycle of the warble fly. (*a*) A, Sites where the warble fly deposits her eggs and where the newly hatched larvae perforate the skin and enter the body; B, first migration of larvae from point of entry to oesophagus; C, larvae encysted around the oesophagus from August to February; D and E, second migration of larvae from oesophagus to back from February to June; (*b*) warbled hide, section and plan views; (*c*) ox warble fly with egg and larvae

bovis maggots pass through the spinal canal on the way to the back. Once in the back each maggot makes a breathing-hole. The swellings which result from the encysted maggots increase in size, soften and burst, discharging the maggot on to the ground, where it pupates and becomes an adult after about 6 weeks (see Figure 8.1).

Symptoms

The economic importance of the warble fly lies in the number of ways in which it causes damage. The flies themselves produce much excitement in stock, causing failure to put on flesh normally and injury to the cattle from their efforts to evade attack. The hides are damaged in their most valuable part where the leather is thickest, and not infrequently large and painful abscesses are formed under the skin. In addition, losses are suffered by the meat industry as the migrating larvae leave greenish-yellow tracts containing a jelly-like material in the tissues, and these have to be trimmed from the carcass.

Resistance to infection

Although it has not yet been definitely ascertained how a degree of resistance is developed by cattle, it seems that certain animals are able to escape the attention of the fly. Calves and young cattle are attacked more frequently and more severely than are older cattle, which appear to have a degree of resistance, due possibly to their tougher hides.

In 1982, it was shown that after natural infection with *H. bovis,* antibody in the sera of 9-month-old calves was found to disappear from the circulation about 14 weeks after the emergence of the warble fly larvae in late June or July, i.e. in October[41]. It seems that the main factors in resistance are the intensity of infection, the number of recurrences and the use of insecticides. Thus, cattle which experienced heavy infection in one year were less susceptible to attack in the following year, whereas those that were lightly infected or had their infection terminated with insecticide experienced heavier infections in the next year[42]. The development of an ELISA (enzyme-linked immunosorbent assay) test for infection has been reported[43].

Incidence of infestation

A survey of warble fly infestation in 164 herds carried out by the Milk Marketing Board in 1973[24] showed that 61.5 per cent of herds were affected, but the infestation was usually slight, with fewer than five warbles per infected animal; only eight herds had 'severe' infestation. The warble fly appeared to be widespread throughout England and Wales and the incidence in heifers did not differ from that in cows. Treatment against warble fly was being given in 60 per cent of the herds and, of the treated herds, 70 per cent were being treated with systemic insecticides.

A survey of the Meat and Livestock Commission of warble-infected fat cattle in 57 selected livestock markets in Great Britain during June 1972 and in May 1973–1984, when 40 000–60 000 cattle were inspected annually, showed that the level of infestation increased from 22 per cent in 1972 to 38 per cent in 1976 and then declined slightly to 34 per cent in 1978. After the introduction of legislation and the eradication scheme in 1978, there was a rapid decrease to 8.6 per cent by 1979 and to <0.02 per cent in 1984[25,44]. The Ministry of Agriculture report a value of <0.01 per cent by 1985[40].

Legislation

Control measures within the European Community vary from the voluntary application of systemic products by individual farmers to comprehensive national eradication schemes by legislation. Denmark has successfully eradicated warble fly and W. Germany, the Republic of Ireland and N. Ireland, like Great Britain, have reduced the incidence to low levels[43].

Control depends upon destroying the larvae before they emerge from the skin of the back. The habits of the fly tend to keep it in a localized area and although benefit may accrue from treatment of cattle on any particular farm, eradication of the fly can only be achieved if treatment is done on every farm in the country.

Under the Warble Fly (Dressing of Cattle) Order 1948 and the 1960 Amendment, it was compulsory to treat cattle visibly infected with the maggots of the warble fly and all cattle in which infection by the maggot could be ascertained by handling the animal. The Order stated that all infected cattle must be treated as follows:

Treatment should start each year between 15 and 22 March or as soon after as the larvae appear under the skin of the backs of the cattle, and should be repeated at intervals of not less than 27 d and not more than 32 d as long as larvae continue to appear, but not after 30 June. The dressing prescribed is 43 g derris resins or 14 g rotenone and a sufficient quantity of dispersing agent to produce a stable emulsion or suspension and to ensure wetting the treated skin. If the derris wash is properly applied and the exudation or scab which forms on the warble holes is removed with a stiff brush to enable the derris to penetrate, every maggot should be killed.

The introduction of systemic organophosphorus insecticides[18–23], which control the maggots before they reach the back, and the realization that the Warble Fly Order could not be adequately enforced, led to the revocation of the Order in 1964. However, if warbles appeared on the back after a systemic insecticide had been used, the cattle had to be dressed externally. The organophosphorus compounds were suitable for beef cattle of all ages, but could not be administered within 60 d of slaughter. They could be given to dairy cows and heifers when dry, but could not be given during lactation or within 60 d of the beginning of lactation. Oral treatment had to be given between the beginning of September and the end of November.

Metriphonate (Trichlorfon) could be given as a single drench by mouth at the rate of 80 mg/kg body weight or as a single dressing early in the spring. The drench, although killing a large proportion of migrating larvae, could cause slight toxic effects in a few animals. This drug was also effective against the nematode parasites causing parasitic gastroenteritis in cattle.

New warble fly regulations were introduced under the Warble Fly (England and Wales) Order 1978 which operated between 15 March and 31 July each year. During this period cattle which were infected with warbles had to be treated with a systemic warble fly dressing, and the movement of warbled cattle (even to a slaughterhouse) was an offence unless they had been treated during this period and were accompanied by an owner's declaration of treatment.

Untreated warbled cattle being moved could be detained and sent either for slaughter or to a specified place of treatment. The owner of the cattle could also be required at his own expense to arrange for his veterinary surgeon to inspect all his cattle over 12 weeks of age and to supervise the treatment of all those cattle found to be affected.

In 1982, warble fly infection was made a notifiable disease. At this time there were 705 affected herds in Great Britain and this decreased to 419 herds in 1985[43]. Under the Warble Fly (England and Wales) Order 1982 and the Warble Fly (Scotland) Order 1982, anyone having in his possession and under his charge, an animal affected or suspected of being affected with warble fly must report the fact immediately to a Divisional Veterinary Officer of the Ministry before attempting any treatment. On confirmation of the existence of warbles, a notice is served on the owner or person in charge of the affected cattle requiring treatment of all cattle over 12 weeks of age on the premises where the infection has occurred and restricting movement of cattle from the premises until treatment has been completed.

Herds known to have been affected in the spring are required to be treated again in the following autumn. Autumn treatment may also be required for herds in proximity to affected herds or which have received cattle from such herds.

Since 1982 'infected areas' have been designated where the disease has persisted or where there is evidence of recurrence, and in these areas all cattle >12 weeks of age must be treated within specified dates[43]. Recent infected areas in Great Britain have covered W. Dorset and neighbouring parts of Somerset and S.W. Devon, including Dartmoor.

Treatment

Treatment by systemic organophosphorus insecticide is either by a liquid 'pour-on' systemic dressing or by injectable systemic solution[40].

Autumn

Although warble fly incidence is low, it is important that cattle are given an autumn preventative treatment to eliminate the serious risk of a resurgence of the disease until eradication is complete.

The best time to carry out this treatment is between 15 September and 30 November, and preferably later within this period to prevent reinfection in late September/early October. Systemic treatment in autumn will kill 95 per cent or more of the larvae in the body before they have caused systemic damage. It is best to avoid treatment during December, January and February, to prevent possible side-effects as a result of destruction of large numbers of larvae while located along the spinal cord or oesophagus. An injectable solution may be used throughout the year.

Systemic dressings are effective against other parasites, e.g. autumn treatment can be used against concurrent mange and lice infection during housing.

Spring

This may only be carried out on instructions from the local Divisional Veterinary Officer after the presence of the disease has been confirmed.

'Pour-on' method

The systemic warble fly dressing is poured along the animal's back from shoulder to tail, the correct quantity in relation to body weight being used. Various formulations are available for treatment, namely coumaphos, famphur, fenthion, phosmet and trichlorfon. All these organophosphorus compounds are extremely potent substances (see also Chapter 7, p. 203).

Instructions for dosage and method of application must be carefully observed. All persons carrying out treatment should wear protective clothing as specified by the manufacturer of the product. Treatment should be in well-ventilated conditions, preferably in the open air, and inhalation of vapour avoided. Sick animals and calves under 12 weeks of age should not be treated, nor should an application be made at the same time as an anthelmintic containing an organophosphorus compound, or levamisole, or diethylcarbamazine citrate is given.

In milking cows the withdrawal period varies between 6 h and 5 d, depending on the product used. Where the product has a 6 h period, treatment should be given immediately after milking and not less than 6 h before the next milking, to avoid contamination of the milk. Where a larger withdrawal period is specified, milk must not be used until the withdrawal period has elapsed. When cattle are intended for slaughter, the appropriate withdrawal period, usually 14–21 d, must be complied with. One product has a withdrawal period of 35 d.

Injectable solution

Currently the only licensed injectable solution contains ivermectin. Cattle must not be treated with this solution within 21 d of slaughter. It cannot be used in cattle producing milk for human consumption or within 28 d of parturition.

Parasitic gastroenteritis

The organisms

This disease is caused by a heavy infestation of the abomasum or small intestine with small round worms (nematodes) belonging to the subfamily Trichostrongylidae. *Ostertagia ostertagi* is most frequently implicated[3].

Nematodirus battus, usually associated with sheep, may also infect calves and cattle, but *N. helvetianus* only infects sheep[89].

One of the most pathogenic trichostrongylids of the water buffalo (*Bubalus bubalis*) is *Paracooperia nodulosa,* which gives rise to the formation of multiple nodules in the intestine causing diarrhoea, anaemia, emaciation and sometimes death[92]. This has been reported in Malaysia in an 8-month-old calf weighing 119 kg[91].

Life cycle

The eggs, which are laid in large numbers by adult worms, pass to the ground with the faeces of the infected animal and hatch into larvae which, after moulting twice, are capable of infecting another animal within 4–5 d, though this period may be prolonged in cold weather. These third-stage larvae leave the faeces and migrate into the herbage, where they await the grazing host. Most cattle harbour some of the parasites, but their health is only affected when a heavy infestation occurs. Thus, experimental infection of calves in Italy with 250 000 third-stage larvae of *O. ostertagi* did not produce much effect on the digestibility of the diet, although growth rate and food conversion were reduced[33].

The main requirements for larval development on the pasture are moisture, warmth and oxygen. Humid weather conditions and long herbage favour larval development, whereas drought and short herbage are less suitable. Nevertheless, a heavy infestation may occur in short dry pasture, as close grazing favours the ingestion of larvae from the lower parts of the herbage. Small paddocks frequently used for the grazing of calves near to the buildings are particularly dangerous.

In N. Nigeria, three peak infestations of third-stage larvae of *Cooperia*, *Haemonchus* and *Trichostrongylus* occurred in the rainy season (April to October) on pastures grazed by beef cattle; the first in the second half of May, the second in late July and the third in October. Although there was a sharp fall in infestation after the onset of the dry season, appreciable numbers of third-stage larvae were still present on herbage in early December[93].

Arrested development of larvae

In the UK two types of clinical infection have been identified, one (type 1) occurring in calves from late July to the end of the autumn grazing season and the other (type 2) being delayed for up to 6 months after grazing an infested pasture from late September onwards. Although whilst grazing there were no clinical signs in this second type of infection, the cattle were heavily infected with *Ostertagia*, which were inhibited in their early fourth larval stage in the gastric glands of the abomasal mucosa, and months later maturation of the larvae caused clinical disease in both housed and outwintered stock[4].

Resistance to infection

Calves are most susceptible, but adult cattle, which are normally resistant, may develop the disease under conditions of poor nutrition, or under the strain of lactation and pregnancy, or if infestation is very severe. High nutritional planes reduce both the number of worms and also the effect of the parasites on the host[34].

The population of gastrointestinal worms in the cow may increase during lactation because the fecundity of the female worms increases and there is inhibition of the lymphocyte-mediated component. It is thought that the cause is endocrinal in origin with prolactin, as well as possibly adrenal glucocorticoids, being involved[32].

Under conventional systems of rearing calves, it is suggested that resistance to the disease starts at the age of 4–6 months and develops to a maximum at 18 months–2 years. However, certain immunological advantages may stem from rearing calves at pasture from birth. In 3 years of experiments[5] calves reared at pasture from birth showed an elevated rectal temperature from 10 to 25 d of age – this pyrexia began a few days after the time when pasture is usually first ingested by the young calf and it was not observed in calves kept indoors for the first 6 weeks of life and then put out to pasture. The faecal egg counts of trichostrongyles in calves on pasture from birth increased rapidly to 13 weeks of age, then at a very much slower rate until 21 weeks and thereafter declined, whereas the counts for calves put out to pasture at 6 weeks of age increased rapidly to very high values at 26 weeks of age. Moreover, in the latter treatment the older the calf when it first became infected, the greater was the increase in egg counts with age. It seems possible, therefore, that the very small increase occurring in output of eggs after 13 weeks, for calves that had been exposed to parasites since birth, was associated with a development of immunity, for it is well known that the immune mechanism may restrict egg production of existing worms[6]. Similarly, a heavy larval challenge with *Ostertagia* to 6-month-old calves that had been suckled at pasture and exposed to a natural field infection from birth failed to produce any lesions, and only small numbers of larvae became established as mature worms[7]. The important point appears to be that calves should be managed so that they ingest a gradually increasing amount of grass from a young age.

Under poor nutritional conditions and when exposed to sudden very heavy infestation, the resistance may break down. The autumn is the most dangerous time, as pastures may be heavily infested and the nutritional quality low. Even if heavily infected store animals are taken into buildings, their condition will gradually deteriorate if they are maintained on a low plane of nutrition.

Symptoms

The symptoms of the disease are a progressive loss of condition resulting in loss of bloom from the coat and in the skin becoming 'hide bound'. There is generally diarrhoea, but not invariably, and the appetite is usually retained until the animal becomes very weak (but see below). Mucous membranes will often appear anaemic. In very severe infestations deaths may occur suddenly.

The anaemia appears to be of a normochromic normocytic type, as there is a reduction in packed cell volume, haemoglobin and red cell concentration, but no change in mean corpuscular volume or mean corpuscular haemoglobin concentration. There is a decline in the serum protein concentration, serum albumin decreasing markedly but serum globulin increasing. In the abomasal fluid the concentration of sodium ions increases and the concentration of potassium and chloride ions decreases, resulting in an increased abomasal pH. Similarly, the plasma pepsinogen (the precursor of pepsin) concentration is raised[4]. In a study of subclinical helminthosis in Brahman cross and Hereford × Shorthorn steers[26] elimination of helminths resulted in a marked reduction in plasma gastrointestinal leakage and an increase in plasma content and total plasma protein concentration. The extent of these changes seemed to be less for the Brahman cross.

In four calves given *H. contortus* larvae, serum pepsinogen rose quickly to 3.5 iu tyrosine/l 14 d after infection. The mean concentration dropped to 1.8 iu/l on day 23 and remained around 2 iu/l until an anthelmintic was given on day 87, when concentrations returned to pre-infection values[85]. Two different forms of pepsinogen were isolated in the serum of calves given third-stage *O. ostertagi* larvae[87] or given adult worms via an abomasal cannula[88].

The adult *O. ostertagi*, as distinct from the larvae, also cause physiological damage. When given via an abomasal cannula, they caused increased concentrations of plasma pepsinogen and, in those calves in which the largest number of worms became established, plasma gastrin concentrations were also increased. However, abomasal pH was unaffected and there was no significant difference in the percentage establishment of adult parasites in those previously infected with *O. ostertagi* third-stage larvae and in those that had been maintained parasite-free before infection with the adult worms[86].

A comparison of the intestinal enzyme activity in infected and uninfected 4–5-month-old Friesian calves in Queensland, in which *Cooperia* spp. predominated, showed no difference in maltase (EC 3.2.1.20), trehalase (EC 3.2.1.28), cellobiase (EC 3.2.1.21), sucrase (EC 3.2.1.48), lactase (EC 3.2.1.108), β-galactosidase (EC 3.2.1.23), or acid (EC 3.1.3.2) and alkaline (EC 3.1.3.1) phosphatase[125] activity. However in heavily infected calves with >20 000 *Cooperia* spp., duodenal maltase and acid phosphatase activities were significantly elevated[31].

The effects of the intensity of parasitic infection on metabolism have been studied in Denmark. A dose of 4000–5000 *O. ostertagi* infective larvae given three times a week for 10 weeks to 4–5-month-old calves had no effect on protein digestibility or nitrogen metabolism. Faecal excretion of eggs was first detected

16–19 d after infection, the diurnal excretion increasing to 10^5–10^6 eggs with great daily variations This lack of clinical effect of spreading an infective dose over a period, may be compared with the marked effect of giving a single dose of 200 000 infective larvae which produced complex clinical reactions[30].

It has been suggested that cobalt-deficient cattle may have impaired immunity to *Ostertagia* infection. Thus, whilst experimental infection with *O. ostertagi* larvae reduced live-weight gains in both cobalt-deficient and animals of normal cobalt status, the prepatent period was shorter, worm egg counts greater and paradoxically plasma pepsinogen levels lower in the cobalt-deficient calves[97].

Prevention

In a survey in 1984 of parasitic worm control in cattle in W. Scotland, grazing management or anthelmintic treatment was used in 92 per cent of herds. The use of anthelmintic drugs was greatest on farms where grazing control was also practised. Benzimidazoles were the most frequently used drugs[98].

Prevention of the condition can thus be afforded by feeding stock adequately, by avoiding overstocking and by the prophylactic use of drugs.

Grazing management

There is no doubt that overstocking is the most important factor in influencing the development of an outbreak of the disease. A few larvae may survive for 12 months in a permanently moist and protected environment, such as deep in grass tufts, but most die within a few days. A period of 6 weeks should be sufficient before a heavily infested pasture is safe for susceptible animals, but where the herbage is long, and in damp cloudy weather, a rest of 5–6 months may be necessary (but see below).

If weaned calves are turned out in spring onto a pasture on which there is an overwintering infestation, the calves will begin to excrete worm eggs about 3 weeks later and by mid-July there will be a large increase in infestation on the pasture. If the calves are left on the pasture, the calves will have acquired sufficient worm burdens to cause retarded growth or clinical disease. Thus, it is usually recommended that calves should be treated with an anthelmintic and moved to a clean pasture in July.

From observations carried out during a 27-year period in S.E. England, it is highly unlikely that the new generation of larvae will reach a significant level before mid-July. Even if due to excessive rainfall in June large numbers of larvae become available at that time, it is of no consequence if calves are moved onto clean pastures in mid-July particularly when accompanied by treatment with an anthelmintic[94].

Young cattle can play an important role in the epidemiology of *N. battus*. Thus, young cattle readily acquire a heavy burden of *N. battus* in the spring and the contamination of pastures with eggs from these infections results in significant populations of larvae on the herbage, which are infective to both calves and lambs grazed on these pastures in the following year. Although the majority of the *N. battus* eggs hatch in the spring, some hatch in the autumn. The calves develop a strong immunity to *N. battus* during the grazing season as shown by the absence of worms at post-mortem examination in the autumn despite the presence of infective larvae on the pasture[90].

Three strategies have been suggested for the control of nematode infection[35].

Preventive strategy
In this calves are either turned out onto clean pastures, or less desirably onto slightly infested pasture, and treated with anthelmintic periodically until the overwintered infection has died down.

Evasive strategy
The calves are moved to a fresh pasture in July before new infestation appears on the original pasture.

Diluting strategy
Susceptible calves, which are a source of contamination, are grazed with resistant adults, which are not a source of contamination because they produce few eggs.

Thus, the infectivity of a pasture previously grazed by adult dairy cows is no greater than that of a matching control pasture, despite the fact that the cows are untreated and carry worm burdens typical of that class of livestock as judged by faecal egg output. Prior grazing of such cows from May 2 to May 30 had no adverse effect on the subsequent protection afforded by use of a morantel sustained-release bolus (see below) in set-stocked calves turned out on May 30[95].

A classification of the cleanliness of pastures in relation to their previous history is given in Table 8.1[35].

However, in Cuba infection in calves at 70 d of age with *Moniezia* (a tapeworm), *Haemonchus* and *Dictyocaulus* was most common in calves reared by multiple suckling to 70 d, whilst *Strongyloides* was more common in calves restricted-suckled to 70 d and *Cooperia* was most common in artificially-reared calves. The proportion infected was 55 per cent for the multiple-suckled calves compared with 26–30 per cent for the other two treatments. It was considered that this finding was due to grazing the cows and calves together in the multiple suckling system, whereas in the restricted suckling the cows did not enter the area designated for the calves. In the artificial rearing system the calves were transferred to pasture at 42 d of age and grazed nine paddocks of 0.2 ha each for about 3–5 d. Calves on all three treatments had access to concentrates[36].

Supplementation of calves on pasture with a 16 per cent crude protein concentrate mixture at the rate of 1 kg/d has prevented clinical signs in calves challenged with larvae, but the changes in the blood chemistry (see above) still occurred owing, presumably, to a subclinical infection[8].

Drugs

When curative treatment is given, all animals in a group should be treated and after treatment should be transferred to a clean pasture[37].

Phenothiazine
For prophylactic purposes young cattle should receive treatment with phenothiazine on one or two occasions during the summer and autumn and a final treatment before being housed in the autumn. The dose varies from 15 to 25 g, depending on the age of the calf. In bright sunny weather it is usually suggested that calves should be kept indoors for 48 h after treatment, as otherwise damage to the eyes may be caused. After treatment the urine becomes pink in colour owing to the presence of a red dye in the product.

Table 8.1 A classification of pastures for calves. (Adapted from Ministry of Agriculture Fisheries and Food Booklet 2154. © Crown Copyright 1989)

Before 15 July	*After 15 July*
Clean: is defined as a pasture that is sufficiently free from infection that susceptible animals grazing on it will not become a source of contamination. If worm-free animals are put on a *clean* pasture it will remain safe for the rest of the season.	
1. New seeds after an arable crop. If grazed by cattle in the previous autumn, these should have been dosed immediately before they were put on.	1. Aftermath not grazed by cattle earlier in the year.
2. Pasture grazed by sheep only in the previous year.	2. Pasture grazed by sheep in the first half of the grazing season.
3. Grassland used for conservation only in the previous year.	
Safe: is defined as a pasture not sufficiently infested to affect the production of susceptible animals that are grazing it but they will become a source of contamination	
1. Pasture grazed in the previous year by cattle in their second or subsequent grazing year.	1. Pasture grazed in the first half of the season by cattle in their third or subsequent grazing year.
2. Pasture grazed in the previous year by beef cows with or without calves at foot.	2. Pasture which was *clean* at the beginning of the season and was grazed thereafter by hand-reared calves straight out of the buildings
3. After 7 May, pastures grazed by calves in the previous year.	
4. Pasture grazed by calves before mid-March or after mid-September of the previous year.	
Acceptable: is defined as usually *safe* unless there are exceptional circumstances or an exceptional year	
1. After 23 April, pasture grazed by calves in the previous year.	1. Pasture grazed by yearling cattle in the first half of the grazing season.
	2. Pasture grazed in the first half of the season by beef cows with calves at foot.
	3. Pasture grazed in the first half of the season by calves dosed every 3 weeks until the end of May.
Potentially unsafe	
1. Before 23 April, pasture grazed by calves in the previous year.	1. Pasture grazed by calves in the first half of the grazing season.

For curative treatment during an outbreak of acute gastroenteritis, two doses of phenothiazine should be given at intervals of a week, followed by one or two further doses at fortnightly intervals. Phenothiazine is less efficient in cattle than in sheep, and alternative drugs which are more efficient against some worms than is phenothiazine have been introduced.

A low dosage of phenothiazine (10.3 mg/kg live weight in the first and 7 mg/kg in the second year) reduced trichostrongylid infection on herbage and in calves grazing a common area of pasture. The effect on herbage infestation was attributed mainly to ovicidal action. The treatment increased weight gain by 22 and 55 per cent in the two seasons indicating a cumulative effect. Since two calves out of 40 in each season developed photosensitivity, i.e. corneal opacity, it is suggested that phenothiazine dosage should be restricted to 6 mg/kg[84].

Organophosphorus compounds
Of the organic phosphorus compounds, haloxon is effective orally against the mature and young adult *Ostertagia* worms, but is less effective against the larvae, especially against the mucosal fourth larval stage. Unlike other organophosphorus compounds, acute toxicity symptoms occur only with very high doses. It is recommended for routine use in young calves at grass in late summer and autumn and is given by drench[9].

Methyridine
Methyridine can be given by mouth or by intraperitoneal or subcutaneous injection and is effective against most worms which cause parasitic gastroenteritis. No other anthelmintic should be given within 24 h of its use, and methyridine should not be used within 2 weeks following the use of an organic phosphorus compound.

Levamisole
This *l*-isomer of tetramisole is highly effective against adult worms and has good activity against the immature stages. It also kills *D. viviparous* worms, responsible for husk.

Experiments in nine different areas of the USA showed no difference in anthelmintic activity of a pour-on formulation of levamisole in warm weather (16° to 36°C) and cold weather in winter (−4° to +7°C) at a level of 10 mg/kg live weight in naturally infected calves when mean worm burdens were compared at slaughter, 7–9 d after treatment. The burdens of *Bunostomum phlebotomum, Cooperia* spp., *H. placei*, *Nematodirus* sp., *Oesophagostomum radiatum, O. ostertagi* and *Trichostrongylus axei* in treated calves were reduced by 83–100 per cent in summer and by 89–100 per cent in winter. Similarly, faecal egg counts were reduced by 90 per cent in summer trials (27°–36°C) and by 94 per cent in winter trials (−18°C to +10°C)[78].

Ivermectin
Two studies have shown the persistent activity of ivermectin (a derivative of avermectin) injected subcutaneously at 200 μg/kg in preventing the establishment of induced infections of *O. ostertagi* and *C. oncophera*. The reduction in mean worm count, compared to that of a control group, were for *O. ostertagi* >99, 45 and 94 per cent with 7, 14 or 21 d interval between treatment with ivermectin and administration of infective larvae respectively in the first trial and >99, >99 and 99 per cent at 7, 10 or 14 d in the second trial. Corresponding values for *C. oncophera* were 99, 0 and 45 per cent at 7, 14 and 21 d respectively in trial 1 and >99, 88 and 31 per cent at 7, 10 and 14 d in trial 2[119].

In a study of 36 autumn-born steers in Hertfordshire on two adjacent sites, treatment with ivermectin 3 and 8 weeks after turn-out in each of 2 years resulted in a reduction in the contamination of pastures compared with that arising from untreated control animals. In both years, dry summers prevented larval challenge building up to high levels until about the time of housing in the autumn. Atypical outbreaks of parasitic gastroenteritis occurred in May and June of the second year in both groups of control calves. Thus, control measures can bring benefits in the early part of the grazing season as well as later in the year[96].

The use of ivermectin (0.2 mg/kg live weight) in 6–16-month-old dairy heifers at 3 and 8 weeks after turn-out to spring pasture in Ohio, USA resulted in a mean adjusted live-weight gain (62.3 kg) that was 62 per cent greater than that of control heifers (38.5 kg) at the time of winter housing. Pasture infectivity was consistently higher in the pastures grazed by the control group with a six-fold difference in the concentration of third stage larvae at the time of winter housing (358 versus 56/kg dry herbage). The increase in weight gain allowed the age at first breeding to be reduced by 1–3 months[124].

Benzimidazoles
Plasma concentrations of five of these anthelmintics have been compared. The compounds were fenbendazole, oxfendazole, febantel, albendazole and thiabendazole given by mouth and oxfendazole also given by intraruminal release bolus[73].

In Louisiana, anthelmintic trials made with albendazole, fenbendazole and ivermectin were found to principally inhibit early fourth larval stages of *O. ostertagi* in naturally infected cattle. The cattle were slaughtered 7–20 d after treatment. *O. ostertagi* was the predominant abomasal nematode recovered with occasional small numbers of *Haemonchus*[76].

Thiabendazole. Thiabendazole is highly effective against adult worms and has good activity against the immature stages. Thiabendazole, having a high limit of toxicity, is suitable for treating heavy mixed infections in debilitated animals. Thiabendazole can also be obtained mixed with another anthelmintic, rafoxamide.

Cambendazole. In the USA, identical twins, a few months old, were used to study the efficacy of cambendazole when given at 2 week intervals to calves on pasture. The calves were on pasture for 4–5 months, then housed for 6 months and returned to pasture for a further 6.5 months. Untreated animals developed moderate parasitic infection of *Cooperia*, *Ostertagia* and *Oesophagostomum* spp. When untreated calves were housed, their hay intake was depressed and they required more nutrients per kg live-weight gain. During the last 111 d of the experiment when the pasture was of poor quality, live-weight gain of the treated animals was considerably higher. Serum albumin concentration and albumin/globulin ratio was lower in the control animals 7 months after the onset of clinical symptoms. It was concluded that gastrointestinal nematode infection affects the performance of calves not only shortly after exposure, but as long as 1 year later especially where affected animals are subjected to adverse conditions[37].

Fenbendazole. In the Republic of Ireland, fenbendazole given daily to calves at 0.4 and 1 mg/kg live weight suppressed faecal output of *Ostertagia* and *Cooperia* spp. eggs and *D. viviparous* larvae in housed calves that had been artificially infected with these parasites; the higher dose rate was more efficacious. In grazing animals fenbendazole at the rate of 1 mg/kg daily, administered intermittently in the drinking water using an automatic device, almost completely suppressed output of trichostrongylid eggs with the result that infection on pasture remained at a low level throughout the grazing season[120].

Albendazole. In Denmark, a comparison was made of albendazole given as an oral drench, through supplementary feed or in the drinking water to four groups of Danish Black Pied heifers grazing separate permanent pasture paddocks in their first season. No clinical disease or weight depression occurred in any of the groups. Although infection level was low, faecal excretion of trichostrongyle eggs, serum pepsinogen activity and larval contamination of the pasture were all markedly reduced in the treated calves. Serum pepsinogen levels were similar for the oral drench and for the supplementary feed groups, at <1 unit tyrosine/l serum, whilst that for the drinking-water group was intermediate between that value and that of the control group. Mixing the drug with a commercial calf concentrate was labour-saving, the pellets containing albendazole being given on 10 June; 8 and 12 July; 8 August and 6 and 30 September, whereas pellets without albendazole were given on 7 and 9 June; 4 and 5 July; 6 August and 4 September[74].

Reticulo-rumen boluses

Morantel (morantel tartrate) sustained-release bolus. The presence of morantel in the faeces of dosed calves prevented the maturation of approximately 99 per cent of *O. ostertagi* eggs to infective larvae between 7 and 84 d after administration of a morantel tartrate bolus and 75 per cent of eggs to infective larvae on day 91. The bolus having a well recognized effect of clearing pastures of *O. ostertagi* can be used as part of a system in which all first year grazing calves are given the device at turn-out onto pasture on which they remain set-stocked throughout the grazing season. During the next 90 d morantel tartrate, released continuously from the bolus residing in the rumen or reticulum, is sufficient to prevent the establishment of infections of *O. ostertagi* and *C. oncophera* arising from infective larvae on the herbage ingested. Moreover, the midsummer rise of larvae on the pastures caused by egg deposition from infections acquired in the spring is eliminated and late summer clinical parasitic gastroenteritis is avoided[71].

A comparison of the intake of two groups of 20 Friesian steer calves grazed from May to September 1984 on adjacent 2 ha paddocks that had been contaminated the previous autumn by cattle infected with *O. ostertagi* showed that the group that was untreated lost 1.25 kg/d in the late autumn. At housing the mean live weight was 277 kg for the group treated with the morantel sustained-release bolus compared to 230 kg for the untreated group. The mean daily organic matter intakes (g/kg $M^{0.75}$) were similar in August at 76 and 77 g respectively but in September the values were 94 and 77 g. Since the apparent digestibility of organic matter did not differ between the groups, the loss in weight of the untreated calves was due to a reduction in appetite rather than an impairment in digestion[38].

However, from a study in S.W. Scotland of the control of parasitic gastroenteritis by the use of the morantel sustained-release bolus, it was clear that considerable numbers of infective larvae of *O. ostertagi* and *C. oncophera* survived on pastures which had not been grazed at all for at least 18 months. Thus, after one or even two seasons with no further contamination from grazing, permanent cattle pastures cannot be considered 'safe'[68].

A further report by these workers studied the efficacy of the bolus in three groups of calves housed from October to May that had previously grazed on permanent pastures contaminated with infective larvae of *Ostertagia* and *Cooperia* spp. During the first grazing season, one group had received 7.5 mg fenbendazole/kg at fortnightly intervals to suppress trichostrongyle infections, the first treatment being administered after 1 d at grass, another group had received a morantel sustained-release bolus before grazing to limit trichostrongyle contamination of the pasture and the control group was only treated (with levamisole and diethylcarbamazine) when heavy infestation in August, after 13 weeks grazing, caused clinical type 1 ostertagiasis and husk. At the same time diethylcarbamazine was given to the morantel-treated group. During the period of housing, digestibility of the whole diet, particularly that of dry matter, crude protein and energy was lower for the control animals, which had increased faecal and urinary nitrogen outputs resulting in a reduced overall nitrogen retention. The effect was most apparent during clinical type 2 ostertagiasis (caused by previously arrested early fourth-stage larvae)[4], which occurred in March in the control group[69].

During their second grazing season, the mean live-weight gains of the fenbendazole-treated, morantel-treated and control groups during 152 d were 105, 131 and 109 kg respectively. The cattle were slaughtered after an indoor fattening period and the dressed carcass weight and killing-out percentage were superior in the treated groups. Rib analyses showed lower total weight, eye muscle weight and area together with a higher bone content in the control cattle[70].

Prior administration of a morantel sustained-release bolus prevented the pathological changes that occurred with daily infections of calves with *C. oncophera* during a 6 week period. The changes included inappetence, weight loss, impaired nitrogen retention and loss of plasma protein into the gut. The infection occurred throughout the length of the small intestine and many larval stages were present in the mucosa. During the first 10 weeks after first infection stunting and thickening of the villi with excessive mucus production was observed. Most of the worms had been expelled by 12 weeks, but a loss of plasma protein to the gut was still occurring[72].

Albendazole pulse-release device. Albendazole was used in an intraruminal pulse-release electronic device in 12 young Charolais cattle of 300 kg live weight. The drug was released three times at the rate of 2 g at 31 d intervals. The bioavailability of the drug appeared to be similar to that from an oral drench.

The electronic device (E bolus) was administered by a balling gun and its high specific gravity allowed its retention in the reticulo-rumen indefinitely. After immersion in the ruminal fluid, two sensory conductive rubber electrodes completed an internal microcircuit contained on a computer chip. After counting for 10 min, programming instructions in a related circuit again looked at the input and if it was still registering, the E bolus was switched-on irreversibly[83].

Oxfendazole pulse-release bolus. The oxfendazole pulse-release bolus (OPRB), provides five therapeutic doses at approximately 3 week intervals after it is given at turn-out The 'front-loaded' bolus (FLOPRB) is similar but provides six doses at 3 week intervals, the first being released within hours of administration; it is useful for farmers who put susceptible cattle on potentially contaminated pasture late in the season[65].

It is claimed that the OPRB reduced pasture infectivity by 96 per cent and that the FLOPRB almost completely prevented the establishment of both gastrointestinal and lungworm populations in calves grazing on contaminated pastures[64]. Trichostrongylid egg output of calves throughout the first grazing season was reduced by 97–99.9 per cent[59]. Not only does the device reduce the accumulation of infective larvae on the pasture but also appears to suppress lungworm infections[60] (but see below).

It has been questioned whether animals thus protected receive sufficient antigen stimulus to induce immunity. However, a comparison of two groups of yearling heifers, of which one group had carried the intraruminal device, showed that there was no difference in immunity when they were given a single massive challenge of *D. viviparous* or *O. ostertagi*[58].

A test of the reliability of OPRB, which is designed to release five doses at 21 d intervals showed that the pulse interval had a range from 16–23 d[67].

A comparison of the OPRB with other anthelmintic strategies has been made. During the first grazing season, a group of calves treated with the device achieved a mean weight gain of 141 kg compared with 107 kg for a group treated once with ivermectin mid-season and 117 kg by a group which received no treatment. The economic advantage was maintained during the period of winter housing and by the end of the second grazing season, during which animals received no anthelmintic medication, the group with the OPRB weighed 20 kg more than did the untreated control, but this difference was not statistically significant. There were no signs of clinical disease in animals given the OPRB or in the undosed control animals. In the first year the treatment with the OPRB greatly reduced the degree of pasture contamination, but in the second year the animals treated in the first year developed higher worm egg counts and thus augmented the levels of pasture contamination compared with the untreated control group. *N. battus* and *N. fillicollis* both produced low grade and fertile infestations in the calves. Thus to maintain the economic advantage gained in the first year protection should be offered in the second year[63].

That comparison of the release of five doses of oxfendazole with one mid-season dose of ivermectin in providing protection against clinical parasitism and consequential pasture contamination over a 4 month period has been criticized[61], since other evidence has shown that strategic doses of ivermectin at 3, 8 and 13 weeks after turn-out gave similar efficacy to that claimed for the bolus[62, 123].

In another study of the OPRB, 71 Friesian steers of 6–9 months were vaccinated against lungworm and were divided into two groups, each rotationally grazed on separate areas of permanent pasture for 6 weeks followed by additional access to silage aftermath. One group received the OPRB at turn-out, the other control group was treated 99 d after turn-out and when housed with 7.5 mg fenbendazole/kg live weight. The faecal worm egg counts and plasma pepsinogen activity were significantly reduced, and live-weight gain in the younger calves increased, in those given the OPRB compared to those of the control calves. Moreover, pasture larval counts were lower in the fields grazed by animals treated with the OPRB[99].

In Belgium, the OPRB was given at the time of turn-out and provided good control of parasitic gastroenteritis, dominated by *Ostertagia*, in grazing calves in their first season. Parasitic gastroenteritis was greatly reduced with low values of serum pepsinogen and *Ostertagia* antibody titres. Similarly, live-weight gain was improved and normal total plasma protein and albumin concentrations were maintained, whereas untreated calves had hypoproteinaemia and hypoalbuminaemia. Most animals exhibited clinical signs of parasitic bronchitis at the end of the grazing season, so the OPRB may not adequately control parasitic bronchitis on all occasions[66].

Netobimin

Netobimin given in drinking water to calves of 118 kg live weight at 2.8 mg/kg, but not at 1.5 mg/kg (assuming a calf drank 4.5 litres water daily during a medication period of 7 d) was sufficient to suppress both trichostrongylids and *D. viviparous* and compared favourably with an oxfendazole regimen of 1.0 mg/kg[75].

Drug resistance
Resistance to fenbendazole by *C. curticei* in lambs has been demonstrated in the Netherlands[79] and in Australia a strain of *T. axei* resistant to oxfendazole in weaned steers has been identified[80].

Levamisole-resistant *O. ostertagi* have been found in cattle fattening farms in Merchtem, Flanders[81, 82].

Parasitic bronchitis (husk)

The organism

This condition is caused by a heavy infestation with a nematode worm, *Dictyocaulus viviparous*.

Life cycle

The larvae of this worm enter the animal's body by way of the mouth from the pasture herbage; they lose their protective sheaths in the rumen and then pass to the small intestine where, after penetrating the mucous membrane, they make their way into the lymphatic system and thence into the bloodstream. The larvae are then carried by the bloodstream to the lungs, where they break through to the alveoli and migrate into the smaller bronchial tubes.

In this site, the larvae grow to maturity and produce eggs about 3–4 weeks after infection has taken place. Pulmonary lesions may be found 7 d after infection[10] (i.e. 2 weeks before the worms become adult and give rise to larvae in the faeces). The eggs hatch into larvae which are coughed up, are swallowed and then pass to the ground in the faeces. On first reaching the ground the first-stage larvae are not infective to other animals, but after passing through several larval stages they become infective after a minimum period of 5 d.

Symptoms

The main symptom of the complaint is a husky cough, which is slight and infrequent when the parasites are few in number. In chronic cases there is emaciation, loss of appetite, a staring coat, diarrhoea, pallor of the mucous membrane and sunken eyes. These symptoms may be manifested for as long as 4–5 months before death or recovery takes place. In acute cases the cough is deeper and more husky, and fits of coughing occur, with great difficulty in breathing. The respiration and pulse rate are raised, the head is extended, the tongue often protrudes, and mucus flows from the nostrils and corners of the mouth. The animal may even fall prostrate as a result of partial suffocation. Coughing fits may occur several times in a day, and single worms or masses of worms and streaks of blood may sometimes be seen in the mucous discharge.

Immunity

Calves are most susceptible to the condition, and yearlings frequently become infected. It is probable that immunity to the condition develops in a similar way to that against parasitic gastroenteritis. Calves sucking their dams at pasture seem less susceptible than bucket-fed calves, and it appears that Channel Island cattle are particularly susceptible to the acute form of the disease.

In herds not vaccinated against husk, 67 per cent had high average titres, 14 per cent had no detectable antibody against lungworm and 20 per cent had a low mean titre by the ELISA test[105]. However, immunity of cattle to *D. viviparous*, whether from vaccination or natural exposure, is not life-long. It requires stimulation by repeated or continuous exposure to the parasite[114].

Grazing management

Husk is associated with damp, low-lying permanent pastures. Nevertheless, the husk worm may also be present on well-drained dry pastures, especially where the herbage is thick and long. There is also considerable danger of infection from contaminated water supplies.

Small fields and orchards adjacent to farm buildings are particularly dangerous places as they tend to be so frequently used by young animals that they become heavily infested. Where a pasture is heavily infested, very considerable benefit will be obtained by grazing with non-susceptible animals such as horses and sheep.

Pastures from which cattle have been completely excluded throughout the winter are usually free from infection in the spring, but this is not necessarily so, for in the climatic conditions of the W. Scotland lungworms may survive for 13 months[11]. Thus, late turn-out to grass should result in safer grazing[107]. In Denmark, it has been found that larvae that survive the winter abruptly disappear during late April to early May[108]. A 14 d delay in turn-out in May resulted in a 10-fold decrease in subsequent larval recoveries from faeces and herbage[109].

Studies of the epidemiology of *D. viviparous* in Denmark suggested that adult lungworms present in calves about 6 weeks after turn-out play an important role in determining the subsequent pattern of disease.

However, outbreaks of infection have been reported in calves turned out onto aftermath[110–112].

D. viviparous can be acquired by housed cattle if they are given fresh grass from contaminated pastures. It was considered that infections by this means in Ecuador could complicate efforts to prevent the disease by vaccination[106].

By ploughing up grassland, the husk worm larvae can be disposed of more effectively than can the larvae responsible for gastroenteritis, which tend to make their way upwards through the soil again. Even so, it is thought that larvae may sometimes persist in the soil to emerge on the grass after a long period[111, 113].

In the Netherlands, a sero-epidemiological survey of *D. viviparous* infection was made on 13 farms by measuring antibody titres with ELISA. Of the herds, 75 per cent were infected; calves on farms where zero-grazing was adopted had a lower level of infection than did calves grazing pastures. Calves had clinical husk on 15 per cent of the farms, whilst 51 per cent of farms had experienced husk in the past. The later the calves were housed in the autumn, the greater was the level of infection. The number of calves grazing together and the stocking rate had a significant positive effect on the level of infection. Surprisingly, there was no significant difference in the occurrence of clinical husk in calves vaccinated or not vaccinated against the disease[123].

Manure

Manure heaps and dirty stockyards where a small amount of herbage is growing are also sources of infection. Manure heaps should be fenced and rain water which has washed the surface of the heap should be removed by proper drainage. Infected manure should not be used on grassland.

Spreading of dung pats by chain harrowing may be dangerous as, by improving the aeration of the dung, the young larvae in the centre of the faecal mass are helped in their development. The use of the chain harrow should therefore be avoided except in dry weather and when the herbage is short and the surface of the ground exposed.

The role of Pilobolus

In 1982, it was suggested from Denmark that *D. viviparous* infection could be transmitted from adjacent fields by projectile spores of the fungus *Pilobolus* sp.[104]. Further work from the Irish Republic showed that *Pilobolus kleinii*, which is specific to ruminant dung, was involved in the dispersal of *D. viviparous* infective third-stage larvae from bovine faeces onto pasture. It was suggested that the sporangia may protect the larvae from death by desiccation[100, 101].

A study of the environmental factors affecting the growth of *Pilobolus* on the surface of dung showed that increases in relative humidity, rainfall and herbage height stimulated growth, whilst increases in hours of sunshine, air temperature and wind speed depressed growth. These variables accounted for 38 per cent of the variation in growth of *Pilobolus* sp. There was a 19-fold increase in numbers of *D. viviparous* third-stage larvae recovered from the herbage surrounding dung pats containing *P. kleinii* that had been infected with first-stage larvae[100, 102].

Other work showed that a substantial proportion of lungworm larvae present in the faeces were transferred by this fungi within 8 d at a temperature of 15°C. Between the beginning of July and middle of September, peak emergence of the sporangia occurred within 1 week and most sporangia emerged within 3 weeks. Emergence of the sporangia was lower from faeces deposited at the end of September to mid-October[103]. The incidence of extensive damage to natural dung pats within 5 d of deposition varied from 0 to 92 per cent of pats depending on their degree of dryness; this may be another factor in dispersal of *D. viviparous* third-stage larvae. Bird activity may also be involved.

Prevention

Nutrition

For the prevention of the condition, it is essential to maintain the cattle on a reasonable level of nutrition and protect them against a heavy infestation. In some localities the risk of severe infestation among calves is so great that it is best to keep the animals in buildings during their first summer. Directly calves out at grass show symptoms of the disease, they should be taken off the infected pasture and housed. When indoors, the calves should be fed on a high plane of nutrition so that they have a good chance of overcoming the infection. Although constant moving of cattle from pasture to pasture is usually recommended as a help towards the limitation of infestation, research work[12] has indicated that, to build up an immunity, calves should be exposed to a low level of herbage infestation and then continue to have constant access to that infected herbage. Where risk of husk infection is great, concentrates should be fed to young stock as soon as the quality of the pasture starts to deteriorate.

Vaccination

An oral vaccine is available as a protection against husk. This vaccine is based on husk worm larvae, irradiated by X-rays to make them sufficiently weak so that they do not migrate further than the mesenteric lymph nodes but sufficiently strong to stimulate immunity[16]. Two vaccinations are necessary. The first can be given after the calf is 6–8 weeks old and the second a month later. Immunity should be developed within 2–4 weeks after the second vaccination. However, even with vaccinated calves, resistance may break down if they are suddenly exposed to a heavy infection.

Drugs

Levamisole
In a trial in England to test a prophylactic treatment, 30 autumn-born Friesian or Friesian-cross bull calves were allowed to graze a 5 ha field for 6 weeks after turn-out and were then divided into two matching groups. One group was given levamisole at this time and again 2 weeks later, whilst the other was a control group. Levamisole treatment resulted in a marked reduction in larval excretion and considerably delayed the build-up of infection on pasture. Moreover, the onset of disease was delayed and the severity of clinical signs reduced. Since disease was not eliminated completely, this prophylactic programme could not be recommended[117].

Ivermectin
A programme of injections of ivermectin at 3, 8 and 13 weeks after turn-out is based on the assumption of early season transmission of *D. viviparous* through susceptible calves and thus aims to limit pasture contamination. The programme does not interfere with the development of immunity in cattle exposed to challenge from pasture[114].

In two studies of the persistent activity of subcutaneous injections of ivermectin at 200 μg/kg live weight in preventing establishment of induced infections of *D. viviparous*, the reductions in worm count with 7, 14 or 21 d intervals between injection of ivermectin and administration of infective larvae were >99, 98 and >99 per cent in trial 1 respectively and 100 per cent for all intervals in trial 2[119].

The use of this ivermectin programme has been questioned by other workers. For instance, it is not known whether treated calves were immune to reinfection in the second grazing season. It was also considered that giving ivermectin three times in the spring and early summer, or four times if the calves were turned out before 1 May, might lead to drug resistance. Because of the spread of infected larvae from farm to farm by *Pilobolus* sp. and the fact that larvae can reappear on pasture after long periods in the soil, it was doubted whether the drug could be completely effective, especially as clinical disease could be produced by as few as several hundred lungworms whereas many thousands were needed to cause clinical ostertagiasis.

It was considered that if eradication of lungworm was achieved in a herd, immunity to reinfection would rapidly wane and in a short time the herd would be very susceptible to reinfection from neighbouring farms, e.g. by *Pilobolus* sp. dissemination[118].

Fenbendazole
In the Republic of Ireland, fenbendazole at 0.4 and 1 mg/kg live weight, given daily in drenches, suppressed faecal output of *D. viviparous* larvae in housed calves that had been artificially infected with the parasite; the higher dose was the most effective. Fenbendazole administered intermittently using an automatic dose dispenser to drinking water at the rate of 1 mg fenbendazole/kg live weight daily suppressed faecal output of *D. viviparous* for most of the grazing season, but calves became infected with the parasite towards the end of the season[77].

Oxfendazole pulse-release bolus
With calves exposed to a heavy early-season challenge with *D. viviparous* on two 1.41 ha paddocks, a front-loaded oxfendazole pulse-release bolus (FLOPRB), in which one dose is given immediately followed by a further five doses at 21 d intervals, was used on five of 11 calves on one paddock as soon as they became clinically ill with parasitic bronchitis 34 d after turn-out. Clinical signs quickly subsided and no further problem occurred despite continued exposure to reinfection. In the other paddock, an ordinary oxfendazole pulse-release bolus was given at turn-out to 11 calves. No patent infections occurred and the infectivity of the pasture in August and September was reduced by 94 per cent. Calves treated at turn-out performed better than those treated with the FLOPRB when clinically ill and had an average live-weight gain advantage of 20.4 kg at housing. However, prepatent disease, i.e. existing in an unobserved state, did occur when susceptible calves were exposed to a heavy challenge after the first anthelmintic dose had been released from the bolus 21 d after treatment. Two calves required treatment then, but otherwise the disease was terminated by the second pulse of oxfendazole[121].

Vaccination versus drugs

On the basis of live-weight gain, respiration rate and faecal egg and larval counts, cattle which had received ivermectin during their first season were resistant to challenge with lungworm and gastrointestinal parasites in their second grazing season. The level of immunity of ivermectin-treated cattle was equivalent to that of second season cattle which had been vaccinated against lungworm before first turn-out. Thus, the use of prophylactic early-season treatment with ivermectin, where calves graze pasture sufficiently contaminated with *D. viviparous,* can allow development of natural immunity to the parasites[115].

In N. Ireland, a comparison of three groups of calves kept on separate paddocks and infected with *D. viviparous* was made. The control group was untreated, whilst one group was immunized with live irradiated lungworm, and the other group was injected at 3, 8 and 13 weeks after turn-out with ivermectin. During the first half of the grazing season all untreated and three of six vaccinated calves excreted *D. viviparous* larvae, whilst none were excreted by the ivermectin-treated group. Parasite-free tracer calves became infected when grazing the paddocks occupied by the control and vaccinated groups, but only one worm occurred in one of 10 tracer calves grazing the paddock occupied by the ivermectin-treated calves[116].

In a comparison of ivermectin injections with vaccination, four groups of 10 4–5-month-old calves were turned out onto permanent pasture seeded with larvae of *D. viviparous*. One group was vaccinated with lungworm vaccine before turn-out and treated with thiabendazole at 3, 8 and 13 weeks after treatment, another group was injected subcutaneously with ivermectin at 3, 8 and 13 weeks after turn-out and the third group was injected with ivermectin at 3 and 8 weeks after turn-out. A severe outbreak of parasitic bronchitis caused the death of three control calves within 5 weeks of turn-out. Parasitic bronchitis and gastroenteritis affected the vaccinated group after about 4 months at pasture. Calves given ivermectin excreted no lungworm larvae and remained free of clinical parasitism. Two treatments, at 3 and 8 weeks, with ivermectin was adequate but it was suggested that with a heavy challenge and a longer grazing season, a third treatment at 13 weeks would be necessary. It was thought that the poor protection afforded by vaccination might be due to poor nutrition as a result of the waterlogged pasture or to heavy burdens of gastrointestinal parasites accumulated in August and September[122].

Treatment

Phenothiazine

Treatment with phenothiazine, similar to that recommended for parasitic gastroenteritis, may be of benefit, as the two conditions are often present at the same time. Thus, recovery from parasitic gastroenteritis may help the calf to recover from a husk infection. Alternatively, injections into the trachea have been found to be effective in some cases. These methods have largely been superseded by other drugs.

Cyanacetohydrazide

This drug may be given orally or by injection. Cyanacetohydrazide causes the worms to become inert and be moved up the trachea by ciliary action[13]. However, it must be borne in mind that deaths of calves may sometimes occur as a result of lesions produced by the larvae during migration. The effect of the drug on immature worms appears to be limited. For subcutaneous injection 15 mg/kg body weight up to a maximum total dose of 5 g can be used, whereas by mouth 18 mg/kg up to the same maximum can be given. A single treatment should be sufficient, but with animals severely infected treatment should be carried out on three consecutive days. Animals should not be excited during treatment.

Diethylcarbamazine

Diethylcarbamazine is highly effective against immature worms, but almost without effect against adult worms. There is some difference of opinion as to whether treatment with this drug reduces immunity to subsequent infection. The drug was given at a level of 300 mg/kg live weight to 7.5–16-week-old calves at 14 d after challenge with larvae; after a second challenge more worms were recovered from the treated calves than from the controls[14]. With a lower dose rate, however, no difference in immunity to a second infection was found[15].

Methyridine and tetramisole

Methyridine and tetramisole given by mouth are effective against all stages of lungworms and also against many of the gastrointestinal helminths which may cause concurrent infections.

Ivermectin

In Belgium, a comparison was made of calves given levamisole (*l*-tetramisole) (10 mg/kg live weight) and ivermectin (200 μg/kg live weight) with a control group in the treatment of 18 3-month-old calves infected 7 d previously with *D. viviparous* third-stage larvae at the rate of 30/kg live weight. The efficacy of ivermectin given at a therapeutic dose against immature *D. viviparous* was greater than that of levamisole given at twice the recommended dose of 5 mg/kg live weight[120].

Ringworm

The organism

This fungal disease in cattle is caused mainly by *Trichophyton verrucosum*.

Symptoms

The very resistant spores become lodged in cracks on the surface of the skin and in the hair follicles. After germination the fungal hyphae grow on the inside or outside of the hair shaft and the hair breaks off. The earliest lesion is a localized raised area on the skin on which the hair stands up. This gradually develops in size, becomes sharply defined and covered with grey or yellowish-grey scales which increase in thickness. The hair on the affected part then breaks off above the scaly crust and leaves a bare patch. The crusts, in turn, fall off naturally in 1 or 2 months, leaving a bald patch, which finally gives place to a new growth of hair. The main sites of the lesions are the head, neck and anal region, and they often give rise to pronounced itching.

Resistance to infection

Calves and yearlings are most susceptible, and the disease may develop under conditions of poor nutrition, cold, wet and worm or lice infestation. The complaint is more prevalent during the winter months, and it is possible that bright sunlight may exert an inhibiting action on the growth of the fungus. Recovery from one ringworm infection usually confers immunity against a subsequent infection.

It has been suggested that cellular immunity modulates the immune response to ringworm and that humoral factors control the duration of infection. In E. Europe a vaccine is available, which appears to confer lasting protection[45].

Where calves in a poor state of nutrition are affected, the disease may last for a year or more, but older animals usually make a speedy recovery when placed out of doors in bright, sunny weather and when they have access to spring grass.

Prevention and treatment

Opinions differ as to the efficacy of various ringworm remedies which may be obtained from any veterinary chemist. By washing the cattle with a dilute solution of formalin (2 per cent) containing soap or a detergent, the spread of infective spores to other parts of the body or to other individuals may be reduced. During this process the crusts may be removed and scraped into a container. It may be necessary to soften the crusts by applying a mixture of equal parts of soft soap and lard, and to remove the crusts on the following day. The lesions or healthy skin around them should not be scrubbed or rubbed hard, as this has been shown to increase the spread of infection. The ringworm ointment should then be applied and a second application made, if necessary, a week later. To prevent the spread of infection, treated animals should be segregated and the premises disinfected. Particular attention should be paid to parts of the buildings against which the infected animals may have rubbed. The premises should be scraped down, scrubbed with hot detergent or soda solution and later flamed with a blow-lamp. Bedding should be burned and all woodwork creosoted.

On account of the transmissibility of the disease in man, overalls and gloves should be worn while dressing the animals and these should be disinfected after use. Similarly, the hands, arms and neck should be washed and disinfected. Grooming equipment should be disinfected by immersion in 2 per cent formalin for 24 h.

Griseofulvin, a metabolic product of a certain species of *Penicillium*, will, when given by mouth, prevent and cure ringworm[17]. This drug is incorporated into freshly formed skin and hair, and inhibits the growth of the fungus in these tissues. The dose rate is about 33–44 mg/kg body weight daily for a period of up to 3 weeks.

References

1. NEWMAN, A. J. *Vet. Rec.* **79**, 240 (1966)
2. FITZGERALD, P. R. *J. Protozool.* **12**, 215 (1965)
3. MICHEL, J. F. *Parasitology* **53**, 63 (1963)
4. ANDERSON, N., ARMOUR, J., JARRETT, W. F. H., JENNINGS, F. W., RITCHIE, J. S. D. and URQUHART, G. M. *Vet. Rec.* **77**, 1196 (1965)
5. ROY, J. H. B. *Vet. Rec.* **76**, 511 (1964)
6. TAYLOR, E. L. *Outl. Agric.* **3**, 139 (1961)
7. ROSS, J. G. and DOW, C. *Br. vet. J.* **120**, 279 (1964)
8. LELAND, S. E. Jr., DRUDGE, J. H. and DILLARD, R. P. *Am. J. vet. Res.* **27**, 1555 (1966)
9. ARMOUR, J. *Vet. Rec.* **76**, 1364 (1964)
10. JARRETT, W. F. H., McINTYRE, W. I. M. and URQUHART, G. M. *J. Path. Bact.* **73**, 183 (1957)
11. JARRETT, W. F. H., McINTYRE, W. I. M. and URQUHART, G. M. *Vet. Rec.* **66**, 665 (1954)
12. MICHEL, J. F. *Vet. Rec.* **69**, 1118 (1957)
13. WALLEY, J. K. *Vet. Rec.* **69**, 815, 880 (1957)
14. KENDALL, S. B. *J. comp. Path.* **75**, 443 (1965)
15. CORNWALL, R. L. *Res. vet. Sci.* **4**, 435 (1963)
16. JARRETT, W. F. H., McINTYRE, W. I. M., JENNINGS, F. W. and MULLIGAN, W. *Vet. Rec.* **69**, 1329 (1957)
17. LAUDER, I. M. and O'SULLIVAN, J. G. *Vet. Rec.* **70**, 949 (1958)
18. HARRISON, I. R. *Vet. Rec.* **70**, 849 (1958)
19. KENNY, J. E. and THORNBERRY, H. *Ir. vet. J.* **12**, 198 (1958)
20. TURNER, E. C. Jr. and GAINES, S. A. *J. econ. Ent.* **51**, 582 (1958)
21. VETERINARY INVESTIGATION SERVICE AND CENTRAL VETERINARY LABORATORY, MINISTRY OF AGRICULTURE, FISHERIES AND FOOD. *Vet. Rec.* **74**, 127 (1962)
22. BROWN, F. G. *Vet. Rec.* **74**, 577 (1962)
23. BROWN, F. G. *Vet. Rec.* **75**, 585 (1963)
24. MILK MARKETING BOARD. *Report of the Breeding and Production Organization* No. 24 (1973–74)
25. MEAT AND LIVESTOCK COMMISSION. *Br. Fmr Stkbreed* **6**(154), 24 (1977)
26. VERCOE, J. E. and SPRINGELL, P. H. *J. agric. Sci., Camb.* **73**, 203 (1969)
27. PAVLASEK, I. *Vet. Med., Praha* **23**, 411 (1978)
28. McDOUGALD, L. R. *Am. J. vet. Res.* **39**, 1748 (1978)
29. LUDVIGSEN, J. B. *3rd EAAP Symposium on Protein Metabolism and Nutrition, Braunschweig*, p. 515 (1980)

30. LUDVIGSEN, J. B. *Proceedings of 7th Symposium on Energy Metabolism of Farm Animals* (eds. M. Vermorel and G. de Bussac), p. 311 (1976)
31. DHARSANA, R. S., FABIYI, J. P. and HUTCHINSON, G. W. *Vet. Parasitol.* **2**, 333 (1976)
32. CONNAN, R. M. *Vet. Rec.* **99**, 476 (1976)
33. CANALE, A., VALENTE, M. E., DOTTA, U. and BALBO, T. *Folia Vet. Lat.* **7**, 82 (1977)
34. GIBSON, T. E. *Proc. Nutr. Soc.* **22**, 15 (1962)
35. BROADBENT, J. S., LATHAM, J. O., MICHEL, J. F. and NOBLE, J. *Grazing Plans for the Control of Stomach and Intestinal Worms in Sheep and in Cattle.* Ministry of Agriculture, Fisheries and Food Booklet 2154 (1982)
36. UGARTE, J., PRIETO, R. and PRESTON, T. R. *Cuban J. agric. Sci.* **8**, 145 (1974)
37. VAN ADRICHEM, P. W. M. and SHAW, J. C. *J. Anim. Sci.* **45**, 417 (1977)
38. BELL, S. L., FROST, A. I., JEFFREY, Y. and THOMAS, R. J. *Anim. Prod.* **42**, 465 (1986)
39. AGRICULTURAL RESEARCH COUNCIL. *ARC News,* Spring, p. 9 (1983)
40. MINISTRY OF AGRICULTURE, FISHERIES AND FOOD. *Warble Fly.* Leaflet 533 (1987)
41. SINCLAIR, I. J., TARRY, D. W. and WASSALL, D. A. *Res. vet. Sci.* **37**, 383 (1984)
42. EVSTAF'EV, M. N. *Parazitologiya* **14**, 197 (1980)
43. WILSON, G. W. C. *Vet. Rec.* **118**, 653 (1986)
44. PETERS, A. R. and MELROSE, D. R. *Vet. Rec.* **117**, 261 (1985)
45. MacKENZIE, D. W. R. *Vet. Rec.* **120**, 348 (1987)
46. DOY, T. G. and HUGHES, D. L. *Res. vet. Sci.* **37**, 219 (1984)
47. BULGIN, M. S., ANDERSON, B. C., HALL, R. F. and LANG, B. Z. *Res. vet. Sci.* 37, 167 (1984)
48. OLDHAM, G. *Res. vet. Sci.* 39, 357 (1985)
49. RAPIC, D., DZAKULA, N., SAKAR, D. and RICHARDS, R. J. *Vet. Rec.* **122**, 59 (1988)
50. STANSFIELD, D. G., LONSDALE, B., LOWNDES, P. A., REEVES, E. W. and SCHOFIELD, D. M. *Vet. Rec.* **120**, 459 (1987)
51. STANSFIELD, D. G. and LOWNDES, P. A. *Vet. Rec.* **118**, 620 (1986)
52. PIPANO, E., KRIGEL, Y., MARKOVICS, A., RUBINSTEIN, E. and FRANK, M. *Vet. Rec.* **117**, 413 (1985)
53. TAYLOR, S. M., ELLIOTT, C. T. and KENNY, J. *Vet. Rec.* **118**, 98 (1986)
54. PIPANO, E., MARKOVICS, A., KRIEGEL, Y., FRANK, M. and FISH, L. *Res. vet. Sci.* **43**, 64 (1987)
55. SHARMA, S. P. and BANSAL, G. C. *Res. vet. Sci.* **37**, 126 (1984)
56. TAYLOR, S. M., KENNY, J., MALLON, T. R., ELLIOTT, C. T., McMURRAY, C. and BLANCHFLOWER, J. *Vet. Rec.* **114**, 454 (1984)
57. TAYLOR, S. M., ELLIOTT, C. T. and BLANCHFLOWER, J. *Vet. Rec.* **116**, 620 (1985)
58. JACOBS, D. E., PITT, S. R., FOSTER, J. and FOX, M. T. *Res. vet. Sci.* **43**, 273 (1987)
59. JACOBS, D. E., FOX, M. T., GOWLING, G., FOSTER, J., PITT, S. R. and GERRELLI, D. *J. vet. Pharm. Ther.* **10**, 30 (1987)
60. JACOBS, D. E., GOWLING, G., FOSTER, J., FOX, M. T. and OAKLEY, G. A. *J. vet. Pharm. Ther.* **9**, 337 (1986)
61. JONES, P. G. H. *Vet. Rec.* **122**, 143 (1988)
62. TAYLOR, S. M., MALLON, T. R. and KENNY, J. *Vet. Rec.* **117**, 521 (1985)
63. HERBERT, I. V. and PROBERT, A. J. *Vet. Rec.* **121**, 536 (1987)
64. MORGAN, D. W. T. and ROWLANDS, D. AP T. *Proc. XIVth Wld Congr. Dis. Cattle, Dublin,* p. 136 (1986)
65. JACOBS, D. E., PILKINGTON, J. G., FOSTER, J., FOX, M. T. and OAKLEY, G. A. *Vet. Rec.* **121**, 403 (1987)
66. VERCRUYSSE, J., DORNY, P., BERGHEN, P. and FRANKENA, K. *Vet. Rec.* **121**, 297 (1987)
67. BOGAN, J. A., ARMOUR, J. A., BAIRDEN, K. and GALBRAITH, E. A. *Vet. Rec.* **121**, 280 (1987)
68. BAIRDEN, K., ARMOUR, J. and McWILLIAM, P. N. *Res. vet. Sci.* **39**, 116 (1985)
69. ENTROCASSO, C. M., PARKINS, J. J., ARMOUR, J., BAIRDEN, K. and McWILLIAM, P. N. *Res. vet. Sci.* **40**, 65 (1986)
70. ENTROCASSO, C. M., PARKINS, J. J., ARMOUR, J., BAIRDEN, K. and McWILLIAM, P. N. *Res. vet. Sci.* **40**, 76 (1986)
71. ROSSITER, L. M., PURNELL, R. E. and SEYMOUR, D. J. *Vet. Rec.* **122**, 81 (1988)
72. ARMOUR, J., BAIRDEN, K., HOLMES, P. H. *et al. Res. vet. Sci.* **42**, 373 (1987)
73. PRICHARD, R. K., HENNESSY, D. R., STEEL, J. W. and LACEY, E. *Res. vet. Sci.* **39**, 173 (1985)
74. JÖRGENSEN, R. J., NANSEN, P., MIDTGAARD, N. and MONRAD, J. *Vet. Rec.* **121**, 468 (1987)
75. DOWNEY, N. E. *Vet. Rec.* **121**, 275 (1987)
76. SNIDER III, T. G., WILLIAMS, J. C., KNOX, J. W., ROBERTS, E. D. and ROMAIRE, T. L. *Vet. Rec.* **116**, 69 (1985)

77. DOWNEY, N. E. and O'SHEA, J. *Vet. Rec.* **116**, 4 (1985)
78. SEIBERT, B. P., GUERRERO, J., NEWCOMB, K. M., RUTH, D. T. and SWITES, B. J. *Vet. Rec.* **118**, 40 (1986)
79. BORGSTEEDE, F. H. M. *Res. vet. Sci.* **41**, 423 (1986)
80. EAGLESON, J. S. and BOWIE, J. Y. *Vet. Rec.* **119**, 604 (1986)
81. BRITT, D. P. *Vet. Rec.* **118**, 467 (1986)
82. GEERTS, S. *Vet. Rec.* **118**, 283 (1986)
83. DELATOUR, P., GYURIK, R. J., BENOIT, E. and GARNIER, F. *Res. vet. Sci.* **43**, 284 (1987)
84. SOMERS, C. J., DOWNEY, H. E. and O'SHEA, J. *Res. vet. Sci.* **43**, 143 (1987)
85. SHOO, M. K. and WISEMAN, A. *Res. vet. Sci.* **41**, 124 (1986)
86. McKELLAR, Q., DUNCAN, J. L., ARMOUR, J., LINDSAY, F. E. F. and McWILLIAM, P. *Res. vet. Sci.* **42**, 29 (1986)
87. ECKERSALL, P. D., MACASKILL, J., McKELLAR, Q. A. and BRYCE, K. L. *Res. vet. Sci.* **43**, 279 (1987)
88. McKELLAR, Q. A., ECKERSALL, P. D., DUNCAN, J. L. and ARMOUR, J. *Res. vet. Sci.* **44**, 29 (1988)
89. JACOBS, D. E. *Vet. Rec.* **121**, 455 (1987)
90. BAIRDEN, K. and ARMOUR, J. *Vet. Rec.* **121**, 326 (1987)
91. SHEIKH-OMAR, A. R., IKEME, M. M. and FATIMAH, I. *Vet. Rec.* **116**, 134 (1985)
92. MOHAN, R. N. *Vet. Bull.* **38**, 735 (1968)
93. CHIEJINA, S. N. and EMEHELU, C. O. *Res. vet. Sci.* **37**, 144 (1984)
94. LANCASTER, M. B. and HONG, C. *Vet. Rec.* **120**, 502 (1987)
95. FOX, M. T., JACOBS, D. E., PITT, S. R. and McWILLIAM, P. N. *Vet. Rec.* **121**, 42 (1987)
96. JACOBS, D. E., FOX, M. T. and RYAN, W. G. *Vet. Rec.* **120**, 29 (1987)
97. MacPHERSON, A., GRAY, D., MITCHELL, G. B. B. and TAYLOR, C. N. *Br. vet. J.* **143**, 348 (1987)
98. GETTINBY, G., ARMOUR, J., BAIRDEN, K. and PLENDERELEITH, R. W. J. *Vet. Rec.* **121**, 487 (1987)
99. MITCHELL, G. B. B. *Vet. Rec.* **121**, 377 (1987)
100. SOMERS, C. J., DOWNEY, N. E. and GRAINGER, J. N. R. *Vet. Rec.* **116**, 657 (1985)
101. SOMERS, C. J. *Res. vet. Sci.* **39**, 124 (1985)
102. SOMERS, C. J. and GRAINGER, J. N. R. *Res. vet. Sci.* **44**, 147 (1988)
103. EYSKER, M. and DE COO, F. A. M. *Res. vet. Sci.* **44**, 178 (1988)
104. JÖRGENSEN, R. J., RÖNNE, H., HELSTED, C. and ISKANDER, A. R. *Vet. Parasit.* **10**, 331 (1982)
105. BAIN, R. K. and SYMINGTON, W. *Vet. Rec.* **118**, 252 (1986)
106. ZURITA, E., VILLALBE, F. P., JARRIN, M. J. H. and SCHILLHORN VAN VEEN, T. W. *Vet. Rec.* **121**, 359 (1987)
107. JACOBS, D. E. and FOX, M. T. *Vet. Rec.* **116**, 75 (1985)
108. JÖRGENSEN, R. J. *Proceedings of Workshop on Epidemiology and Control of Nematodiasis in Cattle,* Copenhagen (eds. P. Nansen, R. J. Jörgensen and E. J. L. Soulsby), Martinus Nijhoff, The Hague, p. 215 (1980)
109. JÖRGENSEN, R. J. *Acta vet. scand.* **21**, 658 (1980)
110. OAKLEY, G. A. *Vet. Rec.* **104**, 460 (1979)
111. DUNCAN, J. L., ARMOUR, J., BAIRDEN, K., URQUHART, G. M. and JÖRGENSEN, R. J. *Vet. Rec.* **104**, 274 (1979)
112. ARMOUR, J., AL SAQUR, I. M., BAIRDEN, K., DUNCAN, J. L. and URQUHART, G. M. *Vet. Rec.* **106**, 184 (1980)
113. OAKLEY, G. A. *Vet. Rec.* **104**, 530 (1979)
114. ANONYMOUS. *Vet. Rec.* **122**, 191 (1988)
115. RYAN, W. G. *Vet. Rec.* **120**, 351 (1987)
116. TAYLOR, S. M., MALLON, T. R. and GREEN, W. P. *Vet. Rec.* **119**, 370 (1986)
117. JACOBS, D. E., FOX, M. T. and KOYO, F. A. *Vet. Rec.* **116**, 492 (1985)
118. BAIN, R. K. and URQUHART, G. M *Vet. Rec.* **118**, 82 (1986)
119. ARMOUR, J., BAIRDEN, K., BATTY, A. F., DAVISON, C. C. and ROSS, D. B. *Vet. Rec.* **116**, 151 (1985)
120. POUPLARD, L., LEKEUX, P. and DETRY, M. *Vet. Rec.* **118**, 557 (1986)
121. JACOBS, D. E., THOMAS, J. G., FOSTER, J., FOX, M. T. and OAKLEY, G. A. *Vet. Rec.* **121**, 221 (1987)
122. ARMOUR, J., BAIRDEN, K., PIRIE, H. M. and RYAN, W. G. *Vet. Rec.* **121**, 5 (1987)
123. BOON, J. H., PLOEGER, H. W. and RAAYMAKERS, A. J. *Vet. Rec.* **119**, 475 (1986)
124. HERD, R. P., REINEMEYER, C. R. and HEIDER, L. E. *Vet. Rec.* **120**, 406 (1987)
125. INTERNATIONAL UNION OF BIOCHEMISTRY. In *Enzyme Nomenclature,* Academic Press, New York and London (1984)

Index of breeds

General index